Golden-winged Warbler Ecology, Conservation, and Habitat Management

STUDIES IN AVIAN BIOLOGY

A Publication of The Cooper Ornithological Society

www.crcpress.com/browse/series/crcstdavibio

Studies in Avian Biology is a series of works published by the Cooper Ornithological Society since 1978. Volumes in the series address current topics in ornithology and can be organized as monographs or multi-authored collections of chapters. Authors are invited to contact the series editor to discuss project proposals and guidelines for preparation of manuscripts.

Volume 49
Studies in Avian Biology
Cooper Ornithological Society

Golden-winged Warbler Ecology, Conservation, and Habitat Management

EDITED BY

Henry M. Streby
University of California
Berkeley, CA

David E. Andersen
University of Minnesota
St. Paul, MN

David A. Buehler
University of Tennessee
Knoxville, TN

CRC Press
Taylor & Francis Group
Boca Raton London New York

CRC Press is an imprint of the
Taylor & Francis Group, an **informa** business

CRC Press
Taylor & Francis Group
6000 Broken Sound Parkway NW, Suite 300
Boca Raton, FL 33487-2742

First issued in paperback 2020

ISBN-13: 978-1-4822-4068-9 (hbk)
ISBN-13: 978-0-367-65833-5 (pbk)

Visit the Taylor & Francis Web site at
http://www.taylorandfrancis.com

and the CRC Press Web site at
http://www.crcpress.com

CONTENTS

Part III · Nonbreeding Season and Migration

Part IV · Synthesis and Future Directions

EDITORS

Henry M. Streby, PhD, is an Assistant Professor in the Department of Environmental Sciences at the University of Toledo and an Adjunct Assistant Professor at the University of Minnesota and at the University of Tennessee. Dr. Streby completed the editorial activities for this volume during his tenure as a National Science Foundation Postdoctoral Research Fellow with the University of Tennessee and as a Visiting Research Scholar at the University of California, Berkeley. He has 15 years of experience studying and publishing extensively on the ecology of birds and their habitats in eastern deciduous and northern hardwood forests of North America. Dr. Streby earned a BS in wildlife biology and an MS in ecology and evolutionary biology at Ohio University, and a PhD in wildlife ecology and management at the University of Minnesota within the Minnesota Cooperative Fish and Wildlife Research Unit. He is broadly interested in the ecology, conservation, and management of migratory and resident birds and the complex and changing landscapes they occupy throughout all stages of their life cycle.

David E. Andersen, PhD, is the Leader of the U.S. Geological Survey Minnesota Cooperative Fish and Wildlife Research Unit at the University of Minnesota. He has more than 35 years of experience in research related to avian population ecology, bird–habitat relations, and conservation of raptors, forest-nesting songbirds, waterfowl, and shorebirds. Dr. Andersen received both an MS and PhD from the University of Wisconsin–Madison, where he worked on population ecology of shortgrass prairie raptor communities in eastern Colorado. Following the completion of his PhD, he worked for the U.S. Fish and Wildlife Service in Colorado, where he oversaw a research program focused on a variety of prairie wildlife, including swift fox, songbirds, raptors, pronghorn, mule deer, and bobcats, in relation to changes in land use. During his time at the Minnesota Cooperative Fish and Wildlife Research Unit, Dr. Andersen led research efforts focused on population ecology of forest-nesting songbirds and raptors, American Woodcocks, and arctic-nesting geese and is currently investigating questions related to migration ecology and bird response to vegetation community manipulation and changing land-use patterns.

David A. Buehler, PhD, is a Professor of wildlife science in the Department of Forestry, Wildlife and Fisheries at the University of Tennessee–Knoxville. He received a BSc and MSc in wildlife ecology from the University of Wisconsin–Madison and a PhD in wildlife science from Virginia Polytechnic Institute and State University. His research program is focused on avian population ecology and developing effective conservation strategies for species of conservation concern. His recent research focused on population viability of Golden-winged and Cerulean Warblers, two of the most rapidly declining passerines breeding in North America. He served as the chair of the Golden-winged Warbler Working Group since its formation in 2005.

CONTRIBUTORS

KYLE R. ALDINGER
West Virginia Cooperative Fish and Wildlife
   Research Unit
West Virginia University
322 Percival Hall
Morgantown, WV 26506
kaldinge@mix.wvu.edu

DAVID E. ANDERSEN
U.S. Geological Survey
Minnesota Cooperative Fish and Wildlife
   Research Unit
200 Hodson Hall
1980 Folwell Avenue
St. Paul, MN 55108
dea@umn.edu

SARA BARKER SWARTHOUT
Conservation Science Program
Cornell Lab of Ornithology
159 Sapsucker Woods Road
Ithaca, NY 14850
sb65@cornell.edu

NICHOLAS J. BAYLY
SELVA: Investigación para la Conservación en el
   Neotropico
Diagonal 42A No. 20-37
Bogotá D.C., Colombia
nick.bayly@selva.org.co

EMILY C. BELLUSH
Department of Biology
Indiana University of Pennsylvania
Weyandt Hall, Room 114
975 Oakland Avenue
Indiana, PA 15705
ebell545@gmail.com

RUTH E. BENNETT
Department of Natural Resources
Cornell University
111A Fernow Hall
Ithaca, NY 14853
reb349@cornell.edu

DAVID A. BUEHLER
Department of Forestry, Wildlife, and Fisheries
University of Tennessee
Knoxville, TN 37996
dbuehler@utk.edu

ROGER D. BULL
Research and Collections
Canadian Museum of Nature
1740 Pink Road
Gatineau, QC J9J 3N7, Canada
rbull@mus-nature.ca

RICHARD B. CHANDLER
Warnell School of Forestry and Natural Resources
University of Georgia
180 E. Green Street
Athens, GA 30602
rchandler@warnell.uga.edu

LILIANA CHAVARRÍA-DURIAUX
Reserva El Jaguar
Managua, Nicaragua
orion.liliana@gmail.com

GABRIEL J. COLORADO
Universidad Nacional de Colombia Sede Amazonia
Kilómetro 2 vía Tarapacá
Leticia, Amazonas, Colombia
gjcoloradoz@unal.edu.co

JOHN L. CONFER
Department of Biology
Ithaca College
Ithaca, NY 14850
confer@ithaca.edu

DOLLY L. CRAWFORD
Department of Biology and Toxicology
Ashland University
401 College Avenue
Ashland, OH 44805
dcrawfo9@ashland.edu

JOSEPH DUCHAMP
Department of Biology
Indiana University of Pennsylvania
Weyandt Hall, Room 114
975 Oakland Avenue
Indiana, PA 15705
jduchamp@iup.edu

GEORGES DURIAUX
Reserva El Jaguar
Managua, Nicaragua
georges.duriaux@gmail.com

PABLO ELIZONDO
INBioparque
Santo Domingo de Heredia, Costa Rica
jpelizondo@pifcostarica.org

DAVID J. FLASPOHLER
School of Forest Resources and Environmental Science
Michigan Technological University
Houghton, MI 49931
djflaspo@mtu.edu

MACK W. FRANTZ
Department of Biology
Indiana University of Pennsylvania
Weyandt Hall, Room 114
975 Oakland Avenue
Indiana, PA 15705

and

West Virginia Cooperative Fish and Wildlife
    Research Unit
West Virginia University
322 Percival Hall
Morgantown, WV 26506
mack.w.frantz@gmail.com

JOHN GERWIN
North Carolina Museum of Natural Sciences
11 West Jones Street
Raleigh, NC 27601
john.gerwin@naturalsciences.org

KEITH A. HOBSON
Science and Technology Branch
Environment and Climate Change Canada
11 Innovation Boulevard
Saskatoon, SK S7N 3H5, Canada
keith.hobson@canada.ca

DAVID I. KING
USFS Northern Research Station
Department of Environmental Conservation
201 Holdsworth Hall
University of Massachusetts
Amherst, MA 01003
dking@fs.fed.us

JEFFERY L. LARKIN
Department of Biology
Indiana University of Pennsylvania
Weyandt Hall, Room 114
975 Oakland Avenue
Indiana, PA 15705

and

American Bird Conservancy
4249 Loudoun Avenue
The Plains, VA 20198-2237
larkin@iup.edu

JOHN P. LOEGERING
Agriculture and Natural Resources Department
    and Extension
University of Minnesota
Crookston, MN 56716
jloegeri@umn.edu

JAMES D. LOWE
29 South Street, PO Box 961
Dryden, NY 13053
jdl6@cornell.edu

TIMOTHY NUTTLE
Civil and Environmental Consultants, Inc.
333 Baldwin Road
Pittsburgh, PA 15205
tnuttle@cecinc.com

KATIE L. PERCY
Department of Forestry, Wildlife, and Fisheries
University of Tennessee
274 Ellington Plant Sciences
Knoxville, TN 37996
kpercy@utk.edu

SEAN M. PETERSON
Minnesota Cooperative Fish and Wildlife
    Research Unit
University of Minnesota
200 Hodson Hall/1980 Folwell Avenue
St. Paul, MN 55108

and

Department of Environmental Science Policy and
    Management
University of California–Berkeley
130 Mulford Hall
Berkeley, CA 94720
sean.michael.peterson@gmail.com

RAUL RAUDALES
Mesoamerican Development Institute
University of Massachusetts
1 University Avenue
Lowell, MA 01854
rraudales@mesoamerican.org

CARLOS G. RENGIFO
Estación Ornitológica La Mucuy
Mérida 5101, Venezuela
crengifo@ula.ve

JEFFREY D. RITTERSON
Department of Environmental Conservation
University of Massachusetts
160 Holdsworth Way
Amherst, MA 01003
jritters@eco.umass.edu

RONALD W. ROHRBAUGH
Conservation Science Program
Cornell Lab of Ornithology
159 Sapsucker Woods Road
Ithaca, NY 14850
rwr8@cornell.edu

KENNETH V. ROSENBERG
Conservation Science Program
Cornell Lab of Ornithology
159 Sapsucker Woods Road
Ithaca, NY 14850
kvr2@cornell.edu

AMBER M. ROTH
School of Forest Resources
and
Department of Wildlife, Fisheries, and
    Conservation Biology
University of Maine
5755 Nutting Hall
Orono, ME 04469
amber.roth@maine.edu

CURTIS SMALLING
Audubon North Carolina
667 George Moretz Lane
Boone, NC 28607
csmalling@audubon.org

HENRY M. STREBY
Department of Environmental Sciences
University of Toledo
Wolfe Hall, Suite 1235
Mail Stop 604, 2801 West Bancroft Street
Toledo, OH 43606
henry.streby@utoledo.edu

THERON M. TERHUNE II
Tall Timbers Research Station and Land
    Conservancy
13093 Henry Beadel Drive
Tallahassee, FL 32312
theron@ttrs.org

WAYNE E. THOGMARTIN
U.S. Geological Survey
Upper Midwest Environmental Sciences Center
2630 Fanta Reed Road
La Crosse, WI 54603
wthogmartin@usgs.gov

SHARNA TOLFREE
Environmental Studies Department
University of North Carolina–Wilmington
601 S College Rd
Wilmington, NC 28403

and

North Carolina Museum of Natural Sciences
11 West Jones Street
Raleigh, NC 27601
sharnatolfree@gmail.com

RICHARD TRUBEY
Mesoamerican Development Institute
University of Massachusetts
1 University Avenue
Lowell, MA 01854
rtrubey@mesoamerican.org

RACHEL VALLENDER
Research and Collections
Canadian Wildlife Service
Environment and Climate Change Canada
351 St. Joseph Boulevard
Gatineau, QC K1A 0H3, Canada

and

Canadian Museum of Nature
1740 Pink Road
Gatineau, QC J9J 3N7, Canada
rachel.vallender@ec.gc.ca

STEVEN L. VAN WILGENBURG
Canadian Wildlife Service
Environment and Climate Change Canada
115 Perimeter Road
Saskatoon, SK S7N 3H5, Canada
steve.vanwilgenburg@canada.ca

ANDREW VITZ
Massachusetts Division of Fisheries and Wildlife
1 Rabbit Hill Road
Westborough, MA 01581
andrew.vitz@state.ma.us

TOM WILL
U.S. Fish and Wildlife Service
Suite 990
5600 American Boulevard West
Bloomington, MN 55437
tom_will@fws.gov

PETRA B. WOOD
U.S. Geological Survey
West Virginia Cooperative Fish and Wildlife
    Research Unit
West Virginia University
Morgantown, WV 26506-6125
pbwood@wvu.edu

PREFACE

Golden-winged Warblers (*Vermivora chrysoptera*) have been the subject of considerable conservation concern over the past 20 years, especially in the Appalachian Mountains region, where they have experienced an approximately 95% reduction in population numbers since the late 1960s. In response to the dramatic decrease in population size in the eastern portion of their breeding distribution, coupled with a relative lack of information about their ecology and conservation in the western Great Lakes region, considerable effort has been directed both toward gaining a better understanding of Golden-winged Warbler ecology and developing conservation strategies to stabilize or increase populations. Issues of particular concern include hybridization with closely related Blue-winged Warblers (*Vermivora cyanoptera*), succession of forest cover types resulting in landscapes lacking stands in seral stages associated with breeding Golden-winged Warblers, land-cover conversion from forest to other uses on the non-breeding grounds, and the potential impacts of global climate change. These and other concerns about Golden-winged Warblers have led to the species being petitioned for listing under the U.S. Endangered Species Act and being afforded special conservation status in Canada and at the state level across much of their breeding distribution.

Several recent collaborative efforts, including the Golden-winged Warbler Working Group, the Golden-winged Warbler Atlas Project, and coordinated survey efforts in nonbreeding areas in Central and northern South America, have begun to shed light on Golden-winged Warbler ecology throughout their annual cycle and provide the basis for targeted conservation activities. However, much of the information resulting from these and other efforts currently exists in unpublished reports and on websites, and has not been subjected to critical scientific review. Moreover, much of the existing information on Golden-winged Warblers derives from efforts in the Appalachian Mountains region, and whether that information adequately represents conditions in the western Great Lakes region, where a majority of the species breeds, is not known. As efforts develop to consider Golden-winged Warblers in land-management strategies and plans, and as some organizations begin to target conservation efforts toward Golden-winged Warblers, it is important for those efforts to be informed by critically reviewed science. To that end, this volume of Studies in Avian Biology is our attempt to compile and critically review recent information on Golden-winged Warbler ecology and conservation, to add this information to the existing peer-reviewed literature, and to make this information widely available to land managers and others in a single volume.

Our aim with this volume is to expand the understanding of Golden-winged Warbler ecology and conservation by compiling current information on breeding and nonbreeding distribution, nesting and postfledging ecology, genetics, nonbreeding ecology, and habitat relations. We note that our volume is not a product of the Golden-winged Warbler Working Group but that many of the authors are members of that group, and we have asked them to share their vision of the future of Golden-winged Warblers and high-priority information needs in Chapter 13 of this volume. The opinions and recommendations expressed in that chapter are solely those of the chapter authors. We close our volume with an overview of the current state of knowledge about Golden-winged Warbler ecology and conservation research, and with our assessment of future directions and information needs.

We are grateful to the peer referees for their thorough and professional reviews of all manuscripts considered for this volume. We thank Nicholas M. Anich, Than J. Boves, Jeffrey D. Brawn, David R. Brown, Lesley P. Bulluck, Andrea Contina, Robert J. Cooper, Randy Dettmers, Duane R. Diefenbach, Laurie A. Hall, Douglas H. Johnson, Andrew W. Jones, Paul M. Kapfer, J. Patrick Kelley, David I. King, Joseph A. LaManna, Jeffrey L. Larkin, Scott R. Loss, Jay McEntee, Christopher E. Moorman, Sean M. Peterson, Matthew E. Reiter, Amanda D. Rodewald, Amber Roth, Kirk W. Stodola, Henry M. Streby, Wayne E. Thogmartin, Frank R. Thompson III, Jason M. Townsend, Rachel Vallender, and Petra Bohall Wood for reviewing manuscripts, and Jeanine M. Refsnider for editorial assistance.

Last, we thank the Studies in Avian Biology Series Editor, Brett K. Sandercock, and editorial board for affording us the opportunity to compile this volume. We acknowledge the efforts of the authors, who by working with us and the reviewers make this volume worthwhile and credible. It is our hope that all those involved felt that they were treated professionally throughout the process of compiling this volume and that their contributions will aid in furthering understanding of the ecology of migratory birds and inform Golden-winged Warbler conservation.

PART ONE

Global Distribution and Status

CHAPTER ONE

Dynamic Distributions and Population Declines of Golden-winged Warblers*

*Kenneth V. Rosenberg, Tom Will, David A. Buehler, Sara Barker Swarthout,
Wayne E. Thogmartin, Ruth E. Bennett, and Richard B. Chandler*

Abstract. Golden-winged Warblers (*Vermivora chrysoptera*) are among the most vulnerable and have one of the most steeply declining populations of North American songbirds, with an estimated breeding population in 2010 of 383,000 adults. The breeding distribution of Golden-winged Warblers has been highly dynamic, expanding and then contracting over the past 150 years in response to regional habitat changes, interactions with closely related Blue-winged Warblers (*V. cyanoptera*), and possibly climate change. To delineate the present-day breeding distribution and to identify population concentrations that could serve as conservation focus areas, we compiled survey data collected during 2000–2006 in 21 states and three Canadian provinces as part of the Golden-winged Warbler Atlas Project (GOWAP), supplemented by data from state and provincial Breeding Bird Atlas projects and more recent observations posted in eBird. Based on >8,000 GOWAP surveys for Golden-winged and Blue-winged Warblers and their hybrids, we also mapped occurrence of phenotypically pure and mixed populations in a roughly 0.5-min latitude–longitude grid across the two species' breeding distributions. Hybrids and mixed populations of Golden-winged and Blue-winged Warblers occurred in a relatively narrow zone across Minnesota, Wisconsin, Michigan, southern Ontario, and northern New York. Phenotypically pure Golden-winged Warbler populations occurred north of the hybrid zone. The region is defined for conservation planning as the *Great Lakes breeding-distribution segment*, but the future of populations in the Great Lakes states and Canada where 90% of the species currently breeds remains highly uncertain because of continued northward expansion of Blue-winged Warblers and the threat of hybridization. A second, now-disjunct range of Golden-winged Warbler populations occurs in the Appalachian Mountains (*Appalachian Mountains breeding-distribution segment*) from southeastern New York to northern Georgia, where they are surrounded at lower elevations by Blue-winged Warbler populations. Concentrations of Golden-winged Warblers persist in the Allegheny Mountains region of West Virginia, the Cumberland Mountains of Tennessee, the Blue Ridge Mountains of western North Carolina, the Allegheny Plateau and Pocono Mountains of Pennsylvania, and the Hudson Highlands of southern New York. High-elevation Appalachian populations have escaped contact with Blue-winged Warblers until recently and represent potentially important refugia for conservation

* Rosenberg, K. V., T. Will, D. A. Buehler, S. B. Swarthout, W. E. Thogmartin, R. E. Bennett, and R. B. Chandler. 2016. Dynamic distributions and population declines of Golden-winged Warblers. Pp. 3–28 in H. M. Streby, D. E. Andersen, and D. A. Buehler (editors). Golden-winged Warbler ecology, conservation, and habitat management. Studies in Avian Biology (no. 49), CRC Press, Boca Raton, FL.

and management. In the 44-year period from 1966 to 2010, the total breeding population of Golden-winged Warblers declined by 66% (−2.3% per year; latest North American Breeding Bird Survey data), with much steeper declines in the Appalachian Mountains Bird Conservation Region (BCR 28; −8.3% per year, 98% overall decline). Despite increasing populations in the northwestern part of the breeding distribution (Minnesota, Manitoba), total population estimates continued to decline by 18% in the 8-year period between 1994 and 2002. If these trends persist, our population projection predicts a further decline to ~37,000 individuals by 2100. In addition, based on historical records and standardized surveys across the nonbreeding distribution, we identified three regions of nonbreeding concentration: mid-elevations and Caribbean slopes from Guatemala and Belize to northwestern Nicaragua, middle elevations (both Caribbean and Pacific slopes) in Costa Rica and western Panama, and an arc of the northern Andes from central Colombia to northern Venezuela. The nonbreeding distribution may be shifting northwest, paralleling shifts in the breeding distribution. Future conservation efforts for Golden-winged Warblers need to include close monitoring of dynamic regional populations and of phenotypic and genetic distributional shifts.

Key Words: abundance, Blue-winged Warbler, conservation, hybridization, Neotropical migrants, population trends.

O ver the last 150 years, the breeding distribution of Golden-winged Warblers (*Vermivora chrysoptera*) in North America has changed dramatically, and the species has experienced one of the steepest population declines of any forest songbird over the past four decades (Sauer et al. 2012). The troubling trends have led to the Golden-winged Warbler being designated as a species of high conservation concern by Partners in Flight (Rich et al. 2004, Berlanga et al. 2010), the U.S. Fish and Wildlife Service (U.S. Fish and Wildlife Service 2008), and the International Union for the Conservation of Nature Red List: Near Threatened (BirdLife International 2012). Golden-winged Warblers have also been petitioned to be listed under protection of the U.S. Endangered Species Act (U.S. Fish and Wildlife Service 2011) and are designated as Threatened in Canada (COSEWIC 2011).

Golden-winged Warblers interbreed with Blue-winged Warblers (*V. cyanoptera*) and produce viable offspring (Parkes 1951, Ficken and Ficken 1968, Gill 1980, Confer 2005). Hybridization and apparent competitive interactions between these species have contributed to the overall population declines of both species (Buehler et al. 2007). Developing effective conservation strategies requires an understanding of limiting factors specific to Golden-winged Warblers, and how limiting factors may be affected by presence of Blue-winged Warblers and hybrids. Breeding distributions of these two species are highly dynamic and shifting, and understanding where hybridization is currently occurring is an important first step in developing effective conservation strategies.

The nonbreeding season is the longest portion of the annual cycle and lasts up to seven months, but until recently little was known about the ecology and conservation status of Golden-winged Warblers during migration or at nonbreeding sites (Chandler and King 2011). Lack of information away from the breeding grounds has hindered conservation efforts because it has not been possible to identify nonbreeding concentration areas or high-quality habitats.

Herein, we summarize the current distribution and status of Golden-winged Warblers at their breeding and nonbreeding grounds. We describe historical data and recently documented changes in species distributions. We also consider the context of simultaneously shifting breeding distribution of Blue-winged Warblers, including present-day geographic areas of overlap and hybridization. Last, we summarize recent population trends, provide estimates of population size, and project future changes to regional populations of Golden-winged Warblers. Most of the earlier topics have been reviewed in part in various publications (Confer 1992, Buehler et al. 2007, Confer et al. 2011), but the synthesis presented here is the first comprehensive review of all current information relating to distribution and population status of Golden-winged Warblers. Our synthesis draws heavily on information compiled for the Golden-winged Warbler Status Review and Conservation Plan (A. M. Roth et al., unpubl. plan), especially Chapter 1 of that document (D. A. Buehler et al. in A. M. Roth et al., unpubl. plan), including results of recent surveys, distribution modeling, extensive literature review, and expert opinion.

METHODS

Historical and Current Breeding Distribution

To examine changes in breeding distributions, we divided the last 140 years into three periods: the late 19th and early 20th Centuries representing historical distribution; the period of extensive change during the 20th Century associated with forest cutting, clearing for agriculture, and subsequent farm-field abandonment up to about 1980 (Confer et al. 2011); and the most recent period of breeding-distribution contraction during the last two decades (Buehler et al. in A. M. Roth et al., unpubl. plan). We base our inference about distribution during the 19th and early 20th Centuries on narrative accounts, whereas changes during the mid-20th Century are better supported by empirical evidence and (since 1966) by the North American Breeding Bird Survey (BBS). Numerous state and provincial breeding-bird atlas projects during the 1980s also helped define the late 20th-Century breeding distribution, as described by the American Ornithologists' Union (1983) and presented in Confer (1992). Only the 21st Century breeding-distribution contraction is well documented by systematic surveys.

In 1999, S. B. Swarthout et al. (unpubl. report) launched the Golden-winged Warbler Atlas Project (GOWAP) to map the current breeding distribution and to understand the geographic extent of present-day interactions between Golden-winged and Blue-winged Warblers. The GOWAP engaged volunteer birders and professional biologists to survey known and potential breeding sites from 1999 to 2006 using three standardized protocols (S. B. Swarthout et al., unpubl. report). More than 200 volunteers and collaborators in 21 states and three provinces conducted GOWAP population surveys to determine locations and numbers of breeding birds, population status, and general habitat characteristics at survey locations, using a combination of passive listening, searching, and standardized playback

of vocalizations of Golden-winged Warblers. The GOWAP also produced an atlas of phenotypic Golden-winged and Blue-winged Warblers and their hybrids, based on >7,200 point counts in 442 DeLorme Atlas (DeLorme Atlas Gazetteers 2014) pages in 17 states and two Canadian provinces from 1999 to 2005. Each page represents a rectangular grid cell, which varied in size across states, but averaged ~4,300 km^2 across most of the Golden-winged and Blue-winged Warbler breeding distributions. Observers selected 20 survey points by locating accessible patches of apparent Golden-winged Warbler habitat distributed across each DeLorme page, with five points in each quarter-page section. Each point count consisted of a 10-min standardized playback sequence. Observers visually confirmed the phenotype of each warbler detected using Type I and Type II songs of Golden-winged and Blue-winged Warblers, because both species sing the same Type II song and can sing each other's Type I song (Confer 1992). Complete description of the GOWAP survey protocols and detailed state-level results and maps can be found in S. B. Swarthout et al. (unpubl. report).

To account for the changing breeding distribution in the years since the GOWAP, we compiled presence and absence records obtained from other ongoing Golden-winged Warbler monitoring programs, locations of recent tissue sampling and other research by Golden-winged Warbler Working Group collaborators, and records in the rapidly increasing citizen science database, eBird (Sullivan et al. 2009, 2014). In addition, second-generation breeding-bird atlas projects completed by many states and Ontario in the past decade have helped document recent breeding distribution; results of the GOWAP will serve as a baseline to index breeding distribution and population size change. Using all of these data sources we define two breeding-distribution segments (see Results and Discussion section) that have been adopted as broad conservation regions by D. A. Buehler et al. (in A. M. Roth et al., unpubl. plan), and refer to populations below as clusters of breeding individuals dispersed, often patchily, within the two breeding-distribution segments.

Estimating Population Size and Trends

Population trends reported in this manuscript are based on the analysis of BBS data from 1966 to

2010 (Sauer et al. 2012). We present distribution-wide and regional population estimates for Golden-winged Warblers based on the Partners in Flight method described by Rosenberg and Blancher (2005) and reviewed by Thogmartin et al. (2006), using BBS annual indices of relative abundance (Sauer et al. 2011) for the decade from 1998 to 2007 (Blancher et al. 2013, Partners in Flight Science Committee 2013). We then compared population estimates with similar estimates using BBS relative abundance for the decade from 1990 to 1999 (Rich et al. 2004). Trends from BBS are now estimated with hierarchical Bayesian count models rather than estimating equations (Sauer and Link 2011, Sauer et al. 2012), and we calculated population size estimates from indices of abundance resulting from Bayesian count models and projected population size estimates to the year 2100.

Nonbreeding Distribution and Survey Methods

Our assessment of the current nonbreeding distribution and migration routes was based on a combination of data sources, including a distribution model based on historical records, an occupancy model based on recent survey data from part of the nonbreeding distribution, specimen records from natural history collections, and additional records in eBird. Museum specimen records and other observational records contributed through Priority Migrant eBird (Barker Swarthout et al. 2008) formed the basis for a preliminary predictive model of the potential nonbreeding distribution of Golden-winged Warblers in the Neotropics (M. Moreno, unpubl. data) using MAXENT (Phillips et al. 2006). In 2008, the Golden-winged Warbler Working Group established survey sites in Honduras, Nicaragua, Costa Rica, Panama, Venezuela, and Colombia using a 100-km^2 grid and selecting locations in grid cells with ≥ 0.08 probability of occurrence for Golden-winged Warblers, based on a preliminary occupancy model (M. Moreno, unpubl. data). The resulting survey was implemented during four field seasons (December to early March of 2009–2012), with 4,856 surveys conducted at 1,499 locations. At each location, trained surveyors conducted 20-min point counts up to three times per field season. Survey effort varied greatly among countries and among field seasons. Venezuela received the least effort, with only 35 locations visited (three times each) during the 2009–2010 field

season. Honduras was only surveyed during the 2010–2011 (47 locations visited three times each) and 2011–2012 field seasons (126 locations visited three times each).

Surveys consisted of a 10-min passive observation period followed by 10-min of broadcasting vocalizations of Golden-winged Warblers. In addition, we divided observation periods during the entire 20-min survey into eight, 2.5-min intervals to estimate probability of occurrence while accounting for variation in probability of detection among observers in each country. The possibility of geographic segregation of male and female Golden-winged Warblers, and the potential for differences in detection probability between sexes, motivated us to model occurrence probability separately for males and females. For each sex, we developed a set of competing models including linear and quadratic terms for effects of precipitation, temperature, elevation, latitude, and longitude. We included linear terms to represent directional changes in resource use and quadratic terms to represent nonlinear trends in resource use. We produced distribution maps by model-averaging predictions of occurrence probability from each model. We fit models using methods described by Chandler et al. (2011) and implemented in the R package unmarked (Fiske and Chandler 2011). Subsequent to this analysis, we conducted 371 additional surveys in Panama in January 2015, using the same protocol, except with a 5-min broadcast of Golden-winged Warbler vocalizations.

To delineate the nonbreeding distribution of Golden-winged Warblers, we combined the results from the surveys and models described earlier with additional records in eBird, and we reviewed published literature for each country with records of Golden-winged Warblers during the nonbreeding season. Although Golden-winged Warblers are not breeding throughout most of the year, the nonbreeding season is defined here and throughout this volume as the period when most Golden-winged Warblers maintain a territory in Central and South America (Chandler 2011). Herein, we define the nonbreeding distribution as that used by Golden-winged Warblers during the nonbreeding season when they are resident in Central and South America, excluding areas only used during migration or by transients. Golden-winged Warblers have been observed in Central and South America from mid-September through early April, but to eliminate including possible transients or migrants we only considered records from 1 November to 15 March to delineate the nonbreeding distribution. Between these dates, most individuals within the nonbreeding distribution maintain a territory (Chandler 2011), and we considered observations of Golden-winged Warblers outside the breeding distribution and outside of this period to be of transients or migrants. In total, we compiled and mapped 3,533 unique georeferenced occurrence records of Golden-winged Warblers during the nonbreeding season (as defined earlier) south of the U.S. Based on this information, we provide our best qualitative assessment of relative abundance and concentration of Golden-winged Warblers within their nonbreeding distribution. We also reviewed records, mostly from eBird, from the spring and fall migration seasons and provide a qualitative assessment of migration timing, routes, and potential stopover areas for Golden-winged Warblers.

RESULTS AND DISCUSSION

Breeding Distribution (1850–1900)

Due to a paucity of records prior to 1850, the pre-European settlement of North America breeding distribution of Golden-winged Warblers remains largely speculative. Most published maps of the historical breeding distribution may underestimate its eastern and northern extent and overestimate its southern extent, and errors have been carried forward to estimates of recent distribution (Confer 1992, Confer et al. 2011). Some records from the late 19th and early 20th Centuries already suggested presence in certain portions of the northeastern U.S., suggesting that even the historical breeding distribution of Golden-winged Warblers may have been dynamic.

In the Upper Midwest, Golden-winged Warbler breeding distribution during the late 1800s and early 1900s likely extended from southwestern Ontario (Speirs 1985, McCracken 1994) across southern Michigan (Berger 1958) and Wisconsin (Robbins 1991) to at least southeastern and central Minnesota (Roberts 1932), and south to northeastern Illinois, northern Indiana, and northern Ohio (Figure 1.1). Barrows (1912) did not report Golden-winged Warblers from the northern part of the Lower Peninsula or from the Upper

Peninsula of Michigan. Although sometimes described as a recent breeding bird in Manitoba (Confer et al. 2011), Golden-winged Warblers were first observed near Winnipeg in 1887 (Batchelder 1890). Five additional sightings from 1905 to 1928 and a small breeding population discovered east of Winnipeg in 1932 (Taylor 2003) suggest that Golden-winged Warblers may have been more widely distributed at the northern edge of their historical breeding distribution than previously reported (Confer 1992; Figure 1.1).

Farther south, Mumford and Keller (1984) claimed that almost every large swamp in northern Indiana had a pair of nesting Golden-winged Warblers during the late 1800s, and they were described as locally common in the early 1900s in northwestern Ohio but rare elsewhere in the state (Peterjohn and Rice 1991). South of our mapped historical breeding distribution (Figure 1.1), Roberts (1932) claimed that Golden-winged Warblers formerly bred south to southern Iowa, although Kent and Dinsmore (1996) reported only three nest records in Iowa, one in southeast (1888) and two in central Iowa (1898). In Missouri, there is one confirmed 1890 nesting record in the southeast and unconfirmed reports in 1884 from the northeast (Robbins and Easterla 1992). In the late 1800s, Ridgway (1889) reported Golden-winged Warblers breeding in southern Illinois, and Butler (1897) reported breeding in the Mississippi River bottoms, but it seems unlikely that Golden-winged Warblers bred regularly in the tallgrass prairie portions of Iowa, Illinois, and Indiana (Graber and Graber 1963). It is unclear, therefore, whether the historical breeding distribution was ever continuous between populations in the Great Lakes region and populations farther east in the Appalachian Mountains region and New England.

Golden-winged Warblers were first documented in southeastern New York in 1867, but breeding was not confirmed until 1897 (Andrle and Carroll 1988). By the early 1900s, Eaton (1914) described the species as common on western and northern Long Island and in the lower Hudson Valley, uncommon or local on southern and eastern Long Island and the central Hudson and Delaware valleys, and extremely rare in central and western New York. Golden-winged Warblers are widely thought to have expanded their distribution into southern New England during the late 1800s and early 1900s (Veit and Peterson 1993). Brewster (1906) first recorded Golden-winged Warblers

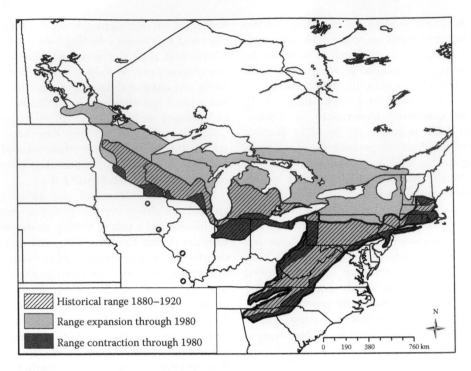

Figure 1.1. Golden-winged Warbler breeding distribution from ~1880 through 1980. Estimated distribution (hash marks) for 1880–1920 was derived largely from state ornithological accounts. Mid-1900s expansion in the north is shown in gray and contraction in the south and east in dark gray. Open circles represent additional early historical records that have previously been included in breeding-distribution maps, but which we regard as questionable (see text).

breeding in eastern Massachusetts in 1874, but J. A. Allen in Bent (1953) already considered the species to be common there by the 1860s. Golden-winged Warblers were first reported in Vermont in 1893 (Stearns and Coues 1893), and in southeastern New Hampshire they occurred as far north and west as Manchester and Concord by 1930 (Foss 1994). The first Connecticut breeding record was in 1875 near the Massachusetts border, followed by observations during the late 1800s of populations scattered throughout northern and southern Connecticut (Gill 1980, Bevier 1994), and in northeastern Rhode Island.

Farther south, the earliest records suggest that Golden-winged Warblers occurred historically throughout most of their current breeding distribution in the Appalachian Mountains region, although some expansion of the breeding distribution may have occurred prior to 1920. During the late 1860s, Golden-winged Warblers were reported as a rare to uncommon breeder throughout Pennsylvania, with most birds concentrated in western Pennsylvania (McWilliams and Brauning 2000). The species was first documented breeding in Maryland in 1895, but

by the early 1900s was reported as an increasingly common breeder in western Maryland (Eifrig 1904, Robbins and Blom 1996). In the central and southern Appalachian Mountains, breeding populations were thought to be limited to scattered summits higher than 850 m prior to European settlement (Mengel 1965). Historical status in Virginia is unclear, but the first description of a nest of a Golden-winged Warbler came from Greenbrier County in 1837 (Bent 1953). Golden-winged Warblers were considered rare by early ornithologists in Tennessee in the late 1800s, occurring in the southern Blue Ridge Mountains and eastern Cumberland Escarpment (Nicholson 1997). Brewster (1885, 1886) reported Golden-winged Warblers to be abundant in the second-growth forests and open oak (*Quercus* spp.) woodlands of western North Carolina at elevations between 600 and 1,200 m. The species was formerly considered a fairly common summer resident in northern Georgia (Burleigh 1958), and was reported from three counties in the mountains of northwestern South Carolina in the late 1800s (Loomis 1890, 1891; Sprunt and Chamberlain 1949; Post and Gauthreaux 1989).

20th-Century Breeding-Distribution Changes (1900–1980)

Golden-winged Warblers experienced a substantial breeding-distribution expansion during the early and mid-1900s in response to extensive forest clearing in the Upper Midwest and a period of extensive farm abandonment in the northeastern U.S. (Figure 1.1). By 1980, however, Golden-winged Warblers had begun to disappear from many portions of their historical and expanded breeding distribution, especially in southern parts of the Upper Midwest, in New England, and at lower elevations in the Appalachian Mountains.

Golden-winged Warblers were rediscovered near Riding Mountain National Park in western Manitoba in 1967 (Artuso 2008) and confirmed as breeding elsewhere in Manitoba in 1978 (Koes 2003). In Minnesota, the species expanded its breeding distribution to the Canadian border sometime after 1930 (Janssen 1987), and in Wisconsin expanded throughout the northern portion of the state but receded from areas where Golden-winged Warblers bred historically in the south (Cutright et al. 2006). Golden-winged Warblers were reported from the Upper Peninsula of Michigan in 1921 (Wood 1951, Payne 2011), and by the mid-1980s bred throughout most of the Upper Peninsula and most of the Lower Peninsula, with the exception of the southern two tiers of counties (Brewer et al. 1991). In general, since 1950, there has been a marked shift in breeding distribution away from the developed southern portions of Minnesota, Wisconsin, and Michigan to the extensively forested, wetland shrub communities farther north (D. A. Buehler et al. in A. M. Roth et al., unpubl. plan). In Ontario, breeding populations of Golden-winged Warblers continued to expand northeastward in the 1930s, were considered fairly common throughout the Bruce Peninsula and northward to Sudbury by the 1970s (Speirs 1985, McCracken 1994), and were common in eastern Ontario by the 1980s (Peck and James 1983).

Breeding-season records of Golden-winged Warblers in northern Illinois continued through the 1980s (Bohlen 1989, Kleen et al. 2004; BBS data), but there are no contemporary records from southern Illinois (Robinson 1996). By the 1980s, Golden-winged Warblers were reported as very rare summer residents in northern Indiana, with

nesting records up until 1983 (J. Castrale, pers. comm.). Breeding populations of Golden-winged Warblers were probably already declining in Ohio by the late 1930s, and there were no confirmed breeding records in the 1980s (Peterjohn 1989, Peterjohn and Rice 1991) or more recently (P. Rodewald, pers. comm.).

In response to extensive farm abandonment, maximum abundance and breeding-distribution extent of Golden-winged Warblers peaked in the northeastern U.S. between 1930 and 1950 (Confer 1992). By 1950, the species was moderately abundant throughout central and western New York but had disappeared from Long Island and the Lower Hudson Valley (Andrle and Carroll 1988). Numbers continued to increase through the 1980s in the Eastern Ontario Plain, Indian River Lakes, and St. Lawrence Plains regions (Confer et al. 1991, McGowan and Corwin 2008). In the mid-1900s, Golden-winged Warblers persisted as rare and local breeders near the southern tip of Lake Champlain and the Connecticut River in Vermont (Laughlin and Kibbe 1985) and in the coastal lowlands of New Hampshire (Foss 1994). Golden-winged Warbler populations increased throughout eastern Massachusetts through the early 1900s (Veit and Peterson 1993). By the 1950s, most breeding records came from eastern Massachusetts (except Cape Cod and adjacent islands) or the Berkshires, where the species was still increasing in abundance (Bailey 1955). After the 1950s, the number of breeding Golden-winged Warblers declined in Massachusetts (Veit and Petersen 1993), with the last breeding populations documented in Essex and Berkshire counties (Petersen and Meservey 2003). Through the 1980s, the breeding distribution continued to shift northward in New York (Confer et al. 1991) and into portions of adjacent Quebec, where Golden-winged Warblers were first detected in 1957 (Gauthier and Aubry 1996).

In Pennsylvania, Golden-winged Warblers remained scarce at higher elevations in the mid-1950s, but bred throughout the Ridge and Valley Region; by 1980 they had expanded through northern Pennsylvania, but were absent at low elevations in the southwest and southeast parts of the state (Brauning 1992). By the mid-1980s, Golden-winged Warblers continued to breed in the mountains of western Maryland (Robbins and Blom 1996) and throughout most of southern and central West Virginia (Buckelew and Hall 1994).

Golden-winged Warblers likely expanded their breeding distribution in southeastern Kentucky and Tennessee following strip-mine reclamation and logging in the early 1900s; Stupka (1963) reported Golden-winged Warblers as fairly common at low and middle elevations in Great Smoky Mountains National Park into the 1950s, and they occurred along the eastern Cumberland Escarpment through the 1980s. A small population persisted through the 1980s in northern Georgia, but disappeared from South Carolina prior to 1980.

Recent Breeding-Distribution Changes and Current Breeding Distribution (1980 to Present)

The 1990s were a period of continued rapid change in breeding distribution and population status of Golden-winged Warblers, leading to their recognition as a species of high conservation concern, especially in the eastern parts of their breeding distribution (Rich et al. 2004, Buehler et al. 2007). Based on GOWAP surveys from 1999 to 2006 (S. B. Swarthout et al.,

unpubl. report), collaborative research and monitoring efforts of the Golden-winged Warbler Conservation Initiative, records and data from the BBS (1966–2012), state and provincial breeding bird atlases, and observations reported in eBird, we derived our estimate of the current (ca. 2013) breeding distribution of Golden-winged Warblers (Figure 1.2). The current distribution map was adopted for use in the Golden-winged Warbler Status Review and Conservation Plan (A. M. Roth et al., unpubl. plan). Golden-winged Warblers have been extirpated from Illinois, Indiana, and Ohio, and no longer breed consistently throughout central New York. Thus, Golden-winged Warblers now occur in two disjunct regions—a Great Lakes and an Appalachian Mountains breeding-distribution segment. The *Great Lakes breeding-distribution* segment extends from the eastern edge of Saskatchewan eastward to the Champlain Valley of northeastern New York and adjacent Vermont. The *Appalachian Mountains breeding-distribution* segment extends from northern Georgia and the Cumberland Mountains of Tennessee to southeastern New York and extreme northwestern

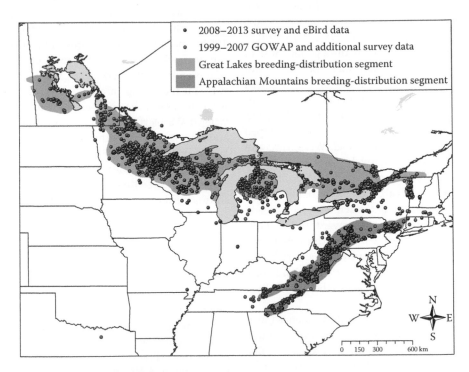

Figure 1.2. Current breeding distribution of Golden-winged Warblers in 2013, as determined by the Golden-winged Warbler Working Group based on all available survey and observational data. The Ontario Breeding Bird Atlas data are not shown on this map but were used to estimate current breeding distribution. Areas with only sporadic occurrence and areas without breeding records since 2008 were not included within the two breeding-distribution segments (shaded regions).

Connecticut (Figure 1.2). These two breeding-distribution segments are treated separately within the Golden-winged Warbler Conservation Plan (A. M. Roth et al., unpubl. plan) and are referred to throughout this volume.

The northward expansion of the Golden-winged Warbler breeding distribution has continued recently into northern New York, eastern and northwestern Ontario, southwestern Manitoba, and even farther west in Canada with one confirmed and several possible breeding records for Saskatchewan (Smith 1996). Monitoring efforts and sightings of Golden-winged Warblers have increased in recent years across Manitoba, with more than 395 territorial birds recorded in 2008 and 2009; observers believed there were >1,000 pairs in Manitoba in the late 2000s. It is unclear, however, whether the number of breeding Golden-winged Warblers in Manitoba is increasing, or whether local populations have only recently been adequately surveyed.

Golden-winged Warblers had virtually disappeared as a breeder by the 1990s from southern Minnesota (Janssen 1987) and southern Wisconsin (Cutright et al. 2006). Recent GOWAP surveys and eBird records show the highest concentrations in north-central and east-central Minnesota and adjacent northwestern Wisconsin. Smaller areas of concentration are located in the Black River State Forest area and in northeastern Wisconsin near the border with Michigan's Upper Peninsula (Figure 1.2). The most recent Michigan Breeding Bird Atlas (BBA) documented continued contraction of the breeding distribution of Golden-winged Warblers in southern Michigan (Payne 2011). The GOWAP surveys found Golden-winged Warblers largely restricted to the northwestern portions of Michigan's Lower Peninsula, and although Golden-winged Warblers have been recorded breeding throughout the Upper Peninsula (Payne 2011), few were detected during the GOWAP. None have been reported at Ottawa National Forest in the western part of the Upper Peninsula, in spite of recent surveys of apparently appropriate habitat (B. Bogaczyk, pers. comm.).

Except for a few sites near Long Point and the Niagara Escarpment, breeding Golden-winged Warblers have vanished from southwestern Ontario. The latest Ontario BBA (2001–2005) found that the recent breeding-distribution expansion had not continued beyond the Canadian Shield region, and the breeding distribution in eastern Ontario had contracted by 27% since the 1980s (Cadman et al. 2007). Scattered breeding locations persist along the Ottawa River, on the Bruce Peninsula, and along the northern shore of Georgian Bay (but not Lake Superior), and a potentially new breeding location has been identified in far western Ontario in the Lake of the Woods area. The breeding distribution in Quebec also appears to have contracted since its peak in the 1980s (Gauthier and Aubry 1996), and was limited to the southwestern corner of the province and the area near Ottawa by 2002.

No breeding Golden-winged Warblers were found in Illinois, Indiana, or Ohio during the GOWAP or more recently. In New York, the largest concentration of Golden-winged Warblers still exists (ca. 2013) in the St. Lawrence Valley, but their breeding distribution is now contracting from the south and does not seem to have expanded farther to the north or east. A few breeding Golden-winged Warblers also persist in the Lake Champlain Valley on both the New York and Vermont sides. Breeding Golden-winged Warblers are now virtually absent from the Finger Lakes region; a few breeding individuals persisted along the south shore of Lake Ontario until about 2005 (McGowan and Corwin 2008), but subsequently this population has also become extirpated.

During the 1990s, Golden-winged Warblers disappeared as a breeding bird from the coastal plain of New Hampshire but continued to breed in the central Connecticut Valley near Hanover. Breeding Golden-winged Warblers have since disappeared from that region and may now be extirpated from New Hampshire (Suomala 2005). Similarly, there have been only four confirmed breeding records in Massachusetts since 1990, and breeding Golden-winged Warblers are now thought to be extirpated in that state. During the 1980s, Bevier (1994) reported confirmed breeding only from northwestern Connecticut but Golden-winged Warblers subsequently disappeared from most of these areas and were known to breed at only three locations by 2009. In Rhode Island, breeding Golden-winged Warblers are now extirpated and are only rarely observed during migration. Note that in this entire zone of recent breeding-distribution contraction, from southern Wisconsin through central New York to New England, reports persist as of 2013 of isolated individuals, including singing males, but no evidence indicates that Golden-winged Warblers consistently breed south or east of the mapped boundary for the Great Lakes breeding-distribution segment (Figure 1.2).

The Appalachian Mountains breeding-distribution segment also has contracted significantly since 1990, and a major focus of the GOWAP was to clearly define its present-day southern boundary. In contrast to the Great Lakes region, Golden-winged Warblers in the Appalachian Mountains have not shifted their breeding distribution geographically through time, but rather have contracted their breeding distribution to refugia, often at higher elevations, as Blue-winged Warblers have expanded their breeding distribution into surrounding lower elevations. At the northern end of the Appalachian Mountains breeding distribution, a population of Golden-winged Warblers persists in the Hudson Highlands of southeastern New York, but the extent of this population's distribution has shrunk by more than 75% since the 1980s (McGowan and Corwin 2008). Golden-winged Warblers may now be restricted to a few sites in and around Sterling Forest State Park. In adjacent New Jersey, an estimated 80–90 Golden-winged Warbler pairs were present in the northwestern part of the state as recently as 2000–2002 (S. Petzinger, pers. comm.). By 2008, however, only a single male remained in the Delaware Water Gap-Kittatinny Ridge area, and the last small cluster of breeding individuals in the state persists in the New Jersey Highlands adjacent to Sterling Forest State Park in New York.

By the 1990s, breeding Golden-winged Warblers had disappeared throughout much of Pennsylvania with the exception of higher elevation sites above 600 m within the Valley and Ridge and Pocono regions, and in more forested landscapes of the mountainous Allegheny High Plateau (Brauning 1992). The GOWAP (1999–2006) and more recent surveys indicated that breeding Golden-winged Warblers remained common in these three, high-elevation regions. Populations in south-central Pennsylvania are contiguous with populations in western Maryland, where the species persists primarily in Garrett County (Ellison 2010), and in the Allegheny Mountains of northeastern West Virginia, which is currently the largest population in that state. In the Allegheny Mountains, breeding Golden-winged Warblers remain at elevations primarily between 850 and 915 m but are virtually absent above 1,200 m,

even in apparently suitable deciduous forest cover types (Canterbury 1997; Gill et al. 2001; P. Wood, pers. comm.; R. Bailey, pers. comm.). In other parts of West Virginia, the number of breeding Golden-winged Warblers has declined rapidly (Canterbury et al. 1993, Canterbury and Stover 1999); relatively large numbers persisted into the 1990s in the coalfields of southern and central West Virginia (McDowell, Wyoming, and Raleigh counties), almost entirely on reclaimed narrow strip benches ≥610 m in elevation, but these populations have nearly disappeared since 2003 (J. Larkin, pers. comm.). A similar pattern has occurred in southeastern Kentucky, where >50 Golden-winged Warbler breeding territories were studied on reclaimed mines as recently as 2004–2005, but breeding Golden-winged Warblers disappeared rapidly from most sites by 2009 (Patton et al. 2010).

Surveys of the historical Golden-winged Warbler breeding distribution in Virginia detected only 50 males and 6 females in 2006, primarily in Bath and Highland counties (Wilson et al. 2007). In Tennessee, >200 pairs continued to breed in the Cumberland Mountains and in several locales in the southern Blue Ridge Mountains in far northeastern Tennessee through the late 2000s. A few breeding pairs also remained at higher elevations on the Cumberland Plateau until about 2000 (Welton 2003). Across western North Carolina, Golden-winged Warblers remained through 2013 as a locally common summer resident at middle and high elevations (600–1,600 m), largely contiguous with the breeding distribution in Tennessee. In 2006, GOWAP surveys located ~100 breeding pairs in North Carolina, and more recent observations through 2013 have come from 13 western counties in that state (C. Smalling, pers. comm.; eBird records). In the mountains of northern Georgia, Golden-winged Warblers remained a rare summer resident through the late 1990s, but after five years (1999–2003) of surveying potential breeding areas on public and private lands within the historical Georgia breeding distribution, Klaus (2004) found <20 territorial males. A small population breeding in regenerating forest following clear-cutting within the Cherokee National Forest reported during the 1990s had disappeared (Klaus 1999), as did a breeding population previously reported from the adjacent Great Smoky Mountains National Park in Tennessee (Nicholson 1997).

Geographic Overlap and Hybridization with Blue-winged Warblers

Changes in the distribution and abundance of Golden-winged Warblers are widely believed to be, at least in part, the result of interactions with Blue-winged Warblers, its closely related sister species. A divergence of 3% (cytochrome B) to 4.5% (NDII) between the mtDNA of Golden-winged and Blue-winged Warblers (Gill 1997, Dabrowski et al. 2005) suggests that isolation of ancestral populations occurred one to two million years before present (BP). In the absence of a fossil record, Short (1963) speculated that the advance and retreat of glaciers during the Wisconsin Glacial Episode (12,000 years BP) led to isolation and left Golden-winged Warblers restricted to the southeastern coastal plain of North America and Blue-winged Warblers to west of the Mississippi Embayment, a water barrier stretching several hundred km from present-day Illinois to the Gulf of Mexico. Following warming climate and retreat of glaciers, breeding distributions of both species likely moved northward but remained isolated until after European settlement (Short 1963). Extensive forest clearing in the Great Lakes region and Appalachian Mountains during the late 1800s and early 1900s undoubtedly increased habitat abundance and distribution for both species in those regions.

As Golden-winged Warblers expanded their breeding distribution northward during the early and mid-20th Century, Blue-winged Warblers also expanded their breeding distribution northward and eastward. Where Blue-winged Warblers colonize areas occupied by Golden-winged Warblers, hybridization is frequent and Golden-winged Warblers have typically been extirpated within 50 years (Gill 1980); species replacement has occurred in the Great Lakes region east to New York and southern New England. Secondary contact has created a dynamic zone of overlap and hybridization that has moved steadily northward as the breeding distributions of the two species shifted (Gill 1980, Gill et al. 2001). In New York, the rate of movement of the hybrid zone was estimated at 3–6 km/year through the mid-20th Century (Confer and Larkin 1998). Blue-winged Warblers also expanded their breeding distribution eastward during the 20th Century, but because of steep elevational relief in the Appalachian Mountains, distributional overlap and interactions with Golden-winged Warblers have been more complex. Remaining mostly at lower elevations, Blue-winged Warblers expanded their breeding distribution north and east along major river valleys and around mountain ridges, extending to central New York, southern New England, eastern Pennsylvania, and New Jersey in the north and to northern Alabama, northern Georgia, and southwestern North Carolina in the south. Blue-winged Warblers remained uncommon, however, on the eastern side of the Appalachian Mountains south of the Maryland Piedmont (Gill et al. 2001).

The GOWAP hybrid atlas surveys from 1999 to 2005 documented a relatively narrow zone of overlap and hybridization between Golden-winged and Blue-winged Warblers extending from central and northern New York west through southwestern Ontario, the middle Lower Peninsula of Michigan, and central Wisconsin, to east-central Minnesota (Figure 1.3). By 2005, the only large phenotypically pure Golden-winged Warbler populations appeared to exist north of this hybrid zone in northern Minnesota and Manitoba. The GOWAP assessment also indicated a second contact zone throughout the Appalachian Mountains, where breeding Golden-winged Warblers persisted, usually at high elevations, surrounded by areas dominated by breeding Blue-winged Warblers (Figure 1.3). Only 4% of the roughly 2,800 individuals detected as part of the GOWAP were phenotypic hybrids, observed in 95 grid cells throughout both zones of overlap. In an additional 80 grid cells, Golden-winged and Blue-winged Warblers were both observed, but no hybrids were detected, suggesting that hybrids are relatively rare even in zones of overlap.

Phenotypic Golden-winged Warbler populations in far northern Minnesota remain outside the expanding breeding distribution of Blue-winged Warblers (Shapiro et al. 2004), but Vallender et al. (2009) found Blue-winged Warbler mtDNA evidence of cryptic hybrids among the phenotypic Golden-winged Warblers in this region. Golden-winged Warblers in Manitoba and eastern Saskatchewan comprise possibly the only nonintrogressed Golden-winged Warbler population anywhere within their breeding distribution (Vallender et al. 2009). The discovery of a Brewster's Warbler (hybrid) in Manitoba in 2008 and the recent identification there of cryptic hybrids indicates that this population may currently be experiencing introgression

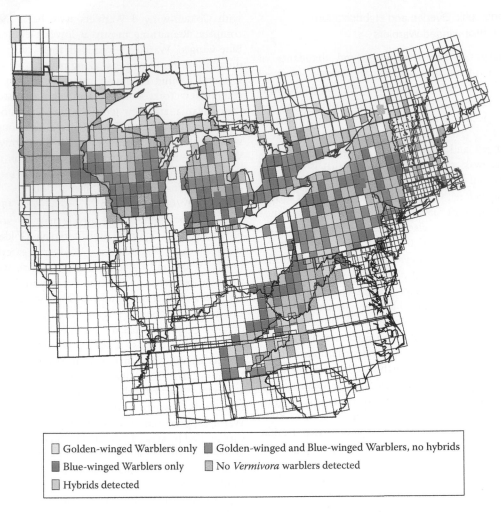

☐ Golden-winged Warblers only	■ Golden-winged and Blue-winged Warblers, no hybrids
■ Blue-winged Warblers only	☐ No *Vermivora* warblers detected
☐ Hybrids detected	

Figure 1.3. Distribution of phenotypic Golden-winged Warblers (GWWA), Blue-winged Warblers (BWWA), and zone of hybridization based on 1999–2005 data from the Golden-winged Warbler Atlas Project (S. B. Swarthout et al., unpubl. report). Each shaded grid cell represents a page from a Delorme Atlas (DeLorme Atlas Gazetteers 2014) ~4,300 km² each.

(Vallender et al. 2009; Chapter 4, this volume), and that genetic monitoring of this population is warranted. Blue-winged Warblers or hybrids were detected during the GOWAP throughout central and eastern Wisconsin, leaving phenotypically pure Golden-winged Warblers only in the northwestern part of that state. Similarly, Blue-winged Warblers and hybrids were found throughout the Lower Peninsula of Michigan, interspersed with a few grid cells where only Golden-winged Warblers were detected. The GOWAP did not sample Ontario, but the second Ontario BBA indicated an expansion of the breeding distribution of Blue-winged Warblers throughout southwestern Ontario, replacing Golden-winged Warblers at all but a few sites (Cadman et al. 2007). Even along the Canadian Shield, where a large population of Golden-winged Warblers existed in the 1990s, Blue-winged Warblers and hybrids began to appear and increased in abundance after 2001. A similar, rapid replacement of Golden-winged Warblers by Blue-winged Warblers occurred in the St. Lawrence Valley of New York, where the first hybrid warblers were detected in the late 1990s. The advancing front of Blue-winged Warbler breeding distribution passed through the well-monitored Fort Drum area by 2005. By 2008, J. Bolsinger (pers. comm.) found a mixed population of 50% Golden-winged Warblers, 32% Blue-winged Warblers, and 17% phenotypic hybrids at that site.

In the Appalachian Mountains, interactions between Golden-winged and Blue-winged Warblers have been more complex, and importantly, several areas exist where both species co-occurred for long periods, mediated by ecological gradients in elevation, vegetation types, or both. For example, in the western portion of the Hudson Highlands in southeastern New York, both species co-occurred for over a century (Eaton 1914, Frech and Confer 1987, Scully 1997, Confer and Tupper 2000). However, Dabrowski et al. (2005) questioned the continuing co-occurrence of these two species in the Hudson Highlands due to increased genetic introgression. Similarly, breeding Golden-winged and Blue-winged Warblers co-occurred in the New Jersey Highlands for almost a century (Confer and Larkin 1998, Confer and Tupper 2000). By 2010, breeding Golden-winged Warblers in that state only persisted in the Newark Watershed within the New Jersey Highlands. Blue-winged Warblers expanded their breeding distribution through the Delaware River Valley in the late 1900s (Gill 1997), and between 2003 and 2008 rapidly replaced breeding Golden-winged Warblers throughout northwestern New Jersey after a brief period of hybridization (S. Petzinger, pers. comm.).

In Pennsylvania, Blue-winged Warblers and hybrids were detected nearly throughout the state during 1999–2006 GOWAP surveys. The breeding distribution of Blue-winged Warblers in Pennsylvania expanded at elevations below 600 m during the late 1900s, but contracted overall in the state since 1990 because of reduction in the extent of early successional forest (Wilson et al. 2012). In adjacent Maryland, the breeding distribution of Blue-winged Warblers also contracted overall since the 1980s and has not expanded to higher elevations in mountainous areas in western Maryland (Ellison 2010). High-elevation, extensively forested areas in Pennsylvania and in the adjacent Allegheny Mountains of western Maryland and West Virginia currently support the largest breeding populations of Golden-winged Warblers within the Appalachian Mountains breeding-distribution segment, and this population of Golden-winged Warblers remains largely isolated geographically from Blue-winged Warblers. High elevation sites therefore represent important conservation areas where management to increase the amount of breeding habitat is especially critical (A. M. Roth et al., unpubl. plan).

A similar situation exists at high elevations in the Blue Ridge and Cumberland mountains of western North Carolina and northern Georgia, where Golden-winged Warblers remain isolated from Blue-winged Warblers. A few hybrids were detected during the GOWAP within areas occupied by Golden-winged Warblers, but with little evidence of extensive hybridization. These areas also currently serve as refugia where management to increase habitat quantity and quality to increase populations is critically needed to avoid extirpation (Percy 2012).

In contrast, the breeding distribution of Blue-winged Warblers expanded rapidly into southeastern Kentucky after 2000, including some high-elevation areas where Golden-winged Warblers had previously been found consistently. Hybrids were commonly observed in these areas after 2005, and breeding Golden-winged Warblers had all but disappeared from Kentucky by 2009 (Patton et al. 2010; S. Vorisek and J. Larkin, pers. comm.). West Virginia supported the largest number of breeding Golden-winged Warblers in the Appalachian Mountains as recently as the 1990s, but GOWAP surveys indicated the presence of both species or hybrids in 20 of 32 grid cells (Figure 1.3), with only Blue-winged Warblers detected in the western third of the state. Since 1995, Blue-winged Warblers replaced Golden-winged Warblers throughout the coalfield region of southern and central West Virginia; only isolated Golden-winged Warbler breeding sites were thought to remain as of 2009. By 2013, Golden-winged Warblers persisted only at middle elevations in the Allegheny Mountains of eastern West Virginia (K. Aldinger, pers. comm.; J. Larkin, pers. comm.). Last, surveys in a 40-county area of western Virginia in 2006 documented the expansion of Blue-winged Warblers into 10 of 11 counties where breeding Golden-winged Warblers were still present; hybrids were detected in 7 of these counties (Wilson et al. 2007).

Breeding Population Trends and Changes in Population Size

Sauer et al. (2012) estimated that the global breeding Golden-winged Warbler population declined from 1966 to 2010 at an average annual rate of −2.6% (95% Credible Limits: −3.5% to −1.6%): one of the steepest declines observed for any North American forest songbird. Extremely steep population declines occurred in the Appalachian Mountains Bird Conservation Region (BCR) (−8.5% per year), whereas more moderate declines occurred in the Prairie Hardwood Transition BCR (−2.9%) and the Boreal-Hardwood Transition BCR (−1.1%).

Using indices of abundance resulting from the Bayesian analysis of BBS data, we estimated that the number of Golden-winged Warblers declined from roughly 1.25 million breeding adults (95% CL: 950,000; 1,700,000) in 1966 to 386,000 (95% CL: 290,000; 520,000) in 2000 and 383,000 (95% CL: 275,000; 565,000) in 2010 (Figure 1.4). Using the population estimate for 2010 and assuming the 45-year BBS trend and the observed variability in that trend will continue, we predict a population of 37,000 breeding adults (95% CL: 27,000; 67,000) by 2100. This evaluation represents a reduction in population size of 1.2 million birds (97%) in 135 years. Note that the apparent leveling off of the overall trend in recent years could be the result of recent stabilization of Golden-winged Warbler populations, which is not suggested by our distributional analysis, or could result from a low detection rate of Golden-winged Warblers on BBS routes in areas where the species has become rare, and therefore the low power of the BBS to detect further changes. If populations have truly begun to stabilize, applying the 45-year trend will result in a lower future population projection than may actually occur.

Based on regional estimates of population size, we estimate that >80% of breeding Golden-winged Warblers currently occur within the Boreal-Hardwood Transition BCR (BCR 12) and nearly 90% of the total population occurs within the area we defined as the Great Lakes breeding-distribution segment (Table 1.1). In contrast, only about 5% of the current breeding population occurs within the Appalachian Mountains BCR (BCR 28), which corresponds with the Appalachian Mountains breeding-distribution segment. Ongoing reduction in Golden-winged Warbler population size in the Appalachian Mountains BCR is especially alarming—a 97.8% population decline from 1966 to 2010 and a 61.7% decline between our average estimate for the 1990s and our average estimate for the 2000s decades (Table 1.1). The change also represents a proportional change, in that the Appalachian Mountains BCR was estimated to support nearly 10% of the total population of breeding Golden-winged Warblers during the decade of the 1990s, but only 5% of the total population during the 2000s. The 1,740% increase in estimated population size in the Boreal Taiga Plains BCR between the 1900s and 2000s is based on relatively few records, such that the credible interval is extremely broad (Table 1.1). Buehler et al. (in A. M. Roth et al., unpubl. plan) used these same estimates of population trends and changes in regional population size to predict

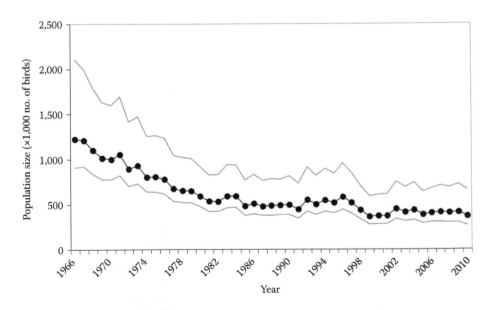

Figure 1.4. Annual breeding population size of Golden-winged Warblers (with 95% credible limits [CL]) estimated with a hierarchical Bayesian analysis of North American Breeding Bird Survey data (1966–2011).

TABLE 1.1

Estimated percent of total Golden-winged Warbler population and change in population size by Bird Conservation Region (BCR) over the entire period (1966–2010) of the North American Breeding Bird Survey (BBS) and between average estimates for the decades of the 1990s vs. the 2000s (see Blancher et al. 2013).

Bird Conservation Region	Percent of population	1966–2010	1990s vs 2000s (average for each decade)
Boreal Hardwood Transition (BCR 12)	80.2	**−27.5 (−30.3, −25.9)**	**−16.3 (−17.8, -15.2)**
Prairie Hardwood Transition (BCR 23)	9.8	**−70.7 (−76.8, −46.3)**	−11.8 (−27.0, 25.9)
Appalachian Mountains (BCR 28)	5.3	**−97.8 (−97.9, −97.7)**	**−61.7 (−63.7, −60.5)**
Great Lakes–St. Lawrence Plain (BCR 13)	4.0	**−16.0 (−18.1, −10.9)**	0.9 (−5.5, 8.2)
Boreal Taiga Plains (BCR 6)	0.6		**1,740 (850, 946,620)**
Total Population Change		**−66.2 (−67.0, −65.5)**	**−17.9 (−20.3, −15.4)**

Trend estimates are derived from the BBS (Sauer et al. 2011). Numbers in parentheses are 95% credible limits; significant change (0.95 probability that the percent change differs from zero) is indicated in bold font.

probability of Golden-winged Warbler population persistence in portions of the species' breeding distribution.

Nonbreeding Distribution

Nonbreeding-season point-count surveys during December–early March 2009–2012 resulted in 543 detections of male Golden-winged Warblers and 76 detections of females. Because of the low number of female detections, we modeled only occurrence of males. Probability of male occurrence was lower in Colombia and Venezuela than in northern Central America (Figure 1.5). Note that these results were based on few survey points in Venezuela; few detections of Golden-winged Warblers in Colombia; and no survey data from Guatemala, Belize, or Mexico. Throughout the modeled distribution, probability of occurrence was greatest at intermediate elevations (700–1,400 m) and was lowest (<0.05) in the lowlands of South America (<500 m), above 3,000 m throughout the estimated nonbreeding distribution, and in the tropical dry forests of Costa Rica, Nicaragua, and Honduras (Figure 1.5). Female Golden-winged warblers were detected

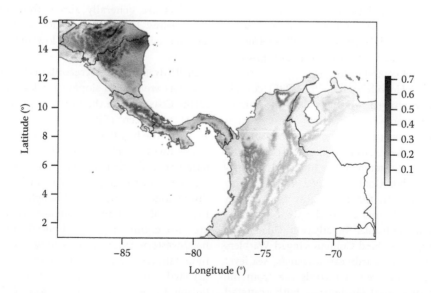

Figure 1.5. Estimated probability of occurrence of male Golden-winged Warblers, based on survey data from December to mid-March, 2009–2012. We modeled probability of occurrence as a function of annual precipitation, temperature, elevation, longitude, and latitude.

throughout the geographic distribution of males, providing no evidence for latitudinal segregation between males and females. Habitat associations for females, however, did suggest some spatial segregation, with females occupying lower elevations and drier areas than males.

Combining our model results with a qualitative assessment of 3,533 specimen and observational records from 1 November to 15 March, we estimated that the nonbreeding distribution (excluding areas used only by migrants or transients) of Golden-winged Warblers extends from southeastern Mexico and Belize south and east through the highlands and Caribbean slope of Guatemala, El Salvador, and Honduras to northwestern Nicaragua, in humid lowlands and middle elevations on both Caribbean and Pacific slopes in Costa Rica and Panama, and in an arc of the northern Andes from northwestern Venezuela to central Colombia (Figure 1.6). Smaller numbers of Golden-winged Warblers regularly occur during the nonbreeding season farther north in eastern Mexico, including the Yucatan Peninsula, in the Santa Marta and Perijá Mountain regions of northern Colombia, and farther south in the western Andes to northwestern Ecuador. Scattered records from the Pacific lowlands of Central America, and in the Greater Antilles, Virgin Islands, and Trinidad may indicate rare but regular occurrence during the nonbreeding season in those regions. Based on high predicted occupancy and frequency of observations, the largest numbers of Golden-winged Warblers during the nonbreeding season likely occur at middle elevations (700–1,400 m) from Honduras to western Panama and possibly in the Andes of central Colombia.

In Mexico, Howell and Webb (1995) considered Golden-winged Warblers an uncommon to fairly common transient and visitor during boreal winter on the Atlantic slope of Chiapas, and rare elsewhere. Nonbreeding-season records in Mexico are clustered in the highlands of eastern Chiapas, but Golden-winged Warblers also regularly occur in the mountains of southern Veracruz and scattered across the wetter parts of the Yucatan Peninsula. In adjacent Guatemala, Land (1970) considered Golden-winged Warblers uncommon in the Caribbean lowlands, sea level to 1,750 m; most recent records are from the highlands of central Guatemala, with scattered additional records in the Petén Department in the north, in the Sierras of the Caribbean region,

and in the lower mountains and volcanoes closer to the Pacific Coast. However, recent surveys in Petén found very few individuals (R. E. Bennett, unpubl. data). Few historical records of Golden-winged Warblers during the nonbreeding season exist for Belize, and Jones (2003:201) described them as a "very uncommon winter visitor;" yet >50 recent eBird records as of December 2015 suggest that Golden-winged Warblers occur during the nonbreeding season throughout the country especially in the low, forested mountainous regions of Belize.

Honduras appears to support a considerable population of Golden-winged Warblers during the nonbreeding season, based on recent surveys and fieldwork. Monroe (1968) considered Golden-winged Warblers uncommon visitors during the nonbreeding season, from sea level to 1,800 m and most frequent on the Caribbean slope and interior highlands. The recent surveys at 503 sites during field seasons of 2010–2011 and 2011–2012 (Bennett 2012), plus additional recent eBird records, indicated that Golden-winged Warblers are widespread and fairly common on the northern Caribbean slope and at middle and upper elevations throughout the interior highlands. Golden-winged Warblers occupy a variety of forest types in Honduras, including humid broadleaf forest, mixed pine-broadleaf forest, and riparian forest corridors within highly disturbed landscapes (Bennett 2012). Golden-winged Warblers are generally absent from the Pacific slope below 1,000 m, and their status is uncertain in the eastern Caribbean lowlands (Mosquita Region).

In El Salvador, the first records of Golden-winged Warblers were from slopes of volcanoes near the Pacific Coast and in the northeast (Thurber et al. 1987), and recent eBird records come from those areas and the highlands near the Honduran border in northwestern El Salvador. It is likely that small numbers of Golden-winged Warblers occur during the nonbreeding season in highland regions throughout the country, although some records may be of migrants. Golden-winged Warblers are also common in the Central Highlands of northwestern Nicaragua, contiguous with the mountainous areas in Honduras; they were detected at 60% of 303 sites visited in that region during three recent field seasons (2009–2011). Martínez-Sánchez and Will (2010), summarizing records through 1993, considered Golden-winged

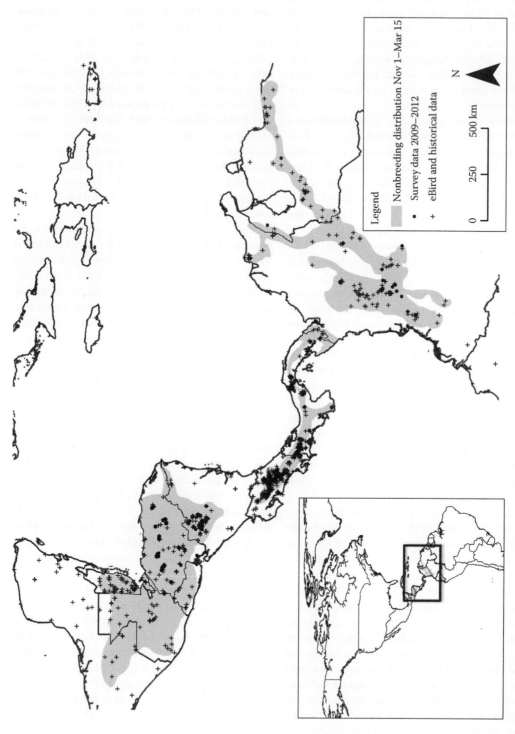

Figure 1.6. Nonbreeding distribution of Golden-winged Warblers in Central and northern South America. Dots represent specific locality records from historical specimens and other published records, standardized surveys conducted during 2009–2012, and additional 1 November to 15 March records from eBird.

Warblers only an uncommon transient and non-breeding-season resident in Nicaragua, with only scattered records from the rest of the country, primarily from volcanoes near the Pacific Coast and the San Juan River region near the border with Costa Rica. As in Honduras, it is uncertain whether Golden-winged Warblers occur during the nonbreeding season in the extensive Caribbean lowlands where no recent surveys have been conducted.

Another major area used by Golden-winged Warblers during the nonbreeding season is in a mid-elevation zone in the highlands of Costa Rica and western Panama. Golden-winged Warblers are most numerous from 700 to 1,400 m on the Caribbean slopes of the Guanacaste, Tilaran, and Central Cordilleras, on both slopes of the Talamanca Cordillera, and in the lower Fila Costeña in southwestern Costa Rica (Slud 1964, Stiles and Skutch 1989; eBird). Smaller numbers of Golden-winged Warblers occur during the non-breeding season regularly where forest remains in the lower foothills of the Caribbean slope, to near sea level in the southwestern Pacific lowlands from Carara to the Osa Peninsula, and possibly in the mountains of the Nicoya Peninsula. During recent surveys (2009–2012) at 486 sites, 184 Golden-winged Warblers were detected at roughly 30% of sites sampled, confirming continued presence at many areas previously used by Golden-winged Warblers during the nonbreeding season throughout Costa Rica although Golden-winged Warblers are apparently absent from the extensive Caribbean lowlands during the non-breeding season.

In Panama, Golden-winged Warblers are most numerous in the foothills and lower Pacific slopes in the westernmost province of Chiriquí (Wetmore et al. 1984, Ridgely and Gwynne 1989; eBird). A large region of potential habitat exists on the Caribbean slopes of the Talamanca Mountains in La Amistad National Park, but few ornithological records exist for this region. In central Panama, records of Golden-winged Warblers are concentrated in several areas, including the Pacific slopes of El Valle volcano, near sea level in the Canal Zone, and in the Cerro Azul-Cerro-Jefe region near Panama City. Golden-winged Warblers also occur regularly in eastern Panama in the mountains of Darien Province. Scattered additional historical and recent records in other parts of Panama suggest that Golden-winged

Warblers may be continuously distributed at lower and mid-elevations throughout the entire country, especially along the Cordillera Central and the San Blas Cordillera in eastern Panama. Recent surveys (2009–2012) at 340 sites in areas that appeared to be suitable for Golden-winged Warblers, however, resulted in only 32 detections (<10% of sites), mostly in Chiriquí and other areas with historical records of occurrence. Our recent surveys in Panama in 2015 revealed that female Golden-winged Warblers consistently occupy areas that receive less annual precipitation than males. Two sites were identified as important for female Golden-winged Warblers, despite the absence of males: the Parque Metropolitano in Panama City and the lowland secondary broadleaf forest around Meteti, Darien.

Most references for South America depict the nonbreeding distribution of Golden-winged Warblers as including much of western Colombia and northwestern Venezuela (Ridgely et al. 2005, Ridgely and Tudor 2009). Our compilation of historical and recent records, however, suggests that the current nonbreeding distribution in these countries is much more narrow and restricted, and may possibly be shrinking in area. In Colombia, Hilty and Brown (1986) described Golden-winged Warblers as fairly common in the western Andes and less numerous eastward, mostly in foothills and lower slopes (700–2,000 m), south to Cauca and Huila in the western Andes and to Meta on the east slope of the eastern Andes. The records we compiled are scattered along the slopes of the Andean mountain chains but are concentrated on slopes facing the interior valleys of the Magdalena and Cauca rivers (Figure 1.5). Fewer records are from farther south in the western Andes including a few south to Mindo, Ecuador, in the more isolated Serranía de San Lucas, Sierra Nevada de Santa Marta, and Sierra de Perijá in northern Colombia, and possibly in the hilly region of the western Chocó. There are few records of Golden-winged Warblers on the eastern slope of the eastern Andes. Many Colombian records are from October, November, and March, and it is difficult to distinguish records of resident birds from those of migrants. During recent surveys in three field seasons (2009–2012) at 421 sites in Colombia, Golden-winged Warblers were detected at only 35 sites (<10%), mostly in the central portions of the central and western Andes; many potentially suitable areas throughout Colombia were not surveyed.

In northwestern Venezuela, Golden-winged Warblers occur during the nonbreeding season in an arc at intermediate elevations (950–2,400 m) along the slopes of the Andes and the Coastal Cordillera, east at least to the Caracas area, in the Sierra de Perijá at the Colombian border, and Sierra de San Luis in the far north (Hilty 2003). A few individuals have been recorded farther east in the Coastal Cordillera (Sucre and Monagas) and in the Tepui region (Sierra de Lema). Most records for Venezuela are from historical specimens collected in the early 20th Century, but it is unclear whether a lack of modern records reflects a lack of recent reports in databases such as eBird or a change in nonbreeding distribution. Of 35 sites surveyed in Venezuela during our 2010–2011 field season, only three Golden-winged Warblers were detected at two sites. We do not have enough recent data, therefore, to make conclusions about the current status of Golden-winged Warblers in Venezuela during the nonbreeding season.

The possibility that Golden-winged Warblers may be relatively less common during the nonbreeding season in South America today than they were historically is intriguing. For a low-density, hard-to-detect species that occupies dense forest canopy during the nonbreeding season, Golden-winged Warblers collected in Central and South America are relatively well represented compared with other migrant warblers in major museum collections, based on general collecting activity in the early 20th Century. Our recent surveys were not extensive, but it was difficult to find Golden-winged Warblers even in areas with high predicted probability of occurrence. In contrast, Golden-winged Warblers may be more numerous than previously thought in the northern part of their nonbreeding distribution in Central America. For example, Monroe (1968) reported only five specimens collected during the period we defined as the nonbreeding season from Honduras, and Belize was not included in many previously published maps or descriptions of the Golden-winged Warblers nonbreeding distribution (American Ornithologists' Union 1983, DeGraaf and Rappole 1995, Birdlife International 2012). Even in Nicaragua, where >300 Golden-winged Warblers were detected on recent surveys, Martínez-Sánchez and Will (2010) cited few definitive records prior to 1990. Considering the extent of the nonbreeding distribution is much greater from west to east than from north to south, and

the breeding distribution has shifted toward the northwest, it is possible that a concomitant shift in the nonbreeding distribution also has occurred.

Distributions of Blue-winged and Golden-winged Warblers during the period we defined as the nonbreeding season did not overlap, with Blue-winged Warblers distributed farther north in northern Central America and southern Mexico (Gill et al. 2001). Substantial overlap of these two species during the nonbreeding season may be limited to portions of Guatemala, Belize, and Honduras (Land 1970, Confer 1992, Gill et al. 2001; eBird records); records of Blue-winged Warblers are rare in Nicaragua (Martínez-Sánchez and Will 2010), Costa Rica (Stiles and Skutch 1989), and Panama (Ridgely and Gwynne 1989), and there are only two records from South America in northern Colombia and northern Venezuela. Golden-winged × Blue-winged Warbler hybrids have been found throughout the Central American nonbreeding distribution of both species, and there are a few records of hybrids from Colombia and Venezuela.

Migration Routes and Timing

Little is known about the migration of Golden-winged Warblers, especially south of the U.S. We examined 1,244 observation and specimen records from south of the U.S. and outside the period we defined as the nonbreeding season (1 November to 15 March). The locations of Golden-winged Warbler eBird records during migration suggest a looped, trans-Gulf of Mexico migration, with spring migration occurring farther west than fall migration. Elliptical migration is a common pattern observed in long-distance migrants breeding in eastern North America (La Sorte et al. 2014). In spring, most individuals cross the western Gulf of Mexico to Texas and Louisiana (or possibly farther north), arriving mostly in late April and early May (earliest, 10 April). The migration period appears to be short, with the latest spring records of Golden-winged Warblers in Nicaragua (17 April), Colombia (19 April), Guatemala (24 April), Honduras (25 April), and Costa Rica (1 May) in late April and early May. Spring migrants have been reported from late March to early May along the Caribbean coast (including offshore islands) from western Panama to Belize and southeastern Mexico (but not on Yucatan Peninsula), indicating that some Golden-winged Warblers may move northward along the Central American coast, or may cross the western Caribbean Sea before crossing the Gulf of Mexico to the U.S. A paucity of spring records from Florida or anywhere in the southeastern Coastal Plain of the U.S. further suggests that some Golden-winged Warblers may fly directly from Central America to breeding grounds in the southern Appalachian Mountains.

Fall records (August–November) of Golden-winged Warblers in eBird also suggest a trans-Gulf of Mexico migration, but farther east, with a much greater concentration of records in Florida in early fall (late August through September) than in spring, but few records from along the Gulf of Mexico coast in Louisiana, Texas, or Mexico. Some fall migrants also likely cross the Caribbean Sea, as evidenced by records from offshore islands along the Central American coast and occasionally from Caribbean islands; a few records from Bermuda suggest that some individuals even may attempt to cross the western Atlantic Ocean to reach South America. Migrant Golden-winged Warblers have been detected in Yucatan and Belize in late August, and the first individuals arrive in Honduras and Nicaragua during the first week of September, and in Costa Rica, Panama, Colombia, and Venezuela by mid-September.

Implications for Conservation

The breeding distribution of Golden-winged Warblers has been highly dynamic, exhibiting among the largest shifts of any North American songbird. Our review of the historical breeding distribution suggests a continual shift northward by Golden-winged Warblers over the past 100 years in the Upper Midwest, but a geographically stable distribution in the Appalachian Mountains, despite steeply declining populations within the Appalachian Mountains breeding-distribution segment. This historical context suggests two different but complementary conservation strategies. Species population stability may best be addressed through land management aimed at creating dynamic forested landscapes (A. M. Roth et al., unpubl. plan) to support source populations in the Great Lakes breeding-distribution segment. At the same time, preservation of the remaining historical breeding distribution and potential genetic variability might best be addressed by focusing on maintenance of Golden-winged

Warbler populations in the Appalachian Mountains breeding-distribution segment.

The three northern states within the Great Lakes breeding-distribution segment currently harbor an estimated 76% of the total global population of breeding Golden-winged Warblers, with 5%, 24%, and 47% of the estimated population in Michigan, Wisconsin, and Minnesota, respectively (Blancher et al. 2013). These three northern states therefore hold the greatest stewardship responsibility for future conservation of Golden-winged Warblers. Wisconsin currently lists Golden-winged Warblers as a Species of Special Concern, and in Minnesota, Golden-winged Warblers were designated a species of greatest conservation need in 2013 based on their high stewardship value (D. Carlson, pers. comm.).

In contrast, the Appalachian Mountains breeding-distribution segment currently includes only 5% of the estimated global population of Golden-winged Warblers. The Appalachian Mountains represent a considerable portion of the historical breeding distribution, and include areas where Golden-winged Warblers remain geographically isolated from Blue-winged Warblers. Golden-winged Warbler populations in the Appalachian Mountains may also be important in terms of genetic diversity in the face of climate change, because these populations are presumably better adapted to warm climates (Hampe and Petit 2005). Within this region, Golden-winged Warblers are listed as endangered, threatened, or special concern by at least six states including Connecticut, New York, New Jersey, Kentucky, Tennessee, and Georgia, but do not have special protection status in the two states with the largest remaining populations: Pennsylvania and West Virginia. Overall, the traditional conservation paradigm of directing resources primarily to jurisdictions where a species is endangered makes it more difficult to focus conservation where larger populations provide greater potential for management success (Wells et al. 2010). The Appalachian Mountains breeding-distribution segment of Golden-winged Warblers has reached levels where BBS is becoming ineffective in tracking changes in distribution and relative abundance, and a new, targeted monitoring approach will be necessary there, such as the spatially balanced sampling design implemented by R. W. Rohrbaugh et al. (unpubl. report).

As with many Neotropical–Nearctic migratory birds, the extent of the nonbreeding distribution is considerably smaller than during the breeding season, and loss of nonbreeding habitat could be an important factor limiting Golden-winged Warbler populations. Within the nonbreeding distribution of Golden-winged Warblers, conservation efforts are likely to be most effective by focusing on the retention of forested landscapes at intermediate elevations in Central American and the northern Andes (Chapter 11, this volume). The probable shift in the nonbreeding distribution to the north and west has important implications for the conservation of Golden-winged Warblers in northern Central America, especially Guatemala, Honduras, and Nicaragua. Conservation of forested, middle-elevation landscapes in these countries will be key to the successful retention of habitat used during the nonbreeding season for a large portion of the Golden-winged Warbler population (Chapter 2, this volume). If Golden-winged Warblers are becoming more numerous at the northern end of their nonbreeding distribution, interactions between Golden-winged and Blue-winged Warblers during the nonbreeding season also may play an increasing role in determining species interactions and hybridization during the breeding season. The recently completed *Wintering Grounds Conservation Plan for Golden-winged Warbler* presents a strategy for prioritized conservation action in each country within the nonbreeding distribution (R. E. Bennett et al., unpubl. report).

The complexity of changes in the breeding distribution, the genetic and ecological interactions with Blue-winged Warblers, and the apparent shifts in distribution and abundance during the nonbreeding season collectively pose unique challenges for the full-life-cycle conservation of Golden-winged Warblers. An aggressive and collaborative conservation strategy, incorporating the latest research results and management guidelines, will be necessary to reverse population declines and maintain healthy numbers of Golden-winged Warblers throughout their present-day distribution.

ACKNOWLEDGMENTS

Much of the information in this chapter was adapted from the Golden-winged Warbler Status Review and Conservation Plan (A. M. Roth et al., unpubl. plan; available at www.gwwa.org), developed and reviewed under the guidance of the Golden-winged Warbler Working Group. Distribution information from the GOWAP presented in this manuscript comes from S. B. Swarthout et al. (unpubl. report), which is available from the authors. We are grateful to the many individuals who contributed significant information on current and historical distribution or conducted GOWAP surveys, as follows: R. J. Adams, K. Aldinger, T. W. Arnold, C. Artuso, R. Baker, M. H. Bakermans, J. C. Bednarz, P. J. Blancher, B. A. Bogaczyk, D. W. Brauning, G. L. Brewer, W. L. Brininger, G. Bryan, L. P. Bulluck, I. Butler, J. S. Castrale, J. E. Cely, J. L. Confer, J. Cornutt, K. Corwin, D. Crawford, C. Croy, A. Darwin, P. J. Delphey, D. W. Demarest, R. Dettmers, C. Dobony, L. Dunn, M. V. d' Entremont, G. Falardeau, R. L. Ferren, J. Fitzgerald, C. Friis, F. B. Gill, K. Hall, J. M. Hanowski, S. Harding, E. Haverlack, T. P. Hodgman, R. Horton, D. Howell, J. Hughes, P. Hunt, D. A. James, M. Johns, A. Jones, S. L. Jones, T. Jones, S. Kearney-McGee, J. Keith, D. I. King, N. A. Klaus, M. Knutson, J. E. Kubel, J. L. Larkin, H. LeGrand Jr., S. Lewis, K. J. Martin, R. McCollum, T. J. Mersmann, L. Moulton, M. D. Nelson, C. P. Nicholson, L. Osterndorf, J. Overcash, S. Parren, L. L. Patton, A. Paulios, B. G. Peterjohn, S. Peterson, S. Petzinger, T. Post, J. T. Price, A. Prince, C. Raithel, R. B. Renfrew, C. C. Rimmer, P. G. Rodewald, M. Roedel, F. Shaffer, L. Shapiro, M. Shumar, C. Smalling, S. Somershoe, K. St. Laurent, H. M. Streby, S. Stucker, C. Thomas, M. Timpf, E. Travis, R. Vallender, S. Van Wilgenburg, W. Vanderschuit, S. Vorisek, S. Warner, M. Welton, J. Wentworth, S. P. Wilds, D. Willard, P. B. Wood, and R. M. Zink.

Surveys of Golden-winged Warblers during the nonbreeding season were coordinated by the American Bird Conservancy and Fundacion Proaves (Colombia) and conducted by M. Arteaga, L. Chavarría-Duriaux, J. Camargo, E. Campos, P. Elizondo, S. Morales, M. Escaño, M. I. Moreno, and A. Naveda. D. I. King made substantial contributions to the design and implementation of surveys during the nonbreeding season. D. I. King and B. Bailey graciously provided data from Honduras, and M. I. Moreno, J. Carlos Verhost, and A. W. Rothman helped to compile and verify nonbreeding-season data. Specimen data referenced in this study were obtained from the Academy of Natural Sciences, American Museum of Natural History, California Academy of Sciences, Canadian Museum of Nature, Cornell University Museum of Vertebrates, Delaware Museum of Natural History, Field Museum of

Natural History, Instituto de Ciencias Naturales National University of Colombia, Los Angeles County Museum of Natural History, Louisiana State University Museum of Zoology, Museum of Comparative Zoology (Harvard University), Moore Laboratory of Zoology, Museum of Vertebrate Zoology, Royal Ontario Museum, UCLA Dickey Collection, University of Michigan Museum of Zoology, National Museum of Natural History, Western Foundation of Vertebrate Zoology, and Yale University Peabody Museum.

Funding for the Golden-winged Warbler Rangewide Conservation Initiative was provided by the National Fish and Wildlife Foundation and U.S. Fish and Wildlife Service, with more than $1 million in matching contributions provided by numerous partner organizations including American Bird Conservancy, Appalachian Mountains Joint Venture, Audubon North Carolina, Cornell Lab of Ornithology, Fundacion Proaves-Colombia, Indiana University of Pennsylvania, Ithaca College, Michigan Technological University, University of Minnesota, University of Tennessee, Wisconsin Department of Natural Resources, Tennessee Wildlife Resources Agency, and the Ruffed Grouse Society. Any use of trade, product, or firm names are for descriptive purposes only and do not imply endorsement by the U.S. Government.

LITERATURE CITED

American Ornithologists' Union. 1983. Check-list of North American birds (6th ed.). American Ornithologists' Union, Washington, DC.

Andrle, R. F., and J. R. Carroll. 1988. The atlas of breeding birds in New York State. Cornell University Press, Ithaca, NY.

Artuso, C. 2008. Golden-winged Warbler surveys in Manitoba. Birdwatch Canada 45:8–11.

Bailey, W. 1955. Birds in Massachusetts: when and where to find them. College Press, Boston, MA.

Barker Swarthout, S., K. V. Rosenberg, T. C. Will, and M. I. Moreno. 2008. A collaborative web-based recording program for housing records of migratory birds during non-breeding periods in Central and South America. Ornitología Neotropical 19(Suppl.):531–539.

Barrows, W. B. 1912. Michigan bird life. Michigan Agricultural College Special Bulletin, East Lansing, MI.

Batchelder, D. G. 1890. Helminthophila chrysoptera [Golden-winged Warbler] in Manitoba. Auk 7:404.

Bennett, R. E. 2012. Habitat associations of the Golden-winged Warbler in Honduras. M.S. thesis, Michigan Technological University, Houghton, MI.

Bent, A. C. 1953. Life histories of North American wood warblers. Dover Publications, Inc., New York, NY.

Berger, A. J. 1958. The Golden-winged-Blue-winged Warbler complex in Michigan and the Great Lakes area. Jack-Pine Warbler 36:37–72.

Berlanga, H., J. A. Kennedy, T. A. Rich, M. C. Arizmendi, C. J. Beardmore, P. J. Blancher, G. S. Butcher, A. R. Couturier, A. A. Dayer, D. W. Demarest, W. E. Easton, M. Gustafson, E. Inigo-Elias, E. A. Krebs, A. O. Panjabi, V. Rodriguez Contreras, K. V. Rosenberg, J. M. Ruth, E. Santana Castellon, R. M. Vidal, and T. Will. 2010. Saving our shared birds: Partners in Flight tri-national vision for landbird conservation. Cornell Lab of Ornithology, Ithaca, NY.

Bevier, D. R. 1994. The atlas of breeding birds of Connecticut. State Geological and Natural History Survey of Connecticut. Hartford, CT.

BirdLife International. [online]. 2012. Vermivora chrysoptera. In IUCN 2013. IUCN red list of threatened species. Version 2013.1. <www.iucnredlist.org> (3 October 2013).

Blancher, P. J., K. V. Rosenberg, A. O. Panjabi, R. Altman, A. R. Couturier, W. E. Thogmartin, and the Partners in Flight Science Committee. [online]. 2013. Partners in Flight population estimates database, version: 2.0. PIF Technical Series No. 6. <www.partnersinflight.org/pubs/23February 2014/>.

Bohlen, H. D. 1989. The birds of Illinois. Indiana University Press, Bloomington, IN.

Brauning, D. W. 1992. Atlas of breeding birds of Pennsylvania. University of Pittsburgh Press, Pittsburgh, PA.

Brewer, R., G. A. McPeek, and R. J. Adams Jr. 1991. The atlas of breeding birds of Michigan. Michigan State University Press, East Lansing, MI.

Brewster, W. 1885. William Brewster's exploration of the southern Appalachian Mountains: the journal of 1885. North Carolina Historical Review 57:43–77.

Brewster, W. 1886. An ornithological reconnaissance of western North Carolina. Auk 3:94–113, 173–179.

Brewster, W. 1906. The birds of the Cambridge region of Massachusetts. Nuttall Ornithological Club, Cambridge, MA.

Buckelew, A. R. Jr., and G. A. Hall. 1994. The West Virginia breeding bird atlas. University of Pittsburgh Press, Pittsburgh, PA.

Buehler, D. A., A. M. Roth, R. Vallender, T. C. Will, J. L. Confer, R. A. Canterbury, S. B. Swarthout, K. V. Rosenberg, and L. P. Bulluck. 2007. Status and conservation priorities of Golden-winged Warbler (*Vermivora chrysoptera*) in North America. Auk 124:1439–1445.

Burleigh, T. D. 1958. Georgia birds. University of Oklahoma Press, Norman, OK.

Butler, A. W. 1897. The birds of Indiana. Pp. 515–1187 in 22nd Annual Report of the Department of Geology and Natural Resources in Indiana, Indianapolis, IN.

Cadman, M. D., D. A. Sutherland, G. G. Beck, D. Lepage, and A. R. Couturier (editors). 2007. Atlas of the breeding birds of Ontario, 2001–2005. Bird Studies Canada, Environment Canada, Ontario Field Ornithologists, Ontario Ministry of Natural Resources, and Ontario Nature, Toronto, ON.

Canterbury, R. A. 1997. Population ecology of Blue-winged Warblers in West Virginia. Proceedings of the West Virginia Academy of Science 69:53–60.

Canterbury, R. A., and D. M. Stover. 1999. The Golden-winged Warbler: an imperiled migrant songbird of the southern West Virginia coalfields. Green Lands 29:44–51.

Canterbury, R. A., D. M. Stover, and T. C. Nelson. 1993. Golden-winged Warblers in southern West Virginia: status and population ecology. Redstart 60:97–106.

Chandler, R. B., and D. I. King. 2011. Habitat quality and habitat selection of Golden-winged Warblers in Costa Rica: an application of hierarchical models for open populations. Journal of Applied Ecology 48:1038–1047.

Chandler, R. B., J. A. Royle, and D. I. King. 2011. Inference about density and temporary emigration in unmarked populations. Ecology 92:1429–1435.

Confer, J. L. 1992. Golden-winged Warbler (*Vermivora chrysoptera*). Pp. 1–16 in A. Poole and F. Gill (editors), The birds of North America, No. 20. The Birds of North America, Inc., Philadelphia, PA.

Confer, J. L. 2005. Secondary contact and introgression of Golden-winged Warblers (*Vermivora chrysoptera*): documenting the mechanism. Auk 123:958–961.

Confer, J. L., D. Coker, M. Armstrong, and J. Doherty. 1991. The rapidly changing distribution of the Golden-winged Warbler (*Vermivora chrysoptera*) in central New York. Kingbird 41:5–11.

Confer, J. L., P. Hartman, and A. Roth. 2011. Golden-winged Warbler (*Vermivora chrysoptera*). In A. Poole (editor), The birds of North America online. Cornell Lab of Ornithology, Ithaca, NY.

Confer, J. L., and J. L. Larkin. 1998. Behavioral interactions between Golden-winged and Blue-winged Warblers. Auk 115:209–213.

Confer, J. L., and S. K. Tupper. 2000. A reassessment of the status of Golden-winged and Blue-winged Warblers in the Hudson Highlands of southern New York. Wilson Bulletin 112:544–546.

COSEWIC. [online]. 2011. Canadian wildlife species at risk. Committee on the Status of Endangered Wildlife in Canada, Gatineau, QC. <http://www.cosewic.gc.ca/eng/sct0/rpt/rpt_csar_e.cfm> (17 October 2011).

Cutright, N. J., B. R. Harriman, and R. W. Howe (editors). 2006. Atlas of the breeding birds of Wisconsin. Wisconsin Society for Ornithology, Inc., Madison, WI.

Dabrowski, A., R. Fraser, J. L. Confer, and I. J. Lovette. 2005. Geographic variability in mitochondrial introgression among hybridizing populations of Golden-winged (*Vermivora chrysoptera*) and Blue-winged (*V. pinus*) Warblers. Conservation Genetics 6:843–853.

DeGraaf, R. M., and J. H. Rappole. 1995. Neotropical migratory birds: natural history, distribution, and population change. Cornell University Press, Ithaca, NY.

DeLorme Atlas Gazetteers. [online]. 2014. Innovative Earthmate GPS, Mapping Software, GIS Solutions, and Data. DeLorme, Yarmouth, ME. <http://www.delorme.com/>.

Eaton, E. H. 1914. Birds of New York, Part 2. New York State Museum Memoir 12. The University of the State of New York, Albany, NY.

Eifrig, G. 1904. Birds of Allegany and Garrett counties, western Maryland. Auk 21:234–250.

Ellison, W. G. 2010. 2nd Atlas of the breeding birds of Maryland and the District of Colombia. The John Hopkins University Press, Baltimore, MD.

Ficken, M. S., and R. W. Ficken. 1968. Reproductive isolating mechanisms in the Blue-winged Warbler-Golden-winged Warbler complex. Evolution 22:166–179.

Fiske, I., and R. B. Chandler. 2011. Unmarked: an R package for fitting hierarchical models of wildlife occurrence and abundance. Journal of Statistical Software 43:1–23.

Foss, C. R. 1994. Atlas of breeding birds in New Hampshire. Audubon Society of New Hampshire, Concord, NH.

Frech, M. H., and J. L. Confer. 1987. The Golden-winged Warbler: competition with the Blue-winged Warbler and habitat selection in portions of southern, central and northern New York. Kingbird 17:65–71.

Gauthier, J., and Y. Aubry (editors). 1996. The breeding birds of Québec: atlas of the breeding birds of southern Québec. Association Québécoise des

Groups d'Ornithologues, Province of Quebec Society for the Protection of Birds, Canadian Wildlife Service, Environment Canada, Québec Region, Montréal, QC.

Gill, F. B. 1980. Historical aspects of hybridization between Blue-winged and Golden-winged Warblers. Auk 97:1–18.

Gill, F. B. 1997. Local cytonuclear extinction of the Golden-winged Warbler. Evolution 51:519–525.

Gill, F. B., R. A. Canterbury, and J. L. Confer. 2001. Blue-winged Warbler (*Vermivora cyanoptera*). In A. Poole (editor), The birds of North America online. Cornell Lab of Ornithology, Ithaca, NY.

Graber, R. R., and J. W. Graber. 1963. A comparative study of bird populations in Illinois, 1906–1909 and 1956–1958. Illinois Natural History Survey Bulletin 28:383–519.

Hampe, A., and R. J. Petit. 2005. Conserving biodiversity under climate change: the rear edge matters. Ecology Letters 8:461–467.

Hilty, S. L. 2003. Birds of Venezuela (2nd ed.). Princeton University Press, Princeton, NJ.

Hilty, S. L., and W. L. Brown. 1986. A guide to the birds of Colombia. Princeton University Press, Princeton, NJ.

Howell, N. G., and S. Webb. 1995. A guide to the birds of Mexico and northern Central America. Oxford University Press, Oxford, UK.

Hunter, W. C., D. A. Buehler, R. A. Canterbury, J. L. Confer, and P. B. Hamel. 2001. Conservation of disturbance-dependent birds in eastern North America. Wildlife Society Bulletin 29:440–455.

Janssen, R. B. 1987. Birds in Minnesota. University of Minnesota Press, Minneapolis, MN.

Jones, H. L. 2003. Birds of Belize. University of Texas Press, Austin, TX.

Kent, T. H., and J. J. Dinsmore. 1996. Birds in Iowa. Published by the authors, Iowa City, IA.

Klaus, N. A. 1999. Effects of forest management on songbird habitat on the Cherokee National Forest, with special emphasis on Golden-winged Warblers. M.S. thesis, University of Tennessee, Knoxville, TN.

Klaus, N. A. 2004. Status of the Golden-winged Warbler in north Georgia, and a nesting record of the Lawrence's Warbler. Oriole 69:1–7.

Kleen, V. M., L. Cordle, and R. A. Montgomery. 2004. The Illinois breeding bird atlas. Illinois Natural History Survey Special Publication No. 26. Illinois Natural History Survey, Champaign, IL.

Koes, R. 2003. The birds of Manitoba. Manitoba Naturalist's Society, Winnipeg, MB.

Land, H. C. 1970. Birds of Guatemala (1st ed.). Livingston Publishing Company, Wynnewood, PA.

La Sorte, F. A., S. H. M. Butchart, W. Jetz, and K. Böhning-Gaese. 2014. Range-wide latitudinal and elevational temperature gradients for the world's terrestrial birds: implications under global climate change. PLoS One 9:e98361.

Laughlin, S. B., and D. P. Kibbe. 1985. The atlas of breeding birds of Vermont. University Press of New England, Hanover, NH.

Loomis, L. M. 1890. Observations of some of the summer birds on the mountain portion of Pickens County, South Carolina. Auk 7:30–39, 124–130.

Loomis, L. M. 1891. June birds of Caesar' Head, South Carolina. Auk 8:322–333.

Martínez-Sánchez, J. C., and T. Will. 2010. Tom Howell's checklist of the birds of Nicaragua as of 1993. Ornithological Monographs 68:1–108.

McCracken, J. D. 1994. Golden-winged and Blue-winged Warblers: their history and future in Ontario. In Ornithology in Ontario. Ontario Field Ornithologists, Hawk Owl Publishing, Burlington, ON.

McGowan, K. J., and K. Corwin (editors). 2008. The second atlas of breeding birds in New York state. Cornell University Press, Ithaca, NY.

McWilliams, G., and D. Brauning. 2000. The birds of Pennsylvania. Cornell University Press, Ithaca, NY.

Mengel, R. M. 1965. The birds of Kentucky. American Ornithologists' Union Monograph No. 3. Allen Press, Lawrence, KS.

Monroe, B. L. Jr. 1968. A distributional survey of the birds of Honduras. American Ornithologists' Union Monograph No. 7. Allen Press, Lawrence, KS.

Mumford, R. E., and C. E. Keller. 1984. The birds of Indiana. Indiana University Press, Bloomington, IN.

Nicholson, C. P. 1997. Atlas of the breeding birds of Tennessee. University of Tennessee Press, Knoxville, TN.

Parkes, K. C. 1951. The genetics of the Golden-winged × Blue-winged Warbler complex. Wilson Bulletin 63:5–15.

Partners in Flight Science Committee. [online]. 2013. Population estimates database, version 2013. <http://rmbo.org/pifpopestimates> (23 February 2014).

Patton, L. L., D. S. Maehr, J. E. Duchamp, S. Fei, J. W. Gassett, and J. L. Larkin. 2010. Do the Golden-winged Warbler and Blue-winged Warbler exhibit species-specific differences in their breeding habitat use? Avian Conservation and Ecology 5:2.

Payne, R. B. [online]. 2011. Golden-winged Warbler (*Vermivora chrysoptera*). *In* A. T. Chartier, J. J. Baldy, and J. M. Brenneman (editors), The second Michigan breeding bird atlas, 2002–2008. Kalamazoo Nature Center, Kalamazoo, MI. <www.MIBirdAtlas.org> (23 February 2014).

Peck, G. K., and R. D. James. 1983. Breeding birds of Ontario, nidiology and distribution. Passerines (vol. 2). The Royal Ontario Museum, Toronto, ON.

Percy, K. L. 2012. Effects of prescribed fire and habitat on Golden-winged Warbler (*Vermivora chrysoptera*) abundance and nest survival in the Cumberland Mountains of Tennessee. M.S. thesis, University of Tennessee, Knoxville, TN.

Peterjohn, B. G. 1989. The birds of Ohio. Indiana University Press, Bloomington, IN.

Peterjohn, B. G., and D. L. Rice. 1991. The Ohio breeding bird atlas. Ohio Department of Natural Resources, Columbus, OH.

Petersen, W. R., and W. R. Meservey. 2003. Massachusetts breeding bird atlas. Massachusetts Audubon Society and University of Massachusetts Press, Amherst, MA.

Phillips, S. J., R. P. Anderson, and R. E. Schapiro. 2006. Maximum entropy modeling of species geographic distributions. Ecological Modelling 190:231–259.

Post, W., and S. A. Gauthreaux Jr. 1989. Status and distribution of South Carolina birds. The Charleston Museum, Charleston, SC.

Rich, T. D., C. J. Beardmore, H. Berlanga, P. J. Blancher, M. S. W. Bradstreet, G. S. Butcher, D. W. Demarest, E. H. Dunn, W. C. Hunter, W. E. Iñigo-Elias, J. A. Kennedy, A. M. Martell, A. O. Panjabi, D. N, Pashley, K. V. Rosenberg, C. M. Rustay, J. S. Wendt, and T. C. Will. 2004. Partners in Flight North American Landbird Conservation Plan. Cornell Lab of Ornithology, Ithaca, NY.

Ridgely, R. S., T. F. Allnutt, T. Brooks, D. K. McNicol, D. W. Mehlman, B. E. Young, and J. R. Zook. 2005. Digital distribution maps of the birds of the Western Hemisphere, version 2.1. NatureServe, Arlington, VA.

Ridgely, R. S., and J. A. Gwynne Jr. 1989. A guide to the birds of Panama (2nd ed.). Princeton University Press, Princeton, NJ.

Ridgely, R. S., and G. Tudor. 2009. Field guide to the songbirds of South America. The passerines. University of Texas Press, Austin, TX.

Ridgway, R. 1889. The ornithology of Illinois, Part 1 (vol. 1). State Laboratory of Natural History, Springfield, IL.

Robbins, C. S., and E. A. T. Blom. 1996. Atlas of breeding birds of Maryland and the District of Columbia. University of Pittsburgh Press, Pittsburgh, PA.

Robbins, M. B., and D. A. Easterla. 1992. Birds of Missouri: their distribution and abundance. University of Missouri Press, Columbia, MO.

Robbins, S. D. Jr. 1991. Wisconsin birdlife: population and distribution—past and present. University of Wisconsin Press, Madison, WI.

Roberts, T. R. 1932. The birds of Minnesota (vol. 2). University of Minnesota Press, Minneapolis, MN.

Robinson, W. D. 1996. Southern Illinois birds. An annotated list and site guide. Southern Illinois University Press, Carbondale, IL.

Rosenberg, K. V., and P. J. Blancher. 2005. Setting numerical population objectives for priority landbird species. Pp. 57–67 *in* C. J. Ralph, and T. D. Rich (editors), Proceedings of the Third International Partners in Flight Conference. USDA Forest Service General Technical Report PSW-GTR-191. USDA Forest Service, Albany, CA.

Sauer, J. R., J. E. Hines, J. E. Fallon, K. L. Pardieck, D. J. Ziolkowski Jr., and W. A. Link. 2011. The North American Breeding Bird Survey, results and analysis 1966–2009. Version 3.23.2011. USGS Patuxent Wildlife Research Center, Laurel, MD.

Sauer, J. R., J. E. Hines, J. E. Fallon, K. L. Pardieck, D. J. Ziolkowski Jr., and W. A. Link. 2012. The North American Breeding Bird Survey, results and analysis 1966–2010. Version 12.07.2011. USGS Patuxent Wildlife Research Center, Laurel, MD.

Sauer, J. R., and W. A. Link. 2011. Analysis of the North American Breeding Bird Survey using hierarchical models. Auk 128:87–98.

Scully, R. L. 1997. A field study of the Golden-winged Warbler in the Pequannock watershed. Proceedings of the Linnaean Society of New York 1977–1995:25–39.

Shapiro, L. H., R. A. Canterbury, D. M. Stover, and R. C. Fleischer. 2004. Reciprocal introgression between Golden-winged Warblers (*Vermivora chrysoptera*) and Blue-winged Warblers (*V. pinus*) in eastern North America. Auk 121:1019–1030.

Short, L. L. Jr. 1963. Hybridization in the wood warblers *Vermivora pinus* and *V. chrysoptera*. Proceedings of the XIII International Ornithological Congress 13:147–160.

Slud, P. 1964. The birds of Costa Rica: distribution and ecology. Bulletin of the American Museum of Natural History 128:5–430.

Smith, A. R. 1996. Golden-winged Warbler. P. 301 *in* Atlas of Saskatchewan birds. Saskatchewan Natural History Society, Nature Saskatchewan, Regina, SK.

Speirs, J. M. 1985. Birds of Ontario (vol. 2). Natural Heritage, Toronto, ON.

Sprunt, A. Jr., and E. B. Chamberlain. 1949. South Carolina bird life. University of South Carolina Press, Columbia, SC.

Stearns, W. A., and E. Coues. 1893. New England bird life, Part I, Oscines (3rd ed. Rev.). Lee and Shepard Publishers, Boston, MA.

Stiles, F. G., and A. F. Skutch. 1989. A guide to the birds of Costa Rica. Cornell University Press, Ithaca, NY.

Stupka, A. 1963. Notes on the birds of Great Smoky Mountains National Park. University of Tennessee Press, Knoxville, TN.

Sullivan, B. L., J. L. Aycrigg, J. H. Barry, R. E. Bonney, N. Bruns, C. B. Cooper, T. Damoulas, A. A. Dhondt, T. Dietterich, A. Farnsworth, D. Fink, J. W. Fitzpatrick, T. Fredericks, J. Gerbracht, C. Gomes, W. M. Hochachka, M. J. Iliff, C. Lagoze, F. A. La Sorte, M. Merrifield, W. Morris, T. B. Phillips, M. Reynolds, A. D. Rodewald, K. V. Rosenberg, N. M. Trautmann, A. Wiggins, D. W. Winkler, W. K. Wong, C. L. Wood, J. Yu, and S. Kelling. 2014. The eBird enterprise: an integrated approach to development and application of citizen science. Biological Conservation 169:31–40.

Sullivan, B. L., C. L. Wood, M. J. Iliff, R. E. Bonney, D. Fink, and S. Kelling. 2009. eBird: a citizen-based bird observation network in the biological sciences. Biological Conservation 142:2282–2292.

Suomala, R. W. 2005. Golden-winged Warbler (*Vermivora chrysoptera*). Pp. A437–A441 in E. Nedeau (editor), New Hampshire wildlife action plan. New Hampshire Fish and Game Department, Concord, NH.

Taylor, P. (editor). 2003. The birds of Manitoba. Manitoba Naturalist's Society, Winnipeg, MB.

Thogmartin, W. E., F. P. Howe, F. C. James, D. H. Johnson, E. Reed, J. R. Sauer, and F. R. Thompson III. 2006. A review of the population estimation approach of the North American landbird conservation plan. Auk 123:892–904.

Thurber, W. A., J. F. Serrano, A. Sermeño, and M. Benítez. 1987. Status of uncommon and previously unreported birds of El Salvador. Proceedings of the Western Foundation of Vertebrate Zoology 3:109–293.

U.S. Fish and Wildlife Service. 2008. Birds of conservation concern. U.S. Department of Interior, Fish and Wildlife Service, Division of Migratory Bird Management, Arlington, VA. <http://www.fws.gov/migratorybirds/> (23 February 2014).

U.S. Fish and Wildlife Service. 2011. 90-day finding on a petition to list the Golden-Winged Warbler as endangered or threatened. Federal Register 76:31920–31926.

Vallender, R., S. L. Van Wilgenburg, L. P. Bulluck, A. Roth, R. Canterbury, J. Larkin, R. M. Fowlds, and I. J. Lovette. 2009. Extensive rangewide mitochondrial introgression indicates substantial cryptic hybridization in the Golden-winged Warbler. Avian Conservation and Ecology 4:4.

Veit, R. R., and W. R. Petersen. 1993. Birds of Massachusetts. Massachusetts Audubon Society, Lincoln, MA.

Wells, J. V., B. Robertson, K. V. Rosenberg, and D. W. Mehlman. 2010. Global versus local conservation focus of U.S. state agency endangered bird species lists. PLoS One 5:e8608.

Welton, M. 2003. Status and distribution of the Golden-winged Warbler in Tennessee. Migrant 74:61–82.

Wetmore, A., R. F. Pasquier, and S. L. Olson. 1984. The birds of the Republic of Panama. Part 4. Passeriformes: *Hirundinidae* (swallows) to *Fringillidae* (finches). Smithsonian Institution, Washington, DC.

Wilson, A. M., D. W. Brauning, and R. S. Mulvihill (editors). 2012. Second atlas of breeding birds in Pennsylvania. Penn State Press, University Park, PA.

Wilson, M. D., B. D. Watts, M. G. Smith, J. P. Bredleau, and L. W. Seal. 2007. Status assessment of Golden-winged Warblers and Bewick's Wrens in Virginia. CCBTR-07-02. Center for Conservation Biology Technical Report Series, College of William and Mary, Williamsburg, VA.

Wood, N. A. 1951. The birds of Michigan. Miscellaneous Publication 75. University of Michigan Museum of Zoology, Ann Arbor, MI.

CHAPTER TWO

Nonbreeding Golden-winged Warbler Habitat*

STATUS, CONSERVATION, AND NEEDS

David I. King, Richard B. Chandler, Curtis Smalling,
Richard Trubey, Raul Raudales, and Tom Will

Abstract. Anecdotal reports and more recent quantitative findings suggest Golden-winged Warblers (*Vermivora chrysoptera*) are most abundant in mid-elevation moist forests of Central America during the nonbreeding season. The species appears to be tolerant of moderate levels of disturbance, inhabiting both primary and secondary forest; however, occupation of agricultural cover types such as shade coffee may be contingent on the presence of adjacent forest. Trends in deforestation in Latin America offer discouraging prospects for the future of habitat for Golden-winged Warblers in the region in the short term. Nevertheless, recent innovations in agroforestry practices offer market-based tools for restoring and maintaining forest for nonbreeding warblers. One example is hybrid solar-biomass coffee driers that eliminate the use of fuelwood for drying coffee and lower the costs of coffee drying by over 80%. Currently, the equivalent of 6,500 ha of forest is harvested annually in Latin America to fuel coffee driers. Another example is Integrated Open Canopy (IOC) coffee, where coffee is grown with sparse or no shade adjacent to forest patches of equivalent or greater size. In addition to promoting the conservation of forest habitat required by Golden-winged Warblers and other species, IOC increases income to farmers by increasing yields. Increased income to farmers is important because alternative agroforestry systems provide a market-based incentive for forest conservation. Future work will be directed at implementing these market-based forest conservation strategies over large areas using co-management agreements as a framework for enhancing communication, cooperation, and policy to decrease rural poverty and the pressure on forest resources for the benefit of both humans and birds alike.

Key Words: agroforestry, coffee, co-management, market-based, renewable energy, *Vermivora chrysoptera*.

* King, D. I., R. B. Chandler, C. Smalling, R. Trubey, R. Raudales, and T. Will. 2016. Nonbreeding Golden-winged Warbler habitat: status, conservation, and needs. Pp. 29–38 in H. M. Streby, D. E. Andersen, and D. A. Buehler (editors). Golden-winged Warbler ecology, conservation, and habitat management. Studies in Avian Biology (no. 49), CRC Press, Boca Raton, FL.

GOLDEN-WINGED WARBLER HABITAT SELECTION

Like many Neotropical migrants, Golden-winged Warblers (*Vermivora chrysoptera*) spend the majority of their annual cycle on their tropical nonbreeding grounds in Central and northern South America, and birds that spend the nonbreeding season in Costa Rica and Nicaragua arrive in late September and stay past the middle of April (Chandler 2011). Anecdotal observations and incidental reports from community-level studies indicate that Golden-winged Warblers may be specialized in their habitat use (Bent 1963, Tramer and Kemp 1980, Blake and Loiselle 2000), apparently restricted to lower- and middle-elevation tropical wet forests (Powell et al. 1992, Robbins et al. 1992, Blake and Loiselle 2000; Chapter 1, this volume), and specializing on invertebrate prey in dead leaf clusters (Morton 1980, Tramer and Kemp 1980, Gradwohl and Greenberg 1982; Chapter 11, this volume).

Habitat specialization could increase the vulnerability of Golden-winged Warblers to habitat destruction or degradation, especially because the wet tropical forests where the species is most abundant have been extensively deforested in recent decades (Sader and Joyce 1988, Myers et al. 2000, Aide 2012). Similarly, dead leaf clusters are patchily distributed, further restricting the habitat available to Golden-winged Warblers (Chandler and King 2011). Until recently, quantitative analyses of habitat selection by Golden-winged Warblers based on systematic surveys across cover-type and land-use gradients on the nonbreeding grounds had not been conducted. Without this information, it is difficult for conservationists to evaluate the current status of habitat availability for this species or develop strategies to preserve or restore appropriate conditions. Here, we briefly review recent findings on habitat selection of Golden-winged Warblers at the nonbreeding grounds, and discuss them relative to current patterns of forest change in the nonbreeding grounds, and review established strategies and new innovations for conserving nonbreeding habitat.

Recent findings in Costa Rica using standardized point-count surveys indicated that Golden-winged Warblers were absent from tropical dry forest, were most abundant in naturally disturbed primary forest and advanced secondary forest, where they were closely associated with intermediate levels of precipitation and canopy height, and with dead-leaf tangles (Chandler and King 2011). These standardized surveys were complemented with radiotelemetry studies that showed similar results, but also demonstrated birds encountered within shade coffee were individuals in transit between forest patches. Within their home ranges, Golden-winged Warblers select microhabitat features such as vine tangles and hanging dead leaves, which occur in greatest abundance in gaps within forest and along edges. Last, recent analyses of data collected using a standardized protocol with 10-min 100-m-radius point counts by the Alianza Alas Doradas over a six-country area indicated these same general patterns held over the entire nonbreeding distribution, with birds most abundant in mid-elevation moist forests with vine tangles (Bennett 2013, Chandler 2013).

Sexual habitat segregation has been reported for a number of Neotropical migrants (Morton 1990, Rappole et al. 1999, Marra and Holmes 2001, Rappole 2013) and has potentially important implications for conservation. For example, if reserves were established based on records for conspicuous males, but females occurred in other regions or habitats, the habitat needs for both sexes might not be met, potentially hampering conservation efforts. Chandler and King (2011) encountered relatively few females on point-count surveys at their study sites in Costa Rica. However, the data they were able to collect on 22 females sighted incidentally in combination with point-count data indicated that males and females occurred in similar habitats within the study area, and in several instances within the same flock (Chandler 2011). The more recent region-wide analyses by Chandler (2013) confirmed that females use the same general forest cover types as males, but females tended to occur at lower elevations and warmer locations.

TRENDS IN DEFORESTATION

Previous analyses of deforestation rates in Latin America have offered little comfort for conservationists concerned about species dependent on primary forests in the humid tropics (Myers 1994). The net amount of forest cover is a function of forest regeneration and deforestation, however, and the association of Golden-winged Warblers with edges, gaps, and secondary forests suggests these disturbed and regenerating habitats may provide an opportunity for conservation. Asner et al. (2009) used a combination of data sources to evaluate the impact of deforestation, selective logging, and forest recovery to derive an estimate of net forest change in the humid forests worldwide and concluded 1.4% of Central American humid forests was impacted by deforestation and logging between 2000 and 2005, and >67% of the forest has <50% tree cover. However, a corresponding level of forest growth (1.2%) had offset losses almost entirely in terms of net change in forest cover. Similarly, Aide et al. (2012) reported a net increase in forest cover in Central America, once increases from forest regeneration were taken into account.

Recent studies suggest no dramatic changes recently in the extent of nonbreeding habitat available for Golden-winged Warblers, because the species tolerates intermediate levels of disturbance, and also uses secondary forest (Chavarría and Duriaux 2009, Chandler and King 2011). Furthermore, forest regeneration was documented throughout the core nonbreeding distribution of Golden-winged Warblers; predominantly within hilly or mountainous terrain and intermediate elevations (Sánchez-Azofeifa et al. 2003, Asner et al. 2009, Aide et al. 2012), which is within the intermediate elevations occupied by Golden-winged Warblers. The trends are encouraging; however, total forest cover is still greatly reduced from historical levels, covering only 41% of the nonbreeding distribution of this species (Honduras, Nicaragua, Costa Rica, and Panama; Chapter 1, this volume) on average (FAO and JRC 2012). Golden-winged Warblers require forest during the nonbreeding period and a reduction in forest cover has almost certainly reduced the carrying capacity of the nonbreeding grounds from previous levels. Furthermore, future populations trends are contingent on the assumption that forests will continue to be allowed to regenerate, which is far from certain given increased interest in biofuels, intensification of agriculture, and increased demand for natural resources associated with an increased standard of living for rural communities (Wunder 2001, Asner et al. 2009). As such, a long-term solution that ensures the protection and restoration of forest cover types associated with Golden-winged Warblers remains a conservation priority.

CONTEMPORARY APPROACHES TO CONSERVING FOREST

Protected Areas

A variety of approaches has been taken to address the issue of tropical deforestation. One of the most established methods is the creation of parks and preserves where extractive uses by people are prohibited or regulated. Approximately 57% of forest within the core nonbreeding distribution of Golden-winged Warblers is encompassed within protected areas, and 43.5% of forest within protected areas consists of lands where biodiversity conservation is the primary purpose, with the remainder of lands within protected areas consisting primarily of areas where sustainable use of natural resources is permitted (FAO and JRC 2012).

Tropical protected areas generally slow the rate of deforestation within their administrative boundaries (Brooks et al. 2009), but deforestation can still occur within protected areas despite statutory protection. For example, worldwide, DeFries et al. (2005) analyzed satellite data for a sample of 198 highly protected areas (IUCN status 1 and 2) throughout the tropics for a 20-year period starting in the early 1980s, and reported that 25% experienced forest loss within their administrative boundaries and a far higher percentage (70%) underwent forest loss within designated buffer zones. The remote sensing data are consistent with our observations that extensive deforestation has occurred and continues in the buffer zones of three Central American protected areas where we work; the BOSAWAS (Bocay River, Mount Saslaya, and Waspuk River) reserve in northern Nicaragua, and Pico Pijol and Texiguat National Parks in Honduras. According to the Honduran Ministry responsible for protected areas (Instituto Nacional de Conservación y Desarrollo Forestal, Áreas Protegidas y Vida Silvestre), Pico Pijol, and particularly Texiguat National Park, are experiencing alarming rates of deforestation due primarily to conversion of cloud forest to intensive coffee production in buffer zones. Similarly, Porter-Bolland et al. (2012) summarized protection outcomes for 40 protected areas and reported that the average annual loss of forest cover was 1.47%, with parks sampled in Costa Rica and Honduras exhibiting annual rates of forest loss of 8.7% and 0.76%, respectively. Last, establishment of protected areas can displace human activities, potentially accelerating deforestation in adjacent areas (Dewi et al. 2013).

Clearly, protected areas have potential to be an effective means of protecting forest, but in practice they appear to be inadequate on their own, judging from studies indicating that forest cover is reduced and continues to be diminished within their boundaries and in adjacent areas. The effectiveness of protected areas may ultimately be contingent on the degree to which the social and economic needs of local inhabitants, and their land tenure rights and local expertise, are recognized (Porter-Bolland et al. 2012). Successful conservation needs to include the development of institutional and administrative frameworks that recognize local governance and seek to promote sustainable livelihoods (Chazdon et al. 2009, Porter-Bolland et al. 2012). If regulatory efforts are responsive to local needs, compliance will be more aligned with self-interest, and enforcement, always the weak link in the top-down paradigm of natural resources conservation, should be less costly and more effective.

Payments for Ecological Services

Payments for ecological services are a strategy by which forest conservation can be accomplished while compensating governments, communities, or individuals for the loss of access to resources extracted from protected forests (Barrett et al. 2013). A well-known example is payment for watershed protection by municipalities (Southgate and Wunder 2009). For example, in Costa Rica, landowners are paid $50/ha to conserve forest, the valuation based on ecosystem services of carbon sequestration and watershed protection (World Resources Institute 2005). The REDD program (Reducing Emissions from Deforestation and Forest Degradation) seeks to create a financial value for the carbon stored in forests to provide financial incentives for conserving forest. In areas where incentive programs have been implemented, they appear to be successful, at least in the short term. Pagiola (2008) suggested that in Costa Rica, most secondary forests still present are the result of financial incentives and conservation regulations.

Programs providing payments for ecological services depend on ability and willingness of entities to pay, but not all ecologically interesting sites have economic value (Pagiola et al. 2004). In addition, the viability of these programs depends on political commitments or economic resources that may change with changes in government policies or economic conditions, and an increased uncertainty can present another impediment to farmer participation in these programs (Chandler 2011). Still, in most cases these agreements are contractual with the rights and obligations of interested parties well defined. Furthermore, the legal framework required for enforcement is well developed, and enforcement can be simpler than policing encroachment on protected areas. The flexibility of payment for ecological services approaches to conserving biodiversity has important advantages over strict regulatory arrangements, so they are certain to remain an important tool for conservation of Golden-winged Warblers in cases where forest provides ancillary values for which parties are willing to pay.

Agroforestry

Agroforestry describes practices that incorporate either planted or retained trees or other woody perennial plants into farming systems (Schroth et al. 2004). Agroforestry systems seek to compensate farmers for losses in yield associated with the retention of trees and other features that enhance the conservation of biodiversity with price premiums and access to specialty markets. With these market-based incentives, and costs of enforcement that are borne by the producer, this approach to conservation is less subject to the limitations inherent in strict protection such as lack of political will or costs of enforcement. Shade coffee is a form of agroforestry where trees are retained or planted over coffee to provide suitable conditions for coffee production in areas with abundant sun and as a means of biodiversity conservation. Shade coffee can enhance biodiversity in landscapes where tree cover is reduced and is clearly favorable to sun coffee in terms of its value for conservation. Some migratory birds are more abundant in shade coffee or other agroforestry habitats compared to primary forests and some appear to maintain or even increase their body condition over the course of the nonbreeding season (King et al. 2007, Bakermans et al. 2009). However, the potential for agroforestry to create incentives for converting native forest could offset the habitat value shade that coffee provides (Rappole et al. 2003).

Golden-winged Warblers are regularly encountered in shade-coffee farms and other agroforestry systems (Komar 2006, King et al. 2007); but, telemetry data from Costa Rica indicate these individuals are birds in transit between forest patches (Chandler 2011). Evidence for transience is supported by observations from Honduras that Golden-winged Warblers are seldom encountered in coffee farms without adjacent suitable forest cover (Chavarría and Duriaux 2009; D. King, unpubl. data). The habitat associations are likely explained by the close association between Golden-winged Warblers and vine tangles, a habitat feature that is typically absent from shade-coffee farms (Chandler and King 2011). It seems unlikely that shade-coffee certification programs could effectively mandate the retention of habitat features such as vine tangles and hanging dead leaves that would potentially make shade coffee suitable for Golden-winged Warblers without placing an unrealistic burden on coffee farmers (Chandler 2011). Agroforestry may still increase population connectivity and perhaps the effective size of forest patches; however, the retention of native forest appears to be key to conserving Golden-winged Warblers during the nonbreeding season.

NEW INNOVATIONS

Current work is directed at developing new approaches and refining existing strategies to conserve forest for Golden-winged Warblers and other bird species in Central and South America. One new approach for conserving forest within a market-based framework is the development of a coffee drier that uses renewable energy. Coffee must be dried prior to shipment, and currently most coffee driers use wood as fuel, which results in the harvest of wood equivalent to 6,500 ha of forest across Latin America annually (Arce et al. 2009). Hybrid solar-biomass coffee driers (Figure 2.1) use solar-thermal and biomass energy to dry coffee, which is supplemented with biofuel produced from oil derived from the fruit of a native Neotropical tree *Jatropha curcas*. The hybrid system completely eliminates the use of wood for fuel, while reducing drying costs by 88% on average (Arce et al. 2009).

Integrated Open Canopy (IOC) coffee represents an example of a refinement of an existing approach to biodiversity conservation (Arce et al. 2009). Like other forms of agroforestry, IOC coffee cultivation systems seek to use financial incentives to influence farmer behavior to maintain

Figure 2.1. Solar-hybrid coffee drying facility in Subirana, Honduras. Panels in foreground collect thermal energy, which is circulated through drying towers located in the building behind.

biodiversity; however, IOC conserves native forest and not just shade trees. Forest conservation is accomplished by reserving forest patches of equivalent size to areas planted with coffee (typically 2–3 ha) grown with a level of shade that promotes the highest yield, which in some cases is sparse or no shade. IOC increases income to farmers by increasing yields relative to shade coffee (>4×/ha on average; Arce et al. 2009). IOC is particularly well suited for carbon trading, as the carbon accounting for farms includes crops and their integrated buffer of existing, regenerating, or both existing and regenerating forest cover, thereby providing substantially more carbon sequestration than other typical agricultural-based carbon credit projects. If combined with solar drying and carbon credits, IOC coffee could increase income for farmers by >150% (Figure 2.2), and thus represents a market-based strategy for conserving forest.

Last, IOC coffee conserves forest patches of equal size to the patches planted in coffee, and forest patches this size are known to support Golden-winged Warblers and forest-dependent birds that do not occur in shade coffee in the absence of forest (Chandler et al. 2013). Thus, the IOC system imposes a lower limit on the extent of forest that will remain in mixed-use landscapes; forest that otherwise may have no explicit protection, thus would be vulnerable to degradation and conversion to other land uses.

IMPLEMENTATION

Implementation is key to the success of even the most carefully developed innovation for conserving forest or other natural resources (DeClerk et al. 2006). Efforts to translate pilot projects to on-the-ground conservation initiatives at meaningful scales are most successful when they are collaborative and reflect the needs and interests of rural communities that rely on forest resources (Hayes 2006, Chazdon et al. 2009). In addition, community integration can be a critical determinant of the success of forest conservation initiatives (Tucker et al. 2005). One potential strategy for accomplishing community integration is through co-management agreements, which describe a condition of shared responsibility between the government and private citizens. Co-management agreements are a vehicle for developing participatory, decentralized, democratic, conservation-oriented solutions that complement and multiply the capabilities and benefits for all parties (Horowitz 1998, Reed 2008).

A co-management agreement approach is being applied to conserve Golden-winged Warbler habitat in regions of central Honduras. The region is currently threatened by the rapid development of strictly high-grown coffee, which is recognized as a principal threat to forest in the region, including forest within the Pico Pijol National Park. The park encompasses 122 km² of forest, and supports large

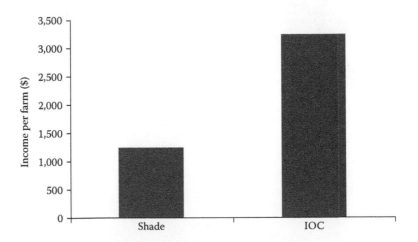

Figure 2.2. Comparison of income for two hypothetical coffee farms, one using shade-coffee cultivation and a conventional wood-fired coffee dryer and the other IOC coffee cultivation and a solar-thermal coffee dryer. On average, coffee yield is 106% higher using IOC, energy costs are reduced by 88% using solar-thermal energy, and sales of carbon credits from forest conservation and solar energy use are 20% of the value of coffee production (Arce et al. 2009).

populations of migrants of heightened conservation concern, including Golden-winged Warblers and Wood Thrushes (*Hylocichla mustelina*). Despite statutory protection, the park is being threatened by extensive degradation of the surrounding buffer zone due to the expansion of commercial production of coffee for export (Figure 2.3). Coffee in this region is typically dried with fuelwood that is harvested locally and can only be grown without shade due to dense cloud cover. The Mesoamerican Development Institute and the COMISUYL (Cooperativa Mixta Subirana Yoro Limitada) Coffee Cooperative have entered into a co-management agreement with the Institute for Forest Conservation (ICF) to manage Pico Pijol National Park using their market-based strategies for conservation and restoration of native forest cover.

A new agreement will serve as a platform for protecting the park through promoting market-based mechanisms for the conservation of native forest biodiversity that serve the common interests of all signatories. For example, ICF has a statutory obligation to protect forest within the buffer zone of the park. The surrounding municipal governments, Yoro, Victoria, El Negrito, and Morazan, depend on the park for water as does a nearby electrical utility company that operates a 12 MW hydropower station, and maintaining forest cover within the park is paramount for maintaining sustainable water flows. The objective of the co-management agreement is to promote the conservation and sustainable management of Pico Pijol National Park through legal and technical implementation of the shared management of the area. Our expectation is that by providing a platform for communication and cooperation among users with similar or compatible interests, the co-management agreement will complement other conventional policies and practices for forest conservation.

FUTURE NEEDS

Future priorities for research should address key gaps in the understanding of the ecology of

Figure 2.3. High-elevation coffee planted in place of forest in the buffer zone of Pico Pijol National Park, Honduras.

Golden-winged Warblers and refining market-based strategies for forest conservation. One key information need for nonbreeding Golden-winged Warblers, which has an important bearing on developing strategies for their conservation, is habitat-specific survival. Nonbreeding migrants are often reported to use suboptimal habitats where survival is lower due to despotic interactions with conspecifics or other mechanisms (Rappole 2013). Thus, it is important to determine whether the patterns of habitat selection observed in Golden-winged Warblers are reflected in patterns of habitat-specific survival, as reported from other species of migratory songbirds, or whether habitat-specific differences in body condition constrain migration or breeding success. Golden-winged Warblers can be cryptic during the nonbreeding season and have large home ranges (~9 ha; Chandler 2011); thus, ordinary mark-resight studies to estimate habitat-specific survival are not practicable for nonbreeding Golden-winged Warblers. Radiotelemetry can overcome these challenges with monitoring, but radios small enough for Golden-winged Warblers do not have long enough battery life to yield sufficient exposure days, and transmitters appear to reduce survival if birds carry them during migration (Chandler 2011). Hierarchical mixture models are an alternative method for estimating survival rates from repeated point counts of unmarked birds (Chandler and King 2011), and research is ongoing to develop and apply mark-resight surveys for cryptic species with large home ranges (Ritterson 2015).

Another important issue related to Golden-winged Warbler conservation is the susceptibility of the species to climate change on the nonbreeding grounds. In Costa Rica, the distribution of Golden-winged Warblers is closely related to moisture, with birds occurring at maximum abundance in areas with ~2.5 m annual rainfall (King et al. 2012). At a nonbreeding-distribution-wide scale, temperature is also an important predictor of local abundance of nonbreeding Golden-winged Warblers (Chandler 2013). Although there is substantial variation, zones of precipitation and temperature within the nonbreeding distribution are linked with elevation, and studies have documented incursions of lower elevation birds into montane areas in the Neotropics (Pounds et al. 1999). The relationship of abundance with moisture raises the possibility that Golden-winged Warblers could be subjected to impacts from climate change, both directly and from increased orographic restriction as their optimal climactic conditions increase in elevation as a result of increased global temperatures. Assessing vulnerability of nonbreeding Golden-winged Warblers to predicted climactic shifts and identification of landscape features associated with potential refugia are research priorities.

Planned refinements of market-based incentives for forest conservation include the development of carbon certification guidelines for small land holders to market carbon credits derived from conservation of forest on their IOC farms. Sales of carbon credits will provide added incentive for farmers to conserve forest, and carbon sales could exceed 20% of the value of the coffee produced. Additionally, the Best Management Practices for IOC coffee specifying the area, shape, and configuration of forest patches must be developed to ensure that farms achieve the goal of conserving biodiversity. Work to develop Best Management Practices is underway based on data on bird abundance versus tree height, patch size, and other forest characteristics from IOC farms in Costa Rica and will also form the basis of a bird-friendly certification that may provide additional income to farmers practicing in IOC coffee cultivation. Last, modeling the relative value of agroforestry landscapes managed using renewable energy and IOC coffee cultivation versus using conventional systems, accounting for changes in income to farmers and its influence on deforestation, will (1) contribute to the growing knowledge base on the socioeconomic aspects of deforestation, (2) help quantify the benefits of this system to support its promotion, and (3) provide the data needed to develop guidelines for agroforesty landscapes implementing these alternative coffee processing and cultivation methods.

ACKNOWLEDGMENTS

We thank the members of the Cooperativa Montes de Oro, particularly V. J. Arce Chavez and A. Morales for their assistance with site selection and logistics, and C. Chandler, S. Beaudreault, C. Orlando Delgado, M. Gonzalez, J. Aguero, J. Wolfe, N. Hazlet, A. Anderson, J. Wells, and J. Ritterson for assistance in the field. Our project was funded by the U.S. Fish and Wildlife Service's Neotropical Migratory Bird Conservation Act program and the U.S. Forest Service International Program. We thank J. P. Kelley, C. Lively, and J. Rappole for comments on an earlier version of this manuscript.

LITERATURE CITED

Aide, T. M., M. L. Clark, H. R. Grau, D. López-Carr, M. A. Levy, D. Redo, M. Bonilla-Moheno, G. Riner, M. J. Andrade-Núñez, and M. Muñiz. 2012. Deforestation and reforestation of Latin America and the Caribbean (2001–2010). Biotropica 45:262–271.

Arce, V. J. C., R. Raudales, R. Trubey, D. I. King, R. B. Chandler, and D. C. Chandler. 2009. Measuring and managing the environmental cost of coffee production in Latin America. Conservation and Society 7:141–144.

Asner, G. P., T. K. Rudel, T. M. Aide, R. S. DeFries, and R. Emerson. 2009. A contemporary assessment of change in humid tropical forests. Conservation Biology 23:1386–1395.

Bakermans, M. H., A. C. Vitz, A. D. Rodewald, and C. G. Rengifo. 2009. Migratory songbird use of shade coffee in the Venezuelan Andes with implications for conservation of Cerulean Warbler. Biological Conservation 142:2476–2483.

Barrett, C. B., E. H. Bulte, P. Ferraro, and S. Wunder. 2013. Economic instruments for nature conservation. Pp. 59–73 in D. W. MacDonald and K. J. Willis (editors), Key topics in conservation biology 2. Wiley Blackwell, Oxford, UK.

Bennett, R. E. 2013. Habitat associations of the Golden-Winged Warbler in Honduras. M.S. thesis, Michigan Technological University, Houghton, MI.

Bent, A. C. 1963. Life histories of North American wood warblers. Dover Publications, Inc., New York, NY.

Blake, J. G., and B. A. Loiselle. 2000. Diversity of birds along an elevational gradient in the Cordillera Central, Costa Rica. Auk 117:663–686.

Brooks, T. M., S. J. Wright, and D. Sheil. 2009. Evaluating the success of conservation actions in safeguarding tropical forest biodiversity. Conservation Biology 23:1448–1457.

Chandler, R. B. 2011. Avian ecology and conservation in tropical agricultural landscapes with emphasis on Vermivora chrysoptera. Ph.D. dissertation, University of Massachusetts, Amherst, MA.

Chandler, R. B. 2013. Analysis of Golden-winged Warbler winter survey data. Report prepared for the Cornell Lab of Ornithology, Ithaca, NY.

Chandler, R. B., and D. I. King. 2011. Habitat quality and habitat selection of Golden-winged Warblers in Costa Rica: an application of hierarchical models for open populations. Journal of Avian Ecology 48:1037–1048.

Chandler, R. B., D. I. King, R. Raudales, R. Trubey, and V. J. Arce. 2013. A small-scale land-sparing approach to conserving biological diversity in tropical agricultural landscapes. Conservation Biology 27:785–795.

Chavarría, L., and G. Duriaux. [online]. 2009. Informe preliminar del primer censo de Vermivora chrysoptera: realizado en Marzo 2009 en la zona norcentral de Nicaragua. <http://www.bio-nica.info/Biblioteca/Chavarria2009FinalGWWA.pdf> (8 June 2014).

Chazdon, R. L., C. A. Harvey, O. Komar, D. M. Griffith, B. G. Ferguson, M. Martínez-Ramos, H. Morales, R. Nigh, L. Soto-Pinto, M. van Breugel, and S. M. Philpott. 2009. Beyond reserves: a research agenda for conserving biodiversity in human-modified tropical landscapes. Biotropica 41:142–153.

DeClerck, F., J. C. Ingram, and C. M. Rumbaitis del Rio. 2006. The role of ecological theory and practice in poverty alleviation and environmental conservation. Frontiers in Ecology and the Environment 4:533–540.

DeFries, R., A. Hansen, A. C. Newton, and M. C. Hansen. 2005. Increasing isolation of protected areas in tropical forests over the past twenty years. Ecological Applications 15:19–26.

Dewi, S., M. van Noordwijk, A. Ekadinata, and J. L. Pfund. 2013. Protected areas within multifunctional landscapes: squeezing out intermediate land use intensities in the tropics? Land Use Policy 30:38–56.

FAO and JRC. 2012. Global forest land-use change 1990–2005, by E. J. Lindquist, R. D'Annunzio, A. Gerrand, K. MacDicken, F. Achard, R. Beuchle, A. Brink, H. D. Eva, P. Mayaux, J. San-Miguel-Ayanz, and H.-J. Stibig. FAO Forestry Paper No. 169. Food and Agriculture Organization of the United Nations and European Commission Joint Research Centre, Rome, Italy.

Gradwohl, J., and R. Greenberg. 1982. The effect of a single species of avian predator on the arthropods of aerial leaf litter. Ecology 63:581–583.

Hayes, T. M. 2006. Parks, people, and forest protection: an institutional assessment of the effectiveness of protected areas. World Development 34:2064–2075.

Horowitz, L. S. 1998. Integrating indigenous resource management with wildlife conservation: a case study of Batang Ai National Park, Sarawak, Malaysia. Human Ecology 26:371–403.

King, D. I., R. B. Chandler, J. H. Rappole, R. Raudales, and R. Trubey. 2012. Community-based agroforestry initiatives in Nicaragua and Costa Rica. Pp. 99–115 in J. A. Simonetti, A. A. Grez, and C. F. Estades (editors), Biodiversity conservation in agroforestry landscapes: challenges and opportunities. Editorial Universitaria, Santiago, Chile.

King, D. I., M. D. Hernandez-Mayorga, R. Trubey, R. Raudales, and J. H. Rappole. 2007. An evaluation of the contribution of cultivated allspice (Pimenta dioca) to vertebrate biodiversity conservation in Nicaragua. Biodiversity and Conservation 16:1299–1320.

Komar, O. 2006. Ecology and conservation of birds in coffee plantations: a critical review. Bird Conservation International 16:1–23.

Marra, P. P., and R. T. Holmes. 2001. Consequences of dominance-mediated habitat segregation in American Redstarts during the nonbreeding season. Auk 118:92–104.

Morton, E. 1990. Habitat segregation by sex in the Hooded Warbler: experiments on proximate causation and discussion of its evolution. American Naturalist. 135:319–333.

Morton, E. S. 1980. Adaptations to seasonal change by migrant land birds in the Panama Canal zone. Pp. 437–553 in A. Keast and E. S. Morton (editors), Migrant birds in the Neotropics: ecology, behavior, distribution, and conservation. Smithsonian Institution Press, Washington, DC.

Myers, N. 1994. Tropical deforestation: rates and patterns. Pp. 27–40 in K. Brown and D. W. Pearce (editors), The economic and statistical analysis of factors giving rise to the loss of the tropical forests. University College London Press, London, UK.

Myers, N., R. A. Mittermeier, C. G. Mittermeier, G. A. da Fonseca, and J. Kent. 2000. Biodiversity hotspots for conservation priorities. Nature 403:853–858.

Pagiola, S. 2008. Payments for environmental services in Costa Rica. Ecological Economics 65:712–724.

Pagiola, S., P. Agostini, J. Gobbi, C. De Haan, M. Ibrahim, E. Murgueitio, E. Ramirez, M. Rosales, and J. P. Ruiz. 2004. Paying for biodiversity conservation services in agricultural landscapes. Environment Department Paper 96. The Word Bank, Washington, DC.

Porter-Bolland, L., E. A. Ellis, M. R. Guariguata, I. Ruiz-Mallén, S. Negrete-Yankelevich, and V. Reyes-García. 2012. Community managed forests and forest protected areas: an assessment of their conservation effectiveness across the tropics. Forest Ecology and Management 268:6–17.

Pounds, J. A., M. P. Fogden, and J. H. Campbell. 1999. Biological response to climate change on a tropical mountain. Nature 398:611–615.

Powell, G. V. N., J. H. Rappole, and S. A. Sader. 1992. Neotropical migrant landbird use of lowland Atlantic habitats in Costa Rica: a test of remote sensing for identification of habitat. Pp. 287–298 in J. M. Hagan and D. W. Johnson (editors), Ecology and conservation of Neotropical migrant landbirds. Smithsonian Institution Press, Washington, DC.

Rappole, J. H. 2013. The avian migrant: the biology of bird migration. Columbia University Press, New York, NY.

Rappole, J. H., D. I. King, and W. C. Barrow. 1999. Winter ecology of the endangered Golden-cheeked Warbler (Dendroica chrysoparia). Condor 101:762–770.

Rappole, J. H., D. I. King, and J. H. Vega Rivera. 2003. Coffee and conservation. Conservation Biology 17:334–336.

Reed, M. S. 2008. Stakeholder participation for environmental management: a literature review. Biological Conservation 141:2417–2431.

Ritterson, J. D. 2015. Generating Best Management Practices for avian conservation in a land-sparing agriculture system, and the habitat-specific survival of a priority migrant. M.S. thesis, University of Massachusetts, Amherst, MA.

Robbins, C. S., J. W. Fitzpatrick, and P. B. Hamel. 1992. A warbler in trouble: Dendroica cerulea. Pp. 549–562 in J. M. Hagan and D. W. Johnson (editors), Ecology and conservation of Neotropical migrant landbirds. Smithsonian Institution Press, Washington, DC.

Sader, S. A., and A. T. Joyce. 1988. Deforestation rates and trends in Costa Rica, 1940 to 1983. Biotropica 20:11–19.

Sánchez-Azofeifa, A., G. C. Daily, A. S. Pfaff, and C. Busch. 2003. Integrity and isolation of Costa Rica's national parks and biological reserves: examining the dynamics of land-cover change. Biological Conservation 109:123–135.

Schroth, G., G. A. B. Fonseca, C. A. Harvey, C. Gascon, H. L. Vasconcelos, and A. M. N. Izac. 2004. Agroforestry and biodiversity conservation in tropical landscapes. Island Press, Washington, DC.

Southgate, D., and S. Wunder. 2009. Paying for watershed services in Latin America: a review of current initiatives. Journal of Sustainable Forestry 28:497–524.

Tramer, E. J., and T. R. Kemp. 1980. Foraging ecology of migrant and resident warblers and vireos in the highlands of Costa Rica. Pp. 285–296 in A. Keast and E. S. Morton (editors), Migrant birds in the Neotropics: ecology, behavior, distribution, and conservation. Smithsonian Institution Press, Washington, DC.

Tucker, C. M., D. K. Munroe, H. Nagendra, and J. Southworth. 2005. Comparative spatial analyses of forest conservation and change in Honduras and Guatemala. Conservation and Society 3:174–200.

World Resources Institute. 2005. World resources 2005: the wealth of the poor–managing ecosystems to fight poverty. World Resources Institute, Washington, DC.

Wunder, S. 2001. Poverty alleviation and tropical forests–what scope for synergies? World Development 29:1817–1833.

PART TWO

Breeding Grounds

CHAPTER THREE

Landscape-Scale Habitat and Climate Correlates of Breeding Golden-winged and Blue-winged Warblers*

Dolly L. Crawford, Ronald W. Rohrbaugh, Amber M. Roth, James D. Lowe, Sara Barker Swarthout, and Kenneth V. Rosenberg

Abstract. Understanding how an animal species uses habitat at a landscape scale is critical in interpreting its ecology and behavior for use in conservation planning. We compiled a dataset of 28,822 location records for Golden-winged Warblers (*Vermivora chrysoptera*; n = 8,266) and Blue-winged Warblers (*V. cyanoptera*; n = 20,556) for a 13-year period (1998–2010) from five sources. We modeled potential habitat of both species as a function of 17 variables related to climate, land cover, and elevation at the distribution-wide, regional, and subregional scales. We used a maximum entropy and ensemble forecasting approach to model the distribution of potential habitat. We evaluated model support with the area-under-the-curve and the true-skill-statistic criteria. We used the best supported models to project the species distribution and identify climate and habitat affinities. We used principal component analysis of dependent variables to further characterize habitats at the subregional scale. At the distribution-wide scale (2.5-km grid-cell size), the occurrence of Golden-winged Warblers was associated with a cool, dry climate, at elevations from ~350 to 1,500 m and habitats comprised of at least 60% deciduous forest cover. Regional correlates of the occurrence of Golden-winged Warblers included breeding season temperature, elevation, land-cover type, and forest type. In the Appalachian Mountains region, an elevation >500 m best predicted the occurrence of Golden-winged Warblers, whereas the distribution of aspen forest (*Populus* spp.) best predicted the occurrence of the species in the Great Lakes region. Models at finer subregional scales demonstrated similar associations; the occurrence of Golden-winged Warblers was associated with land-cover type, tree species composition, elevation, and temperature. The occurrence of Golden-winged Warblers at the subregional scale was negatively associated with agriculture and human disturbance, which were more indicative of the occurrence of Blue-winged Warblers. Across all scales, the relative occurrence of Golden-winged Warblers was associated with deciduous forest at elevations >500 m and inversely associated with disturbed habitats. The opposite pattern was found for Blue-winged Warblers.

Key Words: Blue-winged Warbler, distribution model, ensemble forecasting, Golden-winged Warbler.

* Crawford, D. L., R. W. Rohrbaugh, A. M. Roth, J. D. Lowe, S. B. Swarthout, and K. V. Rosenberg. 2016. Landscape-scale habitat and climate correlates of breeding Golden-winged and Blue-winged Warblers. Pp. 41–66 in H. M. Streby, D. E. Andersen, and D. A. Buehler (editors). Golden-winged Warbler ecology, conservation, and habitat management. Studies in Avian Biology (no. 49), CRC Press, Boca Raton, FL.

tudying how breeding bird species respond to landscape-scale climate variation and the composition and configuration of land cover patterns is key to understanding avian settlement patterns, dispersal, interactions with closely related species, and migration behavior. A given patch might appear suitable at the scale of the breeding territory, but if the patch is not set within an appropriate landscape context or lacks suitable climate during important periods of the breeding cycle, then suitability at the finer scale is lost or compromised (Thogmartin 2007). For example, a grassy field embedded in a forested matrix may not provide habitat for an obligate, grassland-nesting bird species. But if the field is part of a grassland-dominated landscape and is bordered on two sides by forest, the habitat becomes more suitable to the species. The ecological context of suitable breeding habitat is often a function of conditions found at different ecological scales. The only way to quantify such nuances is to undertake landscape-scale research focused on defining multiscale patterns of climate and habitat use (Saab 1999). For example, Johnson (1980) recognized four hierarchical orders of habitat selection that define the nested spatial scales at which organisms select resources needed for survival and reproduction.

Population declines of Golden-winged Warblers (*Vermivora chrysoptera*) have been attributed, in part, to a range-wide fragmentation of suitable habitat for the species, namely, early successional habitat and young forest (Confer and Knapp 1981, Hunter et al. 2001, Buehler et al. 2007, Confer 2008, Confer et al. 2010, Thogmartin 2010). It is also likely that large, contiguous areas of suitable habitat for Golden-winged Warblers are intact but remain unoccupied (Rohrbaugh et al. unpublished report). It is possible that habitat is unoccupied because it lacks an appropriate landscape climatic context or because of competitive exclusion by Blue-winged Warblers (*V. cyanoptera*) that limits or prevents sustainable habitat partitioning between species.

Hybridization between Golden-winged Warblers and Blue-winged Warblers has been documented for over a century (Brewster 1874, Herrick 1874, Parkes 1951, Gill 1980, Sauer et al. 2008). Since the 1990s, however, the incidence of genetically pure Golden-winged Warblers has decreased, while the incidence of Blue-winged Warblers and hybrid phenotypes has increased in the same areas (Gill 2004, Vallender et al. 2007, Sauer et al. 2008;

Chapter 1, this volume). Over two decades, the general trend has been the displacement of Golden-winged Warblers by Blue-winged Warblers from south to north, with a complex pattern of hybridization (Vallender et al. 2007). Golden-winged Warbler subpopulations with the largest proportion of hybrid individuals typically occur in the Appalachian Mountains and eastern Great Lakes region (Vallender et al. 2009). Both climate and landscape context may play a complex role in mitigating hybridization and displacement at a local scale and influencing latitudinal patterns. The first step in understanding the landscape-scale dynamics of hybridization and displacement is quantifying the climatic and landscape variables linked to the distributions of both Golden-winged Warblers and Blue-winged Warblers.

Golden-winged Warbler habitat use is well documented at the territory scale (Confer and Knapp 1981, Patton et al. 2010), but few studies have quantified the habitat affinities of cooccurring Golden-winged Warblers and Blue-winged Warblers (Confer et al. 2010, Patton et al. 2010). Furthermore, development of effective conservation plans requires knowledge of both positive and negative habitat relationships for Golden-winged Warblers and Blue-winged Warblers at multiple spatial scales. Habitat requirements are especially important when developing distribution-wide or regional plans where variation in land cover and physiographic features is likely and the patterns of habitat selection may vary (Chapter 7, this volume).

Golden-winged Warblers, and to a lesser extent Blue-winged Warblers, are patchily distributed throughout portions of their ranges, and monitoring data from the Breeding Bird Survey in North America and state-level breeding bird atlases may not accurately represent the breeding distribution. Incomplete or uneven sampling issues that may result can be mitigated by developing robust predictive models parameterized with empirical data from species presence locations. Model outputs can then be expressed as a predictive map to better visualize the distributions of both species. In this landscape-scale examination of breeding Golden-winged Warblers and Blue-winged Warblers, our specific objectives were to

1. Identify the primary drivers of species-level occurrence patterns, including climate and landscape attributes at the 2.5-km and 500-m scales.

2. Develop predictive species distribution maps representative of current conditions based on the output of climate and landscape attribute models.

METHODS

We examined the environmental and climatic conditions associated with the potential distributions of Golden-winged Warblers and Blue-winged Warblers using a modeling approach. In this approach, location data are projected onto environmental data and the areas predicted to be potential habitats for the species are represented probabilistically (Phillips et al. 2006, Thuiller et al. 2009). A species distribution modeling approach is useful in wildlife and management applications, including reserve design (Araujo et al. 2004), studies of climate impacts on species (Thuiller 2004), and identification of new survey sites to test for the occurrence of the species (Raxworthy et al. 2003). We also applied this approach to suggest management strategies for Golden-winged Warblers (A. M. Roth et al., unpubl. plan). Model development is enhanced through metrics that evaluate model performance (such as the area under the receiver operating characteristic [AUC] and the true skill statistic [TSS]) and evaluate goodness of fit (as the difference in AUC between calibration and evaluation data; Thuiller et al. 2009). The removal of duplicate location data can also enhance model performance and goodness of fit (Phillips et al. 2006, Thuiller et al. 2009). Here, we build on previous work to exploit an additional benefit of this approach, that is, the ability to further optimize predictive output by removing species locations that are predicted to be in suboptimal potential habitat with low probability of occurrence from a priori distribution models (A. M. Roth et al., unpubl. plan). These "optimized" species locations are used in subsequent statistical analyses to quantify the habitat associations of species. We used the output from the model with the highest support to remove species locations from the dataset prior to statistical analyses. Points were excluded if they were located outside of the predicted presence area from the binary presence/absence grid calculated in the best supported model of species distributions.

We compiled a dataset of 42,422 location records for Golden-winged Warblers and Blue-winged Warblers collected from 1931 to 2011 for modeling. We extracted 28,822 "modern" (1998–2011) occurrence records for inclusion in the analyses. We derived the modern period data from presence/absence records for both species from three sources: (1) May eBird records (Golden-winged Warblers = 4,147, Blue-winged Warblers = 14,624; Sullivan et al. 2009), (2) the Cornell Lab of Ornithology Golden-winged Warbler Atlas Project (Golden-winged Warblers = 4,022, Blue-winged Warblers = 4,335), and (3) survey data collected by observers from the Cornell Lab of Ornithology, the Golden-winged Warbler Working Group, and others (Golden-winged Warblers = 97, Blue-winged Warblers = 1,597; Wilson et al. 2007, A. M. Roth et al., unpubl. plan).

Previous studies suggested that environmental variables such as elevation and breeding season temperature are predictive of Golden-winged Warbler habitats (Klaus and Buehler 2001, Hitch and Leberg 2006, Patton et al. 2010, Bakermans et al. 2011). The initial dataset ($n = 28,822$) was used to examine the interaction of these variables. We obtained 10-m elevation data (ELEV), average maximum temperature for May (TMAX5), average precipitation totals in mm for May (PPT5), and average minimum May and June temperatures (TMIN5, TMIN6) at 30-m resolutions and reprojected to 2.5 km and 500 m, respectively, for model development (Supplementary Information A). We projected warbler location data onto the five variables to construct a priori distribution models using a classification tree algorithm in Program R (ver. 2.2.1; R Development Core Team 2011). We removed location records that fell outside of the boundary of the a priori distribution model using a reverse jackknife procedure in DIVA-GIS (ver. 7.5.0; Hijmans et al. 2001, Chapman 2005). The reverse jackknife is suggested for species location data with many observations and selects a subset from the original data that comprise 95% of the original distribution of points, given the abiotic conditions that parameterize the a priori model. In the classification tree analysis, the relative importance of each variable is measured by the Pearson correlation between the standard prediction and the prediction computed when the variable of interest is permuted (Thuiller 2009). A high correlation score (r_s or $r_p > 0.50$) represents a variable with low explanatory power and a low correlation score (r_s or $r_p < 0.50$) indicates a variable with high explanatory power. When we projected the species location records associated with the 95% distribution envelope in Albers Equal Area Conic, numerous locations were outside of the accepted geographic breeding range of the species (A. M. Roth et al., unpubl. plan). We eliminated the extraneous locations (n = 6,052) to increase the accuracy of distribution models (Scheldeman and van Zonneveld 2010).

We modeled distributions for potential habitats of Golden-winged Warblers and Blue-winged Warblers as a function of ecological and climatological parameters across hierarchical spatial extents (area under consideration) and scales (size of each grid cell within the subregion area). We selected spatial extents for analysis that were identified as critical areas for species management by the Golden-winged Warbler Working Group (A. M. Roth et al., unpubl. plan): within the combined regions (Great Lakes and Appalachian Mountains regions), separately within each region, and in subregions within each region (Figure 3.1b). To capture the range of environmental variation expected at coarse to fine scales, we developed distribution models for each species in each extent at 2.5-km and 500-m scales. We projected the presence/absence data into the 2.5-km and 500-m grids to remove duplicate records, which reduced geographical sampling bias and improved the model goodness of fit (Hijmans et al. 2000, Syfert et al. 2013).

Previous studies of Golden-winged Warbler habitat suggested that young deciduous forest cover intermingled with patches of herbaceous or shrub cover are potential habitat areas for breeding Golden-winged Warblers (Klaus and Buehler 2001, Confer et al. 2010, Patton et al. 2010). Distribution models were parameterized with ecological data that reflect characteristics of Golden-winged Warblers' habitat, including land cover data from 2006 (Fry et al. 2011) and 2008 (LANDFIRE 2013) that fell within the timeframe of Golden-winged Warbler and Blue-winged Warbler observations (1998–2010), and aided the development of potential distribution models that were more representative of Golden-winged Warblers' breeding habitat. We obtained spatial data at 30-m resolutions and reprojected to 2.5 km and 500 m, respectively, for model development (Supplementary Information A).

Land cover and climate data were applied in the development of 2.5-km models for Golden-winged Warblers and Blue-winged Warblers. Land cover data included 16 broad vegetation land cover classes at 30-m resolution derived from Landsat satellite data (NLCD 2006). We obtained additional environmental variables, including the percentages of deciduous forest, coniferous forest, and agricultural land, through reclassification and reprojection of Landsat satellite data (NLCD 2006) (Supplementary Information A). As an example, we calculated a grid of percent deciduous forest by averaging the amount of forest across each 30-m area nested within each 2.5-km area. We obtained additional land cover data (EVT) to further characterize warbler habitat. These spatial data represented over 3,000 vegetation cover types for the U.S. in 2008 based on spatial interpolation and ground truthing of Forest Inventory Analysis data at a 30-m resolution (LANDFIRE 2013). EVT data include coverage data for woodland and grassland mosaics, which may reflect the patchy nature of Golden-winged Warbler habitat (Confer et al. 2010). Woodland mosaics were defined as open woodlands with interspersed prairies at southern U.S. latitudes and open woodlands with interspersed shrubby grasslands at northern U.S. latitudes (Stewart 2002). Grassland mosaics were areas with a mixture of grassland or savanna cover in which neither component makes up >60% of the landscape and in which fire is the dominant natural disturbance preventing any cover class from becoming or remaining dominant (Olson 1994). We reclassified EVT data into grids that represent the percentage of grassland and woodland mosaics per 2.5 km using these definitions (Supplementary Information A). Distribution data for 141 tree species in the U.S. were derived from 2002 to 2003 MODIS images in 1-km resolution by the USFS Forest Inventory and Analysis Program and Remote Sensing Applications Center (Ruefenacht et al. 2008). Tree data were used to further refine and characterize the areas predicted as suitable for occupancy by the best supported model by calculating the most common tree species at 2.5 km from the original 1-km grid (TREE; Supplementary Information A). We also applied climate data in the 2.5-km models obtained as average conditions from 1981 to 2010 at an 800-m resolution from PRISM (Daly et al. 2008) and reprojected to 2.5 km. We selected the 30-year data because they reflected climate conditions that were relatively coincident temporally with the warbler location data.

We examined the distribution of potential habitat for Golden-winged Warblers and Blue-winged Warblers at the 500-m scale using land cover data similar to the 2.5-km models. We calculated the percent of agricultural land and deciduous and coniferous forest from NLCD 2006 data (calculations as that obtained earlier). The most common vegetation cover class at 500 m was derived

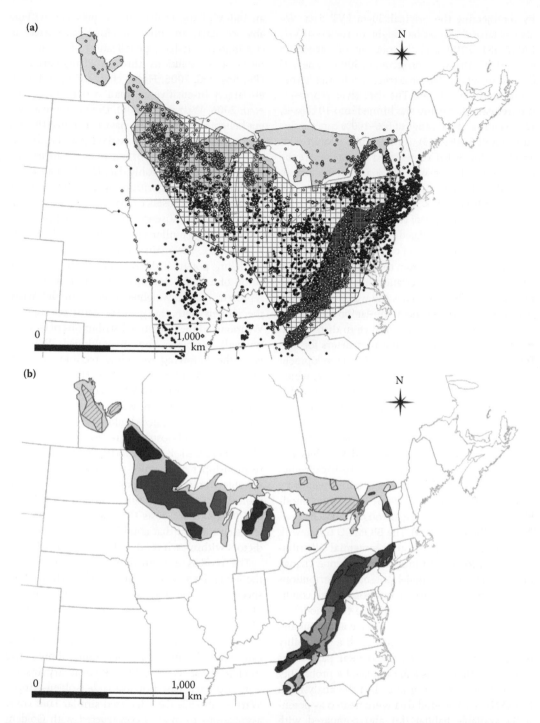

Figure 3.1 (a) The 95% probability distribution (cross-hatched polygon) where Golden-winged Warblers (GWWA) were most likely to occur, as calculated from the data. Warbler locations outside of the polygon boundary were removed from the initial dataset. GWWA and Blue-winged Warbler locations in the initial dataset (point locations) and the location of the Great Lakes and Appalachian Mountains regions for GWWAs (filled polygons; see also [b]) are also shown. (b) Location of the Great Lakes and Appalachian Mountains regions for GWWAs (larger gray polygons) and subregions delimited for GWWA management (smaller colored polygons inside regions). Subregions not included in this analysis are also shown (diagonal hashed polygons).

by reprojecting the original 30-m EVT data. We also obtained vegetation height in meters (EVH; LANDFIRE 2013) and the percent of vegetation cover (EVC) that we reprojected to 500 m. The EVC data were subdivided onto tree, shrub, and herbaceous land-cover types. The data layer representing tree canopy closure was binned into 10 classes, representing a total range of 0%–100% canopy closure (LANDFIRE 2013). For example, a forest with an EVC value of 101 represented an area where the tree canopy closure was ≥10% and <20% and an EVC value of 102 depicted a forest with tree canopy closure ≥20% and <30% (LANDFIRE 2013). To evaluate the influence of canopy closure (Jennings et al. 1999), we calculated a new grid that represented the most common canopy closure value at 101 or 102 per 500 m from the original 30-m data (percent CANOPY). Vegetation height (EVH) was represented as binned values and we prepared these data for modeling using a similar procedure. To reduce the influence of covariance in distribution modeling, we tested all input parameters for correlations in Program R (ver. 2.2.1; R Development Core Team 2011) using the Hmisc package, which calculates a matrix of Pearson's r for all variable pairs. We excluded a variable when a pairwise comparison resulted in r > 0.80.

We used two approaches to model the potential distributions of Golden-winged Warblers and Blue-winged Warblers: (1) a presence-background approach (MAXENT) that uses maximum entropy modeling (Phillips et al. 2006) and (2) six modeling algorithms available through BIOMOD (ver. 2.2; Thuiller et al. 2009). The BIOMOD framework is conducive to forecasting potential distributions as a product of future environmental change but is also used to model potential distributions as a function of current environmental conditions (Jiguet et al. 2010). With the two modeling approaches, we could include both continuous and categorical variables and constructed a probability distribution based on environmental parameters across the study areas and calculated a probability of suitability for each grid cell in the analysis. In MAXENT, background data were treated as potentially available habitat (f_1z) and compared with actual habitat use that was determined based on presence locations (fz). The model algorithm minimizes the distance between density distributions for (f_1z) and (fz) using model fitting rules (Phillips et al. 2006). The features of MAXENT include a modeling algorithm with the capacity to model an index of the probability of presence without absence data, an input of either categorical or continuous variables, penalization of more complex model solutions through L1-regularization (Phillips et al. 2006, Hastie et al. 2009), and models largely insensitive to small sample sizes (Elith et al. 2006, Wisz et al. 2008). Despite these apparent benefits, recent studies have concluded that the main assumption of this method may be flawed, that is, the probability of occurrence can be estimated from presence-only data (Royle et al. 2012).

The modeling algorithms in BIOMOD rely on binary presence/absence data, and pseudo-absence data or artificial absence data are drawn from the background (Thuiller et al. 2009, Angelo-Marini et al. 2010). The main advantage of BIOMOD in modeling potential distributions is in the calculation of a final "consensus" or mean model that represents the unweighted average model across the models with the greatest statistical support. We modeled the potential distributions of Golden-winged Warblers and Blue-winged Warblers with the following modeling techniques inside BIOMOD (Thuiller et al. 2009): (1) classification tree analysis (CTA) with 50-fold cross validation to minimize the deviation, while optimizing the tree topology, (2) generalized additive model (GAM), (3) generalized boosted regression model (GBM), (4) generalized linear model (GLM) with the relationship between the response and predictor variables modeled as a polynomial function, (5) multivariate adaptive regression splines (MARS), and (6) random forests (RF), which is a machine learning method that combines multiple tree predictors within each model (Thuiller et al. 2009).

The models examined within the ensemble-modeling framework require information about species presence and absence. Absence data are often unavailable for many species (Graham et al. 2004) or may represent "false absences" where the species was present but went undetected that can impair the discriminatory ability of predictive models (Lobo et al. 2007). Preliminary inspection of models based on available Golden-winged Warbler absence data depicted similar land cover associations to models constructed with Golden-winged Warbler presence data, suggesting that our absence data may contain an unknown amount of false absences. Several methods can be used to select pseudo-absence data from the study area to represent areas where the species is absent (Fitzgerald et al. 2011, Barbet-Massin et al. 2012).

To mitigate for false absences, we generated a number of pseudo-absence points to replace the false absences. The sample size of the pseudo-absences was the same as the number of false absences. The location of pseudo-absence points was randomly selected from within 2° (~2 km) of a presence location. The use of this radius tends to improve predictive performance (Barbet-Massin et al. 2012). To examine warbler habitat characteristics while minimizing error, we conducted six simulations for each model (n = 7 models; e.g., GLM, RF, MAXENT) and for each species (n = 2) to yield 84 models (6 simulations × 7 models × 2 species) with each change in extent and resolution.

We evaluated the predictive performance of the MAXENT and BIOMOD models by randomly selecting 80% of the data to train the model (calibration data) and we used the remaining 20% for testing (evaluation data). We calculated the mean model as the unweighted average probability distribution across all grid cells for the three BIOMOD models with the greatest mean area under the curve (Araujo and New 2006, Marmion et al. 2009). Model fit was evaluated by comparing the AUC scores between models calculated with calibration versus the evaluation data, where a small difference between AUC values is considered good fit (Thuiller 2003). We employed the same modeling procedure for all spatial extents (region and subregions) and resolutions (500 m and 2.5 km) examined. To ease model interpretation, we transformed the ordinal scores of habitat potential into a more practical grid of presence/absence using the least presence threshold (LPT), which is defined as the lowest suitability value needed for species presence (Allouche et al. 2006). We then used the binary grid to extract environmental parameters from areas predicted to be suitable for Golden-winged Warblers and Blue-winged Warblers. The threshold, which maximizes the agreement between the observed and predicted occurrence distributions, is recommended because of the lower rate of false negatives and false positives (Liu et al. 2005). Several measures of model performance are influenced by prevalence (Allouche et al. 2006). We, therefore, also evaluated model performance using the TSS, which has been shown to be independent of prevalence. The value of TSS ranges from −1 to +1, with +1 representing perfect agreement and TSS ≤ 0 equates to no better than random guessing (Allouche et al. 2006). The relative importance of each predictor

variable is not easy to calculate from each independent model (e.g., GAM, RF) because of varying relationships between response and predictor variables (Thuiller et al. 2009). Therefore, we assessed the relative importance of each variable from the mean model by randomizing each variable and then comparing predictions from the new model versus the calibrated model (Thuiller et al. 2009).

We projected species location onto the binary presence/absence grid from either the MAXENT model or the mean model constructed in BIOMOD with the greatest model accuracy at each spatial extent and resolution (2.5 km and 500 m). Species location data found outside of the predicted presence were trimmed from the dataset for statistical analysis to examine the influence of independent variables on species presence. We examined the effect of the independent variables on the occurrence of Golden-winged Warblers or Blue-winged Warblers using a multivariate analysis of variance (MANOVA) in Program R (ver. 2.12; R Development Core Team 2011). To evaluate ecological factors that influence species habitat at a finer scale, we conducted a principal component analysis (PCA) of the data in Program R to determine the strength of multiple independent factors on Golden-winged Warblers present at the 500-m resolution. We extracted variable coefficients and principal component scores using the prcomp function.

RESULTS

We retained all variables in modeling, although elevation was weakly associated with maximum May and minimum May temperatures at this extent and scale (r = 0.661). Pearson correlation scores (r) obtained from the classification tree analysis of 22,822 warbler locations (Golden-winged Warblers, n = 8,266; Blue-winged Warblers, n = 20,556), identified elevation (r = 0.06), and minimum breeding season temperature (TMIN5, r = 0.17) as the most influential predictors of warbler presence at the distribution-wide extent (Figure 3.2). We used this *a priori* distribution model to remove 523 extraneous locations from the initial dataset (Golden-winged Warblers, n = 112; Blue-winged Warblers, n = 411; Figure 3.1a). We projected the remaining location data onto the extent of the regions at the 2.5-km scale (Figure 3.1b), which removed duplicate records and produced a final set of presence locations for Golden-winged Warblers (n = 2,012) and Blue-winged Warblers (n = 1,077), Golden-winged Warbler–Blue-winged

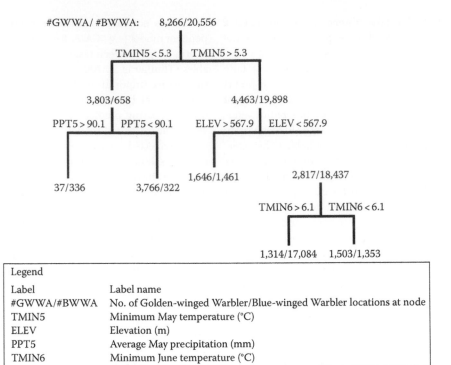

Figure 3.2 Classification tree depicting abiotic variables that were influential in predicting the breeding distribution of Golden-winged Warblers and Blue-winged Warblers at the 2.5-km scale. Environmental conditions that helped to distinguish areas occupied by the species are shown at each node in the tree. The number of warbler locations associated with each condition (Golden-winged Warbler numbers depicted to the left of the bracket and Blue-winged Warblers to the right of the bracket) is shown below in each branch.

Warbler cooccurrence locations (n = 135), Golden-winged Warbler absences (n = 824), Blue-winged Warbler absences (n = 1,246), and pseudo-absence locations for both species (n = 2,070; Supplementary Information B). We also projected location data into a 500-m grid, which removed duplicate records and yielded presence locations for Golden-winged Warblers (n = 1,845) and Blue-winged Warblers (n = 470), Golden-winged Warbler–Blue-winged Warbler cooccurrence of locations (n = 115), 1,367 true absences (Golden-winged Warblers, n = 654; Blue-winged Warblers, n = 713), and pseudo-absences for both species (n = 1,367).

Combined Region Analyses at 2.5-km Scale

We retained all variables in modeling at this scale, although elevation was weakly associated with TMIN5 (r = 0.712). Models of potential habitat demonstrated moderate support for the Great Lakes and Appalachian Mountains regions combined, with AUC > 0.739, and the MAXENT model showed the highest support with AUC > 0.899 for both species (Supplementary Information C). Elevation, breeding season temperatures, and land cover were the most strongly associated variables with occurrence of the species at the 2.5-km extent and resolution (Supplementary Information D). Compared to Blue-winged Warblers, Golden-winged Warblers were associated with higher elevations, cooler breeding season temperatures, and forests composed of aspen (*Populus* spp.). We excluded 86 species locations (Golden-winged Warblers, n = 55; Blue-winged Warblers, n = 31) that fell outside of the area predicted as potential habitat from the best supported model for statistical analyses. Breeding season temperature and elevation explained about 10% of the variation in the data (full MANOVA results not shown), though in opposite directions for species (Table 3.1). Golden-winged Warblers were found at an average elevation of 427 m and a mean maximum May temperature of 20.1°C, whereas Blue-winged Warblers were at 370 m and 21.3°C (Table 3.1). In the combined 2.5-km resolution region analysis,

TABLE 3.1

Mean values (±SE) of explanatory climatic and ecological variables associated with the presence of Golden-winged Warblers and Blue-winged Warblers within the combined Great Lakes and Appalachian Mountains regions at 2.5-km scale

Parameter	Golden-winged Warblers	Blue-winged Warblers	Range (P ≤)
Sample size	n = 2,012	n = 800	
Elevation (m)	427.2 ± 5.1	370.4 ± 3.4	355.2–1,500 (0.001)
Maximum May temperature (°C)	20.1 ± 0.1	21.3 ± 0.04	13.7–15.2 (0.042)
Minimum May temperature (°C)	7.1 ± 0.5	8.8 ± 0.04	3.1–23.3 (0.021)
Percent deciduous forest/area	60.7 ± 6.9	57.3 ±15.3	23.4–64.9 (0.001)
Percent of area that is coniferous	20.1 ± 7.0	15.2 ± 0.8	12.1–19.7 (0.000)
Percent in agriculture	4.5 ± 0.3	10.1 ± 0.5	3.3–12.2 (0.008)
Percent in grassland mosaic	12.7 ± 0.2	8.5 ± 0.2	0.9–8.4 (0.001)
Percent in woodland mosaic	16.4 ± 0.1	7.7 ± 0.4	5.3–22.8 (0.001)

NOTES: Elevation, temperature, and forest type per area values are the averages for each species or group; the values reported for the remaining parameters are the average percent of unique location records for each species or group. Data analysis was constrained to the study extent and duplicate species records were removed by projecting data into a grid at the appropriate scale.

Standard error and P-values are based on MANOVA analysis for Golden-winged Warbler versus Blue-winged Warbler data.

both species were found in roughly equal numbers in grassland mosaic habitats (Golden-winged Warblers, 12.7%; Blue-winged Warblers, 8.5%). However, Golden-winged Warbler locations were more than twice as likely to be found in woodland mosaics (20.2%) compared to Blue-winged Warblers (9.6%; Table 3.1). The greatest percentage of Golden-winged Warbler locations at the 2.5-km extent and resolution were associated with landscapes where quaking aspen (*Populus tremuloides*) was found (39.5%), compared with a smaller percentage of Blue-winged Warbler locations (7.0%; Figure 3.3). A greater percentage of Blue-winged Warbler locations were associated with agricultural landscapes (10.1%) compared to Golden-winged Warbler locations (4.5%).

Great Lakes Region at 2.5-km Scale

The best supported four models with the strongest association with Golden-winged Warbler and Blue-winged Warbler presences/absences at this scale demonstrated AUC values ≥0.721, and the MAXENT model showed the highest support with AUC > 0.835 (Supplementary Information C). Climatic and ecological variables with the strongest association with Golden-winged Warbler presence from the mean model were elevation, summer breeding temperature, landcover type, and the percent of deciduous forest found within a 2.5-km grid cell (Supplementary Information D). We excluded 24 species locations

(Golden-winged Warblers, n = 15; Blue-winged Warblers, n = 9) that fell outside of the area predicted as potential habitat from the best supported model for statistical analyses. Golden-winged Warbler locations had a mean elevation of 335.2 m and a minimum May temperature of 4.7°C, versus a mean of 240.9 m and 7.2°C, respectively, for Blue-winged Warblers (Table 3.2). Twice as many Golden-winged Warbler locations occurred in 2.5-km grid cells that also contained aspen forests compared to Blue-winged Warblers, a large percentage of which were associated with nonforested landscapes (see "No trees," Figure 3.4a,b). A greater percentage of Blue-winged Warbler locations occurred in agricultural landscapes (23.9%) compared to Golden-winged Warblers (6.8%). Golden-winged Warblers demonstrated an affinity for forests with open or broken canopy and with a greater percentage of locations in woodland mosaics (30.7%) compared with grassland mosaics (15.5%). The opposite pattern occurred with Blue-winged Warblers, where 34.6% and 14.3% of locations were in grassland and woodland mosaics, respectively (Table 3.2, Figure 3.5).

Great Lakes Region at 500-m Scale

Pearson correlation scores (r) from pairwise comparisons of explanatory variables were <0.8. Distribution models were well supported; models with the greatest support demonstrated AUC values >0.743 (Supplementary Information C).

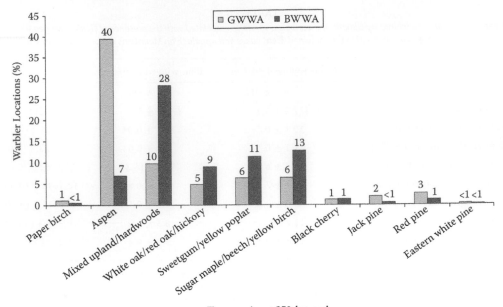

Figure 3.3 The most common tree species found in a 2.5-km area across the Great Lakes and Appalachian Mountains regions associated with occurrences of Golden-winged Warblers and Blue-winged Warblers. The association of each warbler species to each vegetation class is given as a percentage of the total location records for that species (see Table 3.1 for sample sizes).

TABLE 3.2

Mean values (±SE) of explanatory climatic and ecological variables associated with the presence of Golden-winged Warblers and Blue-winged Warblers in the Great Lakes Region at 2.5-km scale

Parameter	Golden-winged Warblers	Blue-winged Warblers	Range (P ≤)
Sample size	n = 1,363	n = 539	
Elevation (m)	335.2 ± 2.7	240.9 ± 7.1	28.5–549.0 (0.001)
Maximum May temperature (°C)	19.1 ± 0.03	19.7 ±0.1	14.1–24.5 (0.080)
Minimum May temperature (°C)	4.7 ± 0.03	7.2 ± 0.1	2.4–8.5 (0.001)
Minimum June temperature (°C)	9.8 ± 0.03	11.4 ± 0.1	7.4–13.4 (0.001)
Percent deciduous forest/area	50.6 ±0.4	48.9 ±1.1	0–80.0 (0.091)
Percent coniferous forest/area	23.6 ± 0.4	19.2 ± 1.0	0–80.0 (0.551)
Percent in agriculture	6.8 ± 0.3	23.9 ±0.5	0–25.2 (0.001)
Percent in grassland mosaic	15.5 ± 2.5	34.6 ± 2.2	8–37.3 (0.002)
Percent in woodland mosaic	30.7 ± 1.1	14.3 ± 1.2	11.2–18.4 (0.001)

NOTES: Elevation, temperature, and forest type values are the averages for each species or group; the values reported for the remaining parameters are the average percent of unique location records for each species or group. Data analysis was constrained to the study extent and duplicate species records were removed by projecting data into a grid at the appropriate scale.

Standard error and P-values are based on MANOVA analysis for Golden-winged Warbler versus Blue-winged Warbler data.

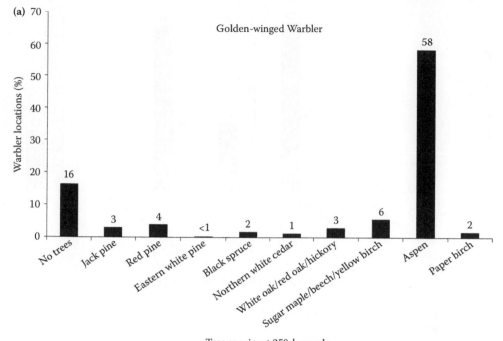

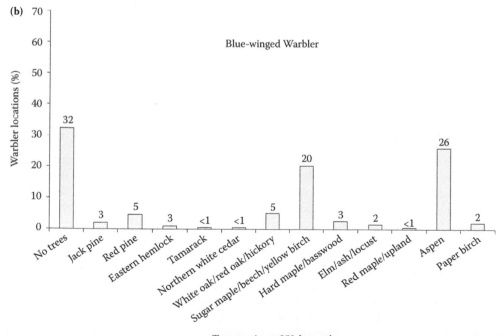

Figure 3.4 The most common tree species found in a 2.5 km area across the Great Lakes region for (a) Golden-winged Warblers and (b) Blue-winged Warblers. The association of each warbler species with each dominant forest cover is given as a percentage of the total location records (see Table 3.2 for sample sizes).

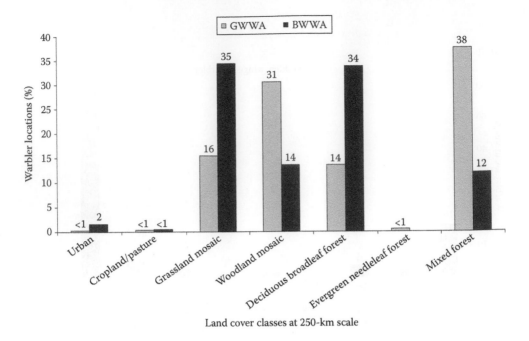

Figure 3.5 The percentage of Golden-winged Warbler (GWWA) and Blue-winged Warbler (BWWA) locations found in each of seven land cover classes in the Great Lakes region at the 2.5-km scale. See Table 3.2 for sample sizes.

The mean model from BIOMOD demonstrated high support (AUC > 0.806) for both species (Supplementary Information C). Variables that were most strongly associated with Golden-winged Warblers and Blue-winged Warblers across the subregions in the Great Lakes region included percent deciduous forest, percent agriculture, and elevation (Supplementary Information D). We excluded 13 species locations (Golden-winged Warblers, n = 11; Blue-winged Warblers, n = 2) that fell outside of the area predicted as potential habitat from the best supported model for statistical analyses. At a 500-m scale, both Golden-winged Warblers and Blue-winged Warblers were strongly associated with yellow birch (*Betula alleghaniensis*)/sugar maple (*Acer saccharum*) forests, although Golden-winged Warblers were also found in woodlands composed of aspen/birch (*Betula* spp.) and jack pine (*Pinus banksiana*), black spruce (*Picea mariana*)/tamarack (*Larix laricina*), or swamp forests (Figure 3.6a). About 13% of Blue-winged Warbler locations were associated with beech (*Fagus* spp.)/sugar maple/basswood (*Tilia* spp.) forests and an additional 11% of locations were found in agricultural land-cover types at lower elevations (Table 3.3; Figure 3.6a).

A PCA on the matrix of climatic and ecological climatic and parameters (Supplementary Information B) for Golden-winged Warbler presence/absence data across subregions at the 500-m scale in the Great Lakes region identified an "eastern" subregion (E) distributed in northern New York and a "western" subregion (W) in Michigan, Minnesota, and Wisconsin (Figure 3.6b). The variables that distinguished Golden-winged Warbler locations across subregions included land-cover type and percent canopy closure (first rotated factor) and elevation (second rotated factor) that explained 40% of the cumulative variation in the data. The first function of canopy closure and land-cover type explained 17% of the total variation. Golden-winged Warblers in both subregions were more commonly associated with yellow birch/sugar maple forests (Figure 3.6b). However, Golden-winged Warblers in the eastern subregion were located in oak (*Quercus* spp.)/pine (*Pinus* spp.) and swamp forests more often than the average occurrence of these forest types for the region as a whole (F = 5.49, df = 2, S.E. = 0.364, P = 0.001; Figure 3.6b). About 10% of Golden-winged Warbler locations were associated with fallow pasture in the eastern subregion (Figure 3.6b). Elevation represented the second function on the PCA and explained 13% of the variance in the data. Golden-winged Warblers in the western subregion occupied a higher average elevation (367 m) compared to the

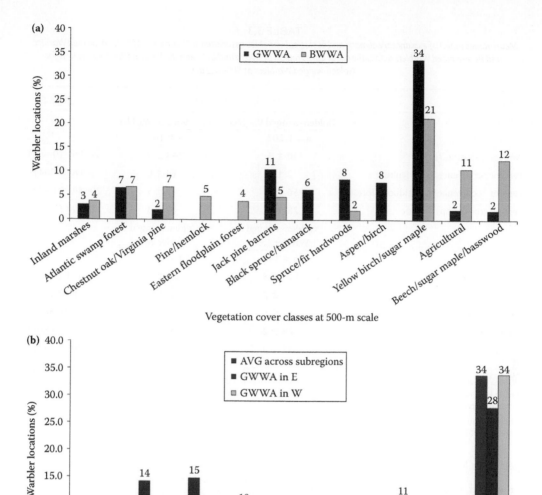

Figure 3.6 The distribution of vegetation classes for (a) Golden-winged Warblers (GWWA) and Blue-winged Warblers (BWWA) across subregions and (b) between Golden-winged Warbler subregions (eastern [E] and western [W] subregions) in the Great Lakes region at the 500-m scale. The association of each warbler species to each vegetation class is given as a percentage of the total location records for that species (see Table 3.3 for sample sizes).

eastern subregion (135 m; Table 3.3). Blue-winged Warbler locations in the same subregions demonstrated lower average elevation values (154 m for the eastern subregion and 303 m for the western subregion), and a large number of Blue-winged Warbler locations were also found in yellow birch/sugar maple forests throughout the subregions, though <1.0% of the species locations were associated with aspen/birch forests (data not shown).

TABLE 3.3

Mean values (±SE) of explanatory climatic and ecological variables associated with presence of Golden-winged Warblers and Blue-winged Warblers across the Great Lakes region at 500 m and between the East and West subregions for Golden-winged Warblers at 500-m scale

Parameter	Species/Region		Range (P ≤)
	Golden-winged Warbler	Blue-winged Warbler	
	n = 1,203	n = 104	
Average elevation (m)	340 ± 3	243 ± 9	34–551 (0.001)
Percent occurrence in agriculture	2 ± 1	10 ± 0	1–18 (0.004)
Percent occurrence in deciduous forest	56 ± 3	51 ± 3	20–80 (0.843)
Percent tree canopy closure	78 ± 1	74 ± 3	1–100 (0.062)
	Golden-winged Warbler East	Golden-winged Warbler West	
	n = 141	n = 104	
Average elevation (m)	135 ± 3	367 ± 2	34–551 (0.001)
Percent occurrence in agriculture	13. ± 2	2 ± 4	0–18 (0.025)
Percent occurrence in deciduous forest	52 ± 6	52 ± 4	22–60 (0.889)
Percent tree canopy closure	78 ± 2	79 ± 1	1–100 (0.867)
Percent in canopy height 10–25 m	65 ± 9	77 ± 10	0–80 (0.417)

NOTES: Elevation values are the averages for each species or region, values reported for remaining parameters are the average percent of unique location records for each species or group. Data analysis was constrained to the study extent and duplicate species records were removed by projecting data into grid at the appropriate scale.

Standard Error and P-values are based on Multivariate Analysis of Variance (MANOVA) for Golden-winged Warbler and Blue-winged Warbler comparison data.

Appalachian Mountains Region at 2.5-km Scale

We removed average maximum May temperature from modeling because of a correlation with elevation ($r = 0.804$; but results for this variable are presented in Table 3.4 for comparison to Blue-winged Warblers). The top four models with the strongest association with Golden-winged Warbler and Blue-winged Warbler presence/absence contained AUC values ≥0.914 and the mean model from BIOMOD demonstrated strong support (AUC > 0.914) for both species (Supplementary Information C). Maximum May temperature, minimum May temperature, elevation, and percent deciduous forest were the top variables associated with Golden-winged Warbler and Blue-winged Warbler presence/absence (Supplementary Information D; Table 3.4).

We removed 22 species locations (Golden-winged Warblers, n = 2; Blue-winged Warblers, n = 20) that fell outside of the area predicted as potential habitat from the best supported model prior to statistical analyses. In the Appalachian Mountains region, Golden-winged Warblers were associated with areas characterized by an average elevation of 659 m compared with an average elevation of 419 m for Blue-winged Warblers (Table 3.4). Golden-winged Warblers and Blue-winged Warblers in the Appalachian Mountains region were found almost exclusively within forested landscapes (% occurrence in deciduous forest in Table 3.4, Figure 3.7a). No single forest type accounted for >20% of occupied sites for either species (Figure 3.7a), but we observed variation in the forest subcanopy associated with each species. About 20% of Golden-winged Warblers were associated with forests dominated by a subcanopy of yellow poplar (*Liriodendron tulipifera*) compared to only 8% for Blue-winged Warblers (Figure 3.7a).

Appalachian Mountains Region at 500-m Scale

Pearson correlation scores (r) from pairwise comparisons of explanatory variables were <0.8. Models for Golden-winged Warbler and Blue-winged Warbler presence/absence in the Appalachian Mountains region at the 500-m scale generally had weaker support when compared to other models, with the range of AUC values from 0.660 to 0.884 (Supplementary Information C). The MAXENT model showed the greatest support

TABLE 3.4

Mean values (± SE) of explanatory climatic and ecological variables associated with presence of Golden-winged Warblers and Blue-winged Warblers in the Appalachian Mountains region at the 2.5-km scale.

Parameter	Golden-winged Warbler	Blue-winged Warbler	Range (P ≤)
	n = 648	n = 538	
Elevation (m)	658 ± 8	419 ± 9	28–1613 (0.000)
Maximum May temperature (°C)	21.2 ± 0.04	21.7 ± 0.1	17.1–25.3 (0.002)
Minimum May temperature (°C)	7.7 ± 0.03	8.8 ± 0.04	5.1–11.1 (0.003)
Minimum June temperature (°C)	12.1 ± 0.03	14.6 ± 0.04	9.7–18.7 (0.003)
Percent deciduous forest/area	62 ± 1	61 ± 1	59–64 (0.018)
Percent in deciduous forest	62.9 ± 0.6	86 ± 1	45–80 (0.018)
Percent in agriculture	3 ± 1	4 ± 1	2–6 (0.064)

NOTES: Elevation, temperature and forest type values are averages for each species or group, values reported for remaining parameters are the average percent of unique location records for each species or group. Data analysis was constrained to the study extent and duplicate species records were removed by projecting data into grid at the appropriate scale.

Standard error and P-values are based on Multivariate Analysis of Variance (MANOVA) analysis for Golden-winged Warbler vs. Blue-winged Warbler data.

for both species, with a range of AUC from 0.832 to 0.863 (Supplementary Information C). Elevation, the percent cover of deciduous forest, the percent of tree canopy closure, the percent cover of agricultural land, and vegetation cover type (EVT) had the strongest associations with Golden-winged Warbler presence across the subregions, based on the mean model (Supplementary Information D). A total of 12 species locations for Golden-winged Warblers that fell outside of the area predicted as potential habitat from the best supported model was excluded from statistical analyses. Golden-winged Warbler locations were reported from deciduous forests comprised mainly of yellow poplar and eastern hemlock (*Tsuga canadensis*) or montane oak forests at an average elevation of 680 m (Figure 3.7b; Table 3.5). In contrast, Blue-winged Warblers were associated with pine/hemlock forests at an average elevation of 435 m (Table 3.5). In addition, a greater percentage of Blue-winged Warbler locations were associated with human-disturbed land-cover classes, such as urban development, quarries, mining, or agricultural activities (Figure 3.7b).

A PCA conducted on the matrix of climatic and ecological parameters for Golden-winged Warbler presence/absence locations across subregions at the 500-m scale (Supplementary Information A) identified a southern subregion in the southern Appalachian Mountains ("Southern" [S], in TN, NC, KY, VA, and WV), a subregion in southeastern New York ("Southeastern New York" [SENY]), and a subregion in northern Pennsylvania

("Northern Pennsylvania" [NPA]; Table 3.5). The subregions were distinguished by elevation (first rotated factor), percent deciduous forest (second rotated factor), and vegetation cover type and percent agricultural land cover (third rotated factor) that explained collectively 46.6% of the variation in the data. The first component in the PCA (19% of the variation) represented differences in the average elevation across the subregions (Table 3.5). The second component (15.4% of the variation) described differences in the vegetation cover type occupied by Golden-winged Warblers across the subregions. Forest tree species composition derived from vegetation cover type for Golden-winged Warblers varied across subregions. Montane oak and yellow poplar/eastern hemlock forests were the predominate tree compositions in the Southern subregion; pine/hemlock, fallow pasture, and yellow poplar/eastern hemlock were the most common vegetation compositions in Southeastern New York, and pine/hemlock and chestnut oak (*Q. prinus*)/Virginia pine (*Pinus virginiana*) were the most common tree compositions in the northern Pennsylvania subregion (28.6%, Figure 3.8a). For the same extent and resolution, the most common forest composition associated with Blue-winged Warbler locations was mixed stands of beech, maple (*Acer* spp.), and basswood (Figure 3.8b). The amount of land cover associated with human disturbance was greatest in the Northern Pennsylvania subregion for Golden-winged Warblers (7.1%; Figure 3.8a), with a larger percentage of Blue-winged Warblers associated

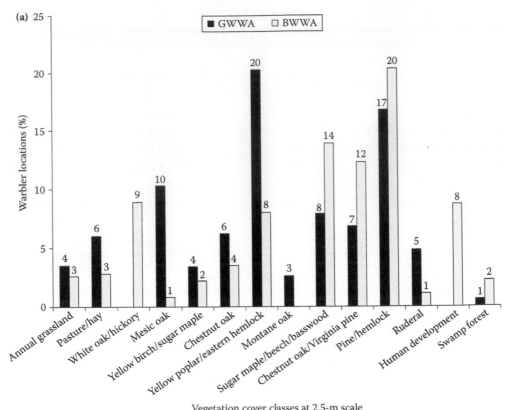

Vegetation cover classes at 2.5-m scale

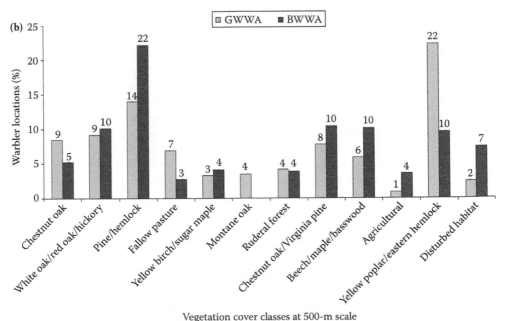

Vegetation cover classes at 500-m scale

Figure 3.7 (a) The distribution of vegetation classes for Golden-winged Warblers and Blue-winged Warblers in the Appalachian Mountains region at the 2.5-km scale. Samples sizes used for the comparison are shown in Table 3.4. (b) Vegetation classes associated with Golden-winged Warblers and Blue-winged Warblers across subregions in the Appalachian Mountains region at the 500-m scale. The association of each warbler species to each vegetation class is given as a percentage of the total location records for that species (see Table 3.5 for sample sizes).

TABLE 3.5

Mean values (±SE) of explanatory climatic and ecological variables associated with presence of (Top) Golden-winged Warblers and Blue-winged Warblers across the Appalachian Mountains region at 500-m scale, and (Bottom) between the Southern Appalachian Mountains (S), southeastern New York (SENY) and northern Pennsylvania (NPA) subregions for Golden-winged Warblers at 500-m scale

Parameter	Species/Subregion			Range (P ≤)
	Golden-winged Warblers	Blue-winged Warblers		
	n = 579	n = 366		
Average elevation (m)	680 ± 12	435 ± 12		226–1788 (0.001)
Percent in agriculture	1 ± 2	4 ± 1		0–7 (0.007)
Percent in deciduous forest	75 ± 1	64 ± 0		60–88 (0.061)
Percent tree canopy closure	84 ± 1	86 ± 1		4–100 (0.045)
	Golden-winged Warblers			
	S	SENY	NPA	
	n = 178	n = 366	n = 98	
Average elevation (m)	919 ± 18	661 ± 12	308.0 ± 11	337–1788 (0.001)
Percent in agriculture	1 ± 0	4 ± 2	3 ± 0	0–10 (0.001)
Percent in deciduous forest	82 ± 3	68 ± 1	75 ± 1	60–92 (0.001)
Percent tree canopy closure	85 ± 2	65 ± 1	90 ± 2	75–100 (0.001)
Percent in canopy height (10–25 m)	55 ± 2	65 ± 2	86 ± 3	0–100 (0.029)

NOTES: Elevation values are the averages for each species or group, values reported for remaining parameters are the average percent of unique location records for each species or group. Data analysis was constrained to the study extent and duplicate species records were removed by projecting data into grid at the appropriate scale.

Standard error and P-values are based on Multivariate Analysis of Variance (MANOVA) analysis for Golden-winged Warbler vs. Blue-winged Warbler data.

with disturbed habitats and agricultural landscapes (Figure 3.8b). Average elevation for Golden-winged Warblers ranged from 308 m for the Northern Pennsylvania subregion to 919 m in the Southern subregion (Table 3.5). Average elevations for Blue-winged Warblers also varied across subregions from a minimum of 250 m in Southeastern New York to 584 m in Northern Pennsylvania to a maximum of 630 m in the Southern subregion (data not shown).

DISCUSSION

Great Lakes and Appalachian Mountains Regions Combined

The results of landscape-scale analyses within the combined regions were difficult to interpret and likely obscured by differences in ecophysiography between the regions. For example, the range in elevation of sites used by Golden-winged Warblers and Blue-winged Warblers in the Great Lakes region was 48–547 m—an elevational range of just 499 m. In contrast, the range of sites used in the Appalachian Mountains region was 1,609 m (range = 28–1,637 m). When modeling data from both regions were combined, observed differences in elevation were largely driven by the availability of high-elevation sites in the Appalachian Mountains region and the lack of an elevation gradient in the Great Lakes region. Furthermore, confounding the problem of interpretation, Golden-winged Warblers are known to use different cover types in each region at the territory scale (A. M. Roth et al., unpubl. plan), potentially making landscape-scale requirements specific to each region.

The model from the combined regions had clear results. Associations between each warbler species and a specific land-cover type that was not evident at smaller spatial scales became apparent at larger spatial scales. For example, mosaic habitats might be too infrequent at the individual region scale to produce statistically meaningful results, but when the two regions were combined,

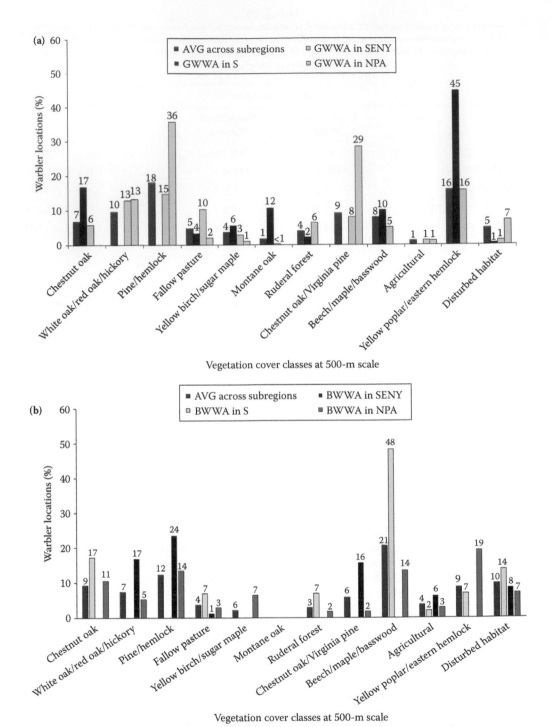

Figure 3.8 Vegetation classes associated with (a) Golden-winged Warblers and (b) Blue-winged Warblers between subregions (Southern [S], Southeastern New York [SENY], and Northern Pennsylvania [NPA]) in the Appalachian Mountains region at the 500-m scale. The association of each warbler species to each vegetation class is given as a percentage of the total location records for that species. Representative sample sizes are shown in brackets.

this land-cover type became a better predictor of explaining a significant amount of variation in the model. This result is likely due to the increased statistical power resulting from the larger samples size when the two regions are taken as a whole.

One commonality in the way that Golden-winged Warblers and Blue-winged Warblers appear to be partitioning habitat across regions at the 2.5-km resolution was in the relative degree of open woodland versus grassland/savanna land-use types. At this scale, Blue-winged Warblers were found in grassland and woodland mosaics in near equal proportions (Table 3.1). Conversely, Golden-winged Warblers were found in woodland mosaics much more frequently than grassland mosaics and Golden-winged Warblers were found in woodland mosaics roughly twice as often as Blue-winged Warblers, while controlling for differences in density (Table 3.1). Also at 2.5-km resolution, Blue-winged Warblers were found in agricultural landscapes more than twice as frequently as Golden-winged Warblers (Table 3.1). Yet, the proportion of forest associated with Golden-winged Warbler sites (deciduous + coniferous = 74.8%) and Blue-winged Warbler sites (deciduous + coniferous = 72.8%) was approximately equal (Table 3.1). Although both species selected landscapes with similar amounts of forest cover, Golden-winged Warblers may have favored less disturbed sites where forest was largely unfragmented by agriculture or human development.

Great Lakes Region

The Great Lakes region holds 95% of the global population of Golden-winged Warblers (A. M. Roth et al., unpubl. plan) and continues to maintain populations with lower hybridization rates compared to the Appalachian Mountains region (Vallender et al. 2009). Thus, habitat protection and management for Golden-winged Warblers that minimizes cooccurrence with Blue-winged Warblers is essential in this region to achieve range-wide population recovery. Golden-winged Warbler and Blue-winged Warbler locations differed in climate and elevation in the Great Lakes region, likely related to the subregional areas occupied by the two species. Golden-winged Warbler was the only species present in roughly the northern two-thirds of the region where temperatures were cooler and elevations higher, whereas in the southern one-third of the region where the two species were sympatric, temperatures were

warmer and there was little elevational variation to influence species-level settlement patterns.

Similar to results for the combined region analysis, at both the 2.5-km and 500-m resolution in the Great Lakes region, Blue-winged Warblers used agricultural landscapes three times more than Golden-winged Warblers (Tables 3.2 and 3.3). The presence of Golden-winged Warblers on agricultural land cover was inconsistent when eastern and western Great Lakes subregions were compared. For example, 13% of Golden-winged Warbler locations in the eastern subregion occurred in agricultural landscapes, whereas only about 3% occurred in this land-use type in the western subregion. The higher percentage of Golden-winged Warblers in less-preferred agricultural landscapes in the eastern subregion is likely an artifact of this land-cover type being dominant in the region. The consequence of Golden-winged Warblers in agricultural areas brings the species into more frequent contact with Blue-winged Warblers, thereby increasing the chance for hybridization. In the western subregion, however, less development in the forested landscape may provide Golden-winged Warblers the choice to avoid agricultural landscapes and subsequent contact with Blue-winged Warblers.

Golden-winged Warblers were most often associated with forested landscapes (mixed, deciduous, and woodland mosaic) in the Great Lakes region, representing over 50% of all occupied sites. Within forests, aspen (58.3% of sites at 2.5-km scale) was by far the most important forest type for Golden-winged Warblers, whereas Blue-winged Warbler sites were distributed across a variety of forest types, including aspen (26.0%), sugar maple/beech/yellow birch (20.3%), and white oak (Q. alba)/red oak (Q. rubra)/hickory (Carya spp., 5.2%). Based on an association of Golden-winged Warblers with forested landscapes in the Great Lakes region, conservation should focus on maintaining or enhancing landscape-scale forest cover, including aspen forest types while minimizing the influence of agriculture (Figures 3.3 and 3.4).

Appalachian Mountains Region

The Appalachian Mountains region supports 5% of the global Golden-winged Warbler population (A. M. Roth et al., unpubl. plan). Active conservation is important to prevent extirpation of key population segments in an effort to reconnect the now geographically isolated Great Lakes and Appalachian Mountains populations. The environmental variables strongly associated with Golden-winged Warbler presence that were identified here can provide valuable insight for directing conservation efforts.

In the Appalachian Mountains region, the primary landscape characteristic upon which Golden-winged Warblers and Blue-winged Warblers partitioned habitat was elevation. Region-wide at the 2.5-km scale, Golden-winged Warblers occupied sites with an average elevation >200 m higher than sites occupied by Blue-winged Warblers. Both species were most commonly found in landscapes characterized by high proportions (>60%) of deciduous and mixed forest types. Unlike the Great Lakes region, however, neither species showed a strong affinity for forests of a specific composition.

As in the Great Lakes region, Golden-winged Warblers occurred less frequently in human-disturbed cover types when compared to Blue-winged Warblers. At the 2.5-km scale, 8.7% of Blue-winged Warbler sites were located in human-disturbed landscapes, whereas no Golden-winged Warblers were found in this land-use type. At the 500-m scale, about 11.0% of Blue-winged Warbler sites were in human-disturbed or agricultural landscapes, but only 3.4% of Golden-winged Warbler sites were found in these land-use types. Conservation for Golden-winged Warblers in the Appalachian Mountains region, as a result, should focus on habitat maintenance and creation at elevations >500 m in forested landscapes with limited agriculture and other sources of human disturbance.

Perhaps the most important conclusion from our work is the confirmation that Golden-winged Warblers are most commonly found in forested landscapes and the species requires substantial amounts of surrounding forest to occupy a given site for breeding. Most field guides and scientific publications describe Golden-winged Warblers as a species of "early successional habitat" or "young forest" (Confer and Knapp 1981, Hunter et al. 2001, Confer 2008, Confer et al. 2010, Thogmartin 2010). The habitat description may be true at the scale of the nest site but omits critical landscape-scale information, because a patch of young forest set in a largely agricultural landscape is unlikely to be occupied by Golden-winged Warblers.

The prevailing view of Golden-winged Warblers requirements for breeding habitat is likely a result of species flexibility toward territory-scale land-cover types, which include abandoned farmland, alder swamps, reclaimed surface mines, and high-elevation pastures—all cover types that are not consistently associated with forest. Adding to the complex relationship between Golden-winged Warblers and habitat use is that not all land-cover types are created equal in space or in time. Habitat selection by a female Golden-winged Warbler is a trade-off between the use of forest edge to maximize nest success and the use of forest interiors to maximize fledgling survival and the nature of this relationship varies over the breeding season (Streby et al. 2014; Chapter 8, this volume). Despite the variable nature of habitat selection and use by the species, forested habitat plays a key role in the breeding success of Golden-winged Warblers. Other researchers have clearly showed a positive relationship between landscape-scale forest and presence of Golden-winged Warblers at the Prairie Hardwood Transition region of the Great Lakes (Thogmartin 2010). Whereas these studies add important insight into the habitat selection of Golden-winged Warblers at local to regional extents, our analysis is the first to quantify and affirm the importance of this relationship in shaping the range-wide breeding distribution of the species.

Performance of Models

We examined output from seven models across 10 combinations of spatial extent and resolution. Most model predictions of warbler distribution across all extents and scales were moderately well supported, with AUC values of 0.722–0.976 where AUC values >0.70 are considered robust model performance (Supplementary Information C; Swets 1988). Several of the modeling approaches consistently ranked among the top four models selected for each warbler extent and resolution. The maximum entropy algorithm in MAXENT was among the top supported models in every analysis. MAXENT AUC values were comparable to those from other top models. The classification trees approach demonstrated consistently poor discriminatory ability and was not selected in any of the 10 models (Supplementary Information C). The CTA tended to overfit during the calibration process (AUC > 0.900) and demonstrated poor fit to the evaluation data with AUC values between 0.634 and 0.926 (Supplementary Information C). CTA models do not incorporate interactive terms, which limits the model's ability to predict species responses to environmental cues (Thuiller et al. 2003). The linear model in the GLM algorithm was also not selected as a top supported model in any of the analyses. The poor discriminatory ability of GLM was unsurprising given the tendency for species to exhibit nonlinear responses to environmental variability (Guisan and Zimmermann 2000). By contrast, both the GAM (selected in 7/10 models) and GBM (selected in 9/10 models) exhibited robust predictive ability. This result was more intuitive, given that both methods are best suited for modeling complex nonparametric relationships (Thuiller et al. 2009).

Sampling bias in species location data is often cited as a confounding variable in species distribution models (Boria et al. 2014). Bias may be due to intense data collection during a specific period such as during a 2005 U.S. birding event or from a specific place due to intense sampling at and near a study site (Hijmans et al. 2000). Not accounting for bias can often lead to spurious conclusions regarding potential habitat, largely as a consequence of overfit of the model to the calibration data, which then yields poor fit of the model to evaluation data (Boria et al. 2014).

Our approach minimized the influence of sampling bias in five different ways. Only recent warbler location data were selected for analysis (1998–2010) and we chose land-cover and climatic data from the same area and period. We removed extraneous locations using a classification tree and reverse jackknife, and we removed duplicate location records by projecting the data into 2.5-km and 500-m extents (grids). As a final treatment, we removed location data that projected outside of an optimally predicted potential distribution from the top supported model. Excluded points were those that projected outside of predicted presence from the binary presence/absence grid calculated as the mean model in BIOMOD or from MAXENT. We applied the resultant data to examine land-cover and climate characteristics of Golden-winged Warblers and Blue-winged Warblers. Generally, models were well supported. Poor performance in some models may be due more to forcing a linear function (e.g., GLM) onto a nonlinear relationship between dependent and independent variables rather than to sampling bias, though additional tests may be needed.

The application of multiple models can also reinforce model performance and model fit (Thuiller et al. 2009). Our conclusions regarding the potential distribution of Golden-winged Warblers and Blue-winged Warblers are reinforced through the evaluation of multiple models, where each model represents a different fit of the predictor variables to the response variable and therefore an independent test of the data. Despite the advantages of this approach, our study was limited in that the MAXENT model algorithm is external to the algorithms in BIOMOD (Thuiller et al. 2009). The separate algorithm environments limited a robust comparison between MAXENT and the models in BIOMOD. Recent changes internalize the maximum entropy algorithm in MAXENT inside BIOMOD2, which allows for better comparison between model output. Future research using BIOMOD2 may help to refine the complex relationship between warbler presence/absence and environmental variables.

SUPPLEMENTARY INFORMATION A

Ecological variables examined in the study

Name	Description	Range	Source
NLCD2006	16 land cover classes	11–95	http://www.mrlc.gov/nlcd06_data.php
EVC	33 vegetation cover classes (2008)	11–129	http://www.landfire.gov/vegetation.php
EVT	Most common of 134 vegetation-type classes per unit area (2008)	11–2559	http://www.landfire.gov/vegetation.php
EVH	22 vegetation height classes (2008)	11.0–112	http://www.landfire.gov/vegetation.php
ELEV	Elevation in meters	0–1614.0	http://ned.usgs.gov/Ned/about.asp
TREE	141 tree species classes	101–995	http://webmap.ornl.gov/wcsdown/
PPT5	Average precipitation (mm) for May from 1981 to 2010	31.0–173.0	http://www.prism.oregonstate.edu/normals/

(Continued)

Name	Description	Range	Source
TMAX5	Average maximum temperature for May from 1981 to 2010	14–28.0	http://www.prism.oregonstate.edu/normals/
TMIN5	Average minimum temperature, for May from 1981 to 2010	2.0–16.1	http://www.prism.oregonstate.edu/normals/
TMAX6	Average maximum temperature for June from 1981 to 2010	3.0–23.2	http://www.prism.oregonstate.edu/normals
TMIN6	Average minimum temperature, June from 1981 to 2010	7.0–20.2	http://www.prism.oregonstate.edu/normals/
% CONIFER	% coniferous forest from NLCD 2006	0–100.0	http://www.mrlc.gov/nlcd06_data.php
% DECID	% deciduous forest from NLCD 2006	0–100.0	http://www.mrlc.gov/nlcd06_data.php
% AGRIC	% agricultural land from NLCD 2006	0–100.0	http://www.mrlc.gov/nlcd06_data.php
% CANOPY	% tree canopy closure from EVC 2008 (500 m)	1.0–100.0	http://www.landfire.gov/vegetation.php
% WOOD	% woodland mosaic from EVT 2008	0–100.0	http://www.landfire.gov/vegetation.php
% GRASS	% grassland mosaic from EVT 2008	0–100.0	http://www.landfire.gov/vegetation.php

SUPPLEMENTARY INFORMATION B

Sample sizes for unique species locations for Golden-winged Warblers (GWWA) and Blue-winged Warblers (BWWA), locations of sympatry (GWBW) and species-absence locations (ABS) across the regions (Great Lakes [GL] and Appalachian Mountains [Apps]), within each region (GL, Apps) at the 2.5-km and 500-m scales.

Species	GL and Apps, 2.5 km	GL, 2.5 km	GL, 500 m	Apps, 2.5 km	Apps, 500 m
Golden-winged Warbler	2,012	1,363	1,203	648	642
Blue-winged Warbler	1,077	539	104	538	366
GWBW	135	42	32	93	83
ABS	2,070	824	559	1,242	808
Total	5,017	2,768	1,898	2,521	1,899

SUPPLEMENTARY INFORMATION C

Area under the receiver operating characteristic (AUC) values for each model, the mean AUC for the best four models (AVG), and model accuracy scores for the top performing model (LPT/TSS).

Model	Species	MXT	CTA	GAM	GBM	GLM	MARS	RF	AVG	LPT[a]/TSS[b]
A₁	GWWA	0.900	0.718	0.735	0764	0.728	0.743	0.808	0.804	0.712/0.902
	BWWA	0.936	0.729	0.728	0.782	0.724	0.740	0.797	0.814	0.696/0.802
B₁	GWWA	0.836	0.715	0.740	0.773	0.741	0.748	0.789	0.786	0.696/0.813
	BWWA	0.952	0.702	0.722	0.761	0.718	0.727	0.787	0.791	0.678/0.780

(Continued)

Model	Species	MXT	CTA	GAM	GBM	GLM	MARS	RF	AVG	LPT[a]/TSS[b]
B₂	GWWA	**0.802**	0.737	0.722	**0.812**	0.721	**0.744**	**0.975**	0.833	0.602/0.733
	BWWA	**0.800**	0.736	**0.781**	**0.807**	0.775	**0.779**	0.775	0.792	0.723/0.636
C₁	GWWA	**0.974**	0.926	**0.935**	0.888	0.909	**0.943**	**0.957**	0.953	0.835/0.889
	BWWA	**0.976**	0.685	**0.915**	**0.916**	0.710	0.700	**0.946**	0.938	0.682/0.776
C₂	GWWA	**0.832**	0.634	**0.661**	**0.771**	0.657	0.657	**0.884**	0.787	0.562/0.666
	BWWA	**0.863**	0.713	**0.761**	**0.764**	0.757	**0.757**	0.705	0.786	0.675/0.551

NOTE: The results are shown for each species at the 2.5-km resolution for (A₁) across regions, (B₁) in the Great Lakes region, and (C₁) in the Appalachian Mountains region and at the 500-m resolution for the same extents (Models B₂, C₂). The four models with the strongest support at each extent and scale are shown (bold). Model abbreviations are reported in text.

[a] The LPT, or the lowest value of the prediction for any of the presence locations.
[b] True skill statistic, a measure of the predictive performance of each model independent of prevalence.

SUPPLEMENTARY INFORMATION D

Top model variables and measure of variable importance (in brackets) extracted from the BIOMOD mean model that differentiates GWWA and BWWA habitats at each spatial extent and scale (2.5-km and 500-m scales).

Range-Wide, 2.5 km	GL, 2.5 km	GL, 500 m	Apps, 2.5 km	Apps, 500 m
ELEV (0.191)	ELEV (0.102)	ELEV (0.112)	ELEV (0.357)	ELEV (0.556)
TMAX5 (0.740)	TMAX5 (0.711)	% AGRIC (0.443)	% DECID (0.325)	% DECID (0.124)
TMIN5 (0.172)	TMIN5 (0.811)	% DECID (0.088)	% TMAX5 (0.402)	% AGRIC (0.111)
% DECID (0.185)	% DECID (0.176)		% TMIN5 (0.428)	% CANOPY (0.112)
% WOOD (0.343)	% AGRIC (0.341)			EVT (0.108)
% GRASS (0.220)	TMIN6 (0.265)			
TREE (0.223)	NLCD (0.189)			

NOTE: Model abbreviations are reported in text.

ACKNOWLEDGMENTS

We greatly appreciate the assistance provided by the Golden-winged Warbler Working Group and the participants of eBird and the Golden-winged Warbler Atlas Project for collecting population survey data. Financial assistance for data collection and analysis was provided by the National Fish and Wildlife Foundation and U.S. Fish and Wildlife Service (Regions 3 and 5).

LITERATURE CITED

Allouche, O., A. Tsoar, and R. Kadmon. 2006. Assessing the accuracy of species distribution models: prevalence, kappa and the true skill statistic (TSS). Journal of Applied Ecology 43:1223–1232.

Angelo-Marini, M., M. Barbet-Massin, J. Martinez, N. P. Prestes, and F. Jiguet. 2010. Applying ecological niche modeling to plan conservation actions for the Red-spectacled Amazon Parrot (*Amazona pretrei*). Biological Conservation 143:102–112.

Araujo, M. B., and M. New. 2006. Ensemble forecasting of species distributions. Trends in Ecology and Evolution 22:42–47.

Araújo, M. B., M. Cabeza, W. Thuiller, L. Hannah, and P. H. Williams. 2004. Would climate change drive species out of reserves? An assessment of existing reserve-selection methods. Global Change Biology 10:1618–1626.

Bakermans, M. H., J. L. Larkin, B. W. Smith, T. M. Fearer, and B. C. Jones. 2011. Golden-winged Warbler habitat Best Management Practices for forestlands in Maryland and Pennsylvania. American Bird Conservancy, The Plains, VA.

Barbet-Massin, M., F. Jiguet, C. H. Albert, and W. Thuiller. 2012. Selecting pseudo-absences for species distribution models: how, where and how many? Methods in Ecology and Evolution 3:327–338.

Boria, R. A., L. E. Olson, S. M. Goodman, and R. P. Anderson. 2014. Spatial filtering to reduce sampling bias can improve the performance of ecological niche models. Ecological Modelling 275:73–77.

Brewster, W. 1874. A new species of North American warbler. American Sportsmen 5:33.

Buehler, D. A., A. M. Roth, R. Vallender, T. C. Will, J. L. Confer, R. A. Canterbury, S. B. Swarthout, K. V. Rosenberg, and L. P. Bulluck. 2007. Status and conservation priorities of Golden-winged Warbler (*Vermivora chrysoptera*) in North America. Auk 124:1439–1445.

Chapman, A. D. 2005. Principles of data quality, version 1.0. Report for the Global Biodiversity Information Facility, Copenhagen, Denmark.

Confer, J. L. 2008. Golden-winged Warbler (*Vermivora chrysoptera*). Pp. 468–469 in K. J. McGowan and K. Corbin (editors), The second atlas of the breeding birds of New York State. Cornell University Press, Ithaca, NY.

Confer J. L., K. W. Barnes, and E. C. Alvey. 2010. Golden- and Blue-winged Warblers: distribution, nesting success, and genetic differences in two habitats. Wilson Journal of Ornithology 122:273–278.

Confer, J. L., and K. Knapp. 1981. Golden-winged Warblers and Blue-winged Warblers: the relative success of a habitat specialist and a generalist. Auk 98:108–114.

Daly, C., M. Halbleib, J. I. Smith, W. P. Gibson, M. K. Doggett, G. H. Taylor, J. Curtis, and P. P. Pasteris. 2008. Physiologically sensitive mapping of climatological temperature and precipitation across the conterminous United States. International Journal of Climatology 28:2031–2064.

Elith, J., C. H. Graham, R. P. Anderson, M. Dudik, S. Ferreir, A. Guisan, R. J. Hijmans, F. Huettmann, J. R. Leathwick, A. Lehmann, J. Li, L. G. Lohmann, B. A. Loiselle, G. Manion, C. Moritz, M. Nakamura, Y. Nakazawa, J. M. Overton, A. T. Peterson, S. J. Phillips, K. S. Richardson, R. Scachetti-Pereira, R. E. Schapire, J. Soberon, S. Williams, M. S. Wisz, and N. E. Zimmerman. 2006. Novel methods improve prediction of species' distributions from occurrence data. Ecography 29:129–151.

Fitzgerald, M. C., N. J. Sanders, S. Ferrier, J. T. Longino, M. D. Weiser, and R. Dunn. 2011. Forecasting the future of biodiversity: a test of single- and multi-species models for ants in North America. Ecography 34:836–847.

Fry, J., G. Xian, S. Jin, J. Dewitz, C. Homer, L. Yang, C. Barnes, N. Herold, and J. Wickham. 2011. Completion of the 2006 National Land Cover database for the conterminous United States. Photogrammetric Engineering and Remote Sensing 77:854–864.

Gill, F. B. 1980. Historical aspects of hybridization between Blue-winged and Golden-winged Warblers. Auk 97:1–18.

Gill, F. B. 2004. Blue-winged Warblers (*Vermivora pinus*) versus Golden-winged Warblers (*V. chrysoptera*). Auk 4:1014–1018.

Graham, C. H., S. R. Ron, J. C. Santos, C. J. Schneider, and C. Moritz. 2004. Integrating phylogenetics and niche models to explore speciation mechanisms in dendrobatid frogs. Evolution 58:1781–1793.

Guisan, A., and N. E. Zimmermann. 2000. Predictive habitat distribution models in ecology. Ecological Modelling 135:147–186.

Hastie, T., R. Tibshirani, and J. H. Friedman. 2009. The elements of statistical learning: data mining, inference, and prediction. 2nd edition. Springer-Verlag Publishing, New York, NY.

Herrick, H. 1874. Description of a new species of *Helminthophaga*. Proceedings of the Academy of Natural Sciences, Philadelphia 26:220.

Hijmans, R. J., K. A. Garrett, Z. Huamán, D. P. Zhang, M. Schreuder, and M. Bonierbale. 2000. Assessing the geographic representativeness of Genebank collections: the case of Bolivian wild potatoes. Conservation Biology 14:1755–1765.

Hijmans, R. J., L. Guarino, M. Cruz, and E. Rojas. 2001. Computer tools for spatial analysis of plant genetic resources data: 1. DIVA-GIS. Plant Genetic Resources Newsletter 127:15–19.

Hitch, A. T., and P. L. Leberg. 2006. Breeding distributions of North American bird species moving north as a result of climate change. Conservation Biology 21:534–539.

Hunter, W. C., D. A. Buehler, R. A. Canterbury, J. L. Confer, and P. B. Hamel. 2001. Conservation of disturbance-dependent birds in eastern North America. Wildlife Society Bulletin 29:440–455.

Jennings, S. B., N. D. Brown, and D. Sheil. 1999. Assessing forest canopies and understory illumination: canopy closure, canopy cover and other measures. Forestry 72:59–73.

Jiguet, F., M. Barbet-Massin, and P. Henry. 2010. Predicting potential distribution of two rare allopatric sister species, the globally threatened *Doliornis* cotingas in the Andes. Journal of Field Ornithology 81:325–339.

Johnson, D. H. 1980. The comparison of usage and availability measurements for evaluating resource preference. Ecology 61:65–71.

Klaus, N. A., and D. A. Buehler. 2001. Golden-winged Warbler breeding habitat characteristics and nest success in clearcuts in the southern Appalachian mountains. Wilson Bulletin 113:297–301.

LANDFIRE. [online]. 2013. LANDFIRE Existing Vegetation Type layer. U.S. Department of Interior, Geological Survey. http://landfire.cr.usgs.gov/viewer/ (8 May 2012).

Liu, C., P. M. Berry, T. P. Dawson, and R. G. Pearson. 2005. Selecting thresholds of occurrence in the prediction of species distributions. Ecography 28:385–393.

Lobo, J. M., A. Jiménez-Valverde, and R. Real. 2007. AUC: a misleading measure of the performance of predictive distribution models. Global Ecology and Biogeography 17:145–151.

Marmion M., M. Parviainen, M. Luoto, R. K. Heikkinen, and W. Thuiller. 2009. Evaluation of consensus methods in predictive species distribution modelling. Diversity and Distributions 15:59–69.

Olson, J. S. 1994. Global ecosystem framework-definitions. U.S. Geological Survey EROS Data Center Internal Report, Sioux Falls, SD.

Parkes, K. C. 1951. The genetics of the Golden-winged and Blue-winged Warbler complex. Wilson Bulletin 63:5–15.

Patton, L. L., D. S. Maehr, J. E. Duchamp, S. Fei, J. W. Gassett, and J. L. Larkin. 2010. Do the Golden-winged Warbler and Blue-winged Warbler exhibit species-specific differences in their breeding habitat use? Avian Conservation and Ecology 5:2.

Phillips, S. J., R. P. Anderson, and R. E. Schapire. 2006. Maximum entropy modeling of species geographic distributions. Ecological Modeling 190:231–259.

R Development Core Team. 2011. R: a language and environment for statistical computing. R Foundation for Statistical Computing, Vienna, Austria.

Raxworthy, C. J., E. Martinez-Meyer, N. Horning, R. A. Nussbaum, G. E. Schneider, M. A. Ortega-Huerta, and A. T. Peterson. 2003. Predicting distributions of known and unknown reptile species in Madagascar. Nature 426:837–841.

Royle, J. A., R. B. Chandler, C. Yackulic, and J. D. Nichols. 2012. Likelihood analysis of species occurrence probability from presence-only data for modelling species distributions. Methods in Ecology and Evolution 3:545–554.

Ruefenacht, B., M. V. Finco, M. D. Nelson, R. Czaplewski, E. H. Helmer, J. A. Blackard, G. R. Holden, A. J. Lister, D. Salajanu, D. Weyermann, and K. Winterberger. 2008. Conterminous U.S. and Alaska forest type mapping using Forest Inventory and Analysis data. Photogrammetric Engineering and Remote Sensing 74:1379–1388.

Saab, V. 1999. Importance of spatial scale to habitat use by breeding birds in riparian forests: a hierarchical analysis. Ecological Applications 9:135–151.

Sauer, J. R., J. E. Hines, and J. Fallon. 2008. The North American Breeding Bird Survey, results and analysis 1966–2007. Version 5.15.2008. USGS, Patuxent Wildlife Research Center, Laurel, MD.

Scheldeman, X., and M. van Zonneveld. 2010. Training manual on spatial analysis of plant diversity and distribution. Biodiversity International, Rome, Italy.

Stewart, O. C. 2002. Pp. 65–217 in Lewis, H. T., and M. K. Anderson (editors), Forgotten fires: Native Americans and the transient wilderness. University of Oklahoma Press, Norman, OK.

Streby, H. M., J. M. Refsnider, S. M. Peterson, and D. E. Andersen. 2014. Retirement investment theory explains patterns in songbird nest-site choice. Proceedings of the Royal Society of London B 281:20131834.

Sullivan, B. L., C. L. Wood, M. J. Iliff, R. E. Bonney, D. Fink, and S. Kelling. 2009. eBird: a citizen-based bird observation network in the biological sciences. Biological Conservation 142:2282–2292.

Swets, J. A. 1988. Measuring the accuracy of diagnostic systems. Science 240:1285–1293.

Syfert, M. M., M. J. Smith, and D. A. Coomes. 2013. The effects of sampling bias and model complexity on the predictive performance of MaxEnt species distribution models. PloS One 8:e55158.

Thogmartin, W. E. 2007. Effects at the landscape scale may constrain habitat relations at finer scales. Avian Conservation and Ecology 2:6.

Thogmartin, W. E. 2010. Modeling and mapping Golden-winged Warbler abundance to improve regional conservation strategies. Avian Conservation and Ecology 5:12.

Thuiller, W. 2004. Patterns and uncertainties of species range shifts under climate change. Global Change Biology 10:2020–2027.

Thuiller, W. B. 2003. BIOMOD—optimizing predictions of species distributions and projecting potential future shifts under global change. Global Change Biology 9:1353–1362.

Thuiller, W., M. Araújo, and S. Lavorel. 2003. Generalized models vs. classification tree analysis: predicting spatial distributions of plant species at different scales. Journal of Vegetation Science 14:669–680.

Thuiller W., B. Lafourcade, R. Engler, and M. Araujo. 2009. BIOMOD—a platform for ensemble forecasting of species distributions. Ecography 32:369–373.

Vallender, R., R. J. Robertson, V. L. Friesen, and I. J. Lovette. 2007. Complex hybridization dynamics between Golden-winged and Blue-winged Warblers (*Vermivora chrysoptera* and *Vermivora pinus*) revealed by AFLP, microsatellite, intron and mtDNA markers. Molecular Ecology 16:2017–2029.

Vallender, R., S. L. Van Wilgenburg, L. P. Bulluck, A. Roth, R. Canterbury, J. Larkin, R. M. Fowlds, and I. Lovette. 2009. Extensive range-wide mitochondrial introgression indicates substantial cryptic hybridization in the Golden-winged Warbler. Avian Conservation and Ecology 4:4.

Wilson, M. D., B. D. Watts, M. G. Smith, J. P. Bredlau, and L. W. Seal. 2007. Status assessment of Golden-winged Warblers and Bewick's Wrens in Virginia. Center for Conservation Biology Technical Report Series, CCBTR-07-02. College of William and Mary, Williamsburg, VA.

Wisz, M. S., R. J. Hijmans, A. T. Peterson, C. H. Graham, A. Guisan, and NCEAS Predicting Species Distributions Working Group. 2008. Effects of sample size on the performance of species distribution models. Diversity and Distributions 14:763–773.

CHAPTER FOUR

Genetic Insights into Hybridization between Golden-winged and Blue-winged Warblers*

Rachel Vallender and Roger D. Bull

Abstract. Hybridization between some pairs of closely related taxa is increasing due to anthropogenic factors such as habitat destruction and fragmentation, as well as an introduction of exotics. Outcomes of hybridization vary according to the species involved. With the advancement in molecular techniques in recent years, it is now possible to examine hybridization with far greater precision and to more fully understand hybridization systems and relationships between species. A better understanding of patterns and process can be used to inform appropriate conservation actions. Hybridization between Golden-winged Warblers (*Vermivora chrysoptera*) and Blue-winged Warblers (*V. cyanoptera*) has been ongoing for at least the past 140 years and is thought to be a primary cause of recent declines of Golden-winged Warblers in populations throughout their breeding distribution. Here, we summarize the current knowledge of the genetics of this hybridization system. Specifically, recent analyses demonstrated extensive introgression in populations of Golden-winged Warblers throughout their breeding distribution, and bidirectional gene flow in almost all surveyed populations. To date, the use of mitochondrial genetic markers has provided the best resolution for understanding the warbler hybridization system, and a need for informative nuclear markers remains. We conclude with suggested avenues for future genetic research to increase understanding of hybridization, and to guide conservation management for Golden-winged Warblers.

Key Words: conservation, introgression, species replacement, *Vermivora chrysoptera*, *Vermivora cyanoptera*.

Hybrid zones, the physical areas where two taxa meet and interbreed, represent natural laboratories where evolutionary processes can be studied in situ (Barton and Hewitt 1985, Hewitt 1988, Rohwer and Wood 1998). Hybridization in plants is common and frequently results in the formation of new species (Ehrlich and Wilson 1991). Although less common in animal taxa, hybridization events occur with an approximate 10% frequency in nonmarine avian species (Grant and Grant 1992, Allendorf et al. 2001, McCarthy 2006). In many cases, hybridization events occur as isolated incidences (Parkes 1978, McCarthy 2006, MacDonald et al. 2012), whereas in others they occur with great frequency (Gill 1980, Morrison and Hardy 1983). Because

* Vallender, R. and R. D. Bull. 2016. Genetic insights into hybridization between Golden-winged and Blue-winged Warblers. Pp. 67–80 in H. M. Streby, D. E. Andersen, and D. A. Buehler (editors). Golden-winged Warbler ecology, conservation, and habitat management. Studies in Avian Biology (no. 49), CRC Press, Boca Raton, FL.

many hybrid zones contain a variety of phenotypes it is possible to address questions of sexual selection (Møller and Birkhead 1994, Higashi et al. 1999, Sheldon and Ellegren 1999), genetic control of phenotypic characters (Gill and Murray 1972, Hewitt 1988), fitness of parental species (Rohwer and Wood 1998), and forces that separate the hybridizing species (Barton and Hewitt 1985, Brumfield et al. 2001). As hybridization increases in frequency due to human-caused habitat change and other factors (Gill 1997), studies of these systems become important for conservation (Haig 1998, Allendorf et al. 2001, Randler 2002), and maintenance of species (Barton and Hewitt 1985, Avise 1994, Brumfield et al. 2001, Yuri et al. 2009).

Worldwide decreases in biodiversity tend to be attributed primarily to habitat modification or destruction, unsustainable harvest, competition with introduced species, and chains of extinction in which the extinction of one species contributes to the extinction of one-or-more ecologically dependent taxa (Rhymer and Simberloff 1996). Less attention has been paid to the effects of hybridization, even though mixing of gene pools can threaten the existence of rare species and is problematic for undertaking some conservation initiatives (Barton and Hewitt 1985, Rhymer and Simberloff 1996, Randler 2002). Understanding the evolutionary consequences of hybridization is key to prevention of further extinctions for species engaging in such genetic admixture.

Over the last 20 years, molecular techniques have become important tools for studying evolutionary dynamics of hybrid zones. Gill (1997) found that asymmetries in gene flow due to a bias in mating behavior can have a pronounced influence on the genetic architecture of a hybrid population: an effect that can go undetected by morphological analyses alone (Avise 1994, Rhymer and Simberloff 1996, Curry 2005, Vallender et al. 2009). More recently, molecular studies documented that extreme cases of hybridization can lead to reductions in genetic differentiation between species and an eventual breakdown of species integrity (Mank et al. 2004).

Whereas molecular analyses can be uniquely informative, it is important to ensure careful selection of genetic markers and techniques when examining hybridization systems (van Dongen et al. 2012). For example, several studies documented discordance between plumage characters

and genetic analyses as an indication of the location of a hybrid zone (Brumfield et al. 2001; Vallender et al. 2007b, 2009), and variation in selection pressures throughout the genome may lead to underestimates of introgression (Brumfield et al. 2001, Yuri et al. 2009). Despite these challenges, genetic analysis of hybridization dynamics has the potential to add greater understanding of evolutionary relationships among groups and individuals.

Ultimately, hybridization is a question of mate choice. Why mate with a heterospecific when a conspecific is available? Genetic techniques can be especially informative in answering this question. Several well-studied avian hybridization systems include four species pairs: Collared (*Ficedula albicollis*) and Pied Flycatchers (*F. hypoleuca*), Black-capped (*Poecile atricapillus*) and Carolina Chickadees (*P. carolinensis*), Townsend's (*Setophaga townsendi*) and Hermit Warblers (*S. occidentalis*), and between Mallards (*Anas platyrhynchos*) and American Black Ducks (*A. rubripes*). In all four of these systems, analysis of genetic markers revealed clear patterns of mate choice. Collared and Pied Flycatchers paired with heterospecifics sought extra-pair copulations with conspecifics, presumably to minimize the fitness costs of heterospecific pairing (Veen et al. 2001). Furthermore, the sex ratio of young was biased in accordance with "Haldane's Rule" when offspring were produced through a heterospecific pairing (Veen et al. 2001). "Haldane's Rule" suggests that hybrids of the heterogametic sex are more likely to be infertile (Haldane 1922).

Female Black-capped and Carolina Chickadees both appeared to prefer Carolina Chickadee males for their extra-pair partners in a zone of sympatry (Reudink et al. 2006), perhaps due to the dominance of Carolina Chickadee males over Black-capped Chickadee males (Bronson et al. 2003; Reudink et al. 2006, 2007). Interestingly, in a hybridization system between Black-capped and Mountain Chickadees (*P. gambeli*), extra-pair copulations between Mountain Chickadee females and Black-capped Chickadee males appear to result from Black-capped Chickadee males being behaviorally dominant over Mountain Chickadee males (Grava et al. 2012).

A similar phenomenon is thought to occur in the Mallard and American Black Duck system where females of both species preferentially select more dominant Mallard males, perhaps because these males are more able to protect their mates from harassment by intruding males (including forced copulations), through predator alerting mechanisms, or by acquisition of superior feeding sites (Brodsky et al. 1988). Likewise, in the Townsend's Warbler and Hermit Warbler hybridization system, females of both species socially prefer the more aggressive, behaviorally dominant Townsend's Warbler males (Rohwer and Wood 1998, Pearson 2000), although this has not yet been quantified using genetic parentage analysis.

GOLDEN-WINGED × BLUE-WINGED WARBLERS

In eastern North America, there is a widespread mosaic hybrid zone between Blue-winged Warblers (*Vermivora cyanoptera*) and Golden-winged Warblers (*V. chrysoptera*). Hybrids of these two species were documented as early as 1870 (Herrick 1874) and hybridization is common (Gill 1980). The taxa likely occupied largely allopatric distributions prior to the early 1900s, after which the widespread abandonment of agricultural fields in eastern North America led to the growth of early-successional vegetation favored by both species, thereby putting them into secondary contact (Gill 1980). The breeding distribution of Blue-winged Warblers subsequently advanced northward into the Golden-winged Warbler range and, although their wide contact zone has steadily moved northwestward, hybridization remains extensive in areas where both species occur. Typically, Golden-winged Warbler breeding populations are negatively affected by hybridization and their local extirpation occurs within 50 years of the appearance of the first Blue-winged Warblers (Gill 1987, 1997). Two classic hybrid phenotypes known as "Brewster's" and "Lawrence's" warblers depicted in most popular field guides, were mistakenly ascribed new species names following their initial discovery (Parkes 1951). Subsequent examination of hybrid phenotypes revealed much more variation in plumage characters than was originally assumed (Short 1963, 1969; Vallender et al. 2009). Unlike other well-studied avian hybrid systems (Veen et al. 2001; Bronson et al. 2003, 2005; Grava et al. 2012), work examining mate-choice patterns is still largely lacking for the Blue-winged × Golden-winged Warbler hybridization system. Only one analysis of genetic mate choice has been carried out in this system (Vallender et al. 2007a), even though heterospecific extra-pair

fertilizations, which override assortative mating preferences, can play a significant role in hybridization and genetic introgression (Hartman et al. 2012). Vallender et al. (2007a) concluded that female Golden-winged Warblers do not appear to avoid hybrid males as either social partners or extra-pair sires, but their work was conducted in a region with a general paucity of hybrids and Blue-winged Warblers and they were unable to fully elucidate any pattern in mate-choice decisions by females of either species.

Mate-choice preferences are not fully understood, but a general pattern of the progression of hybridization has been recognized for this species pair: advancement of Blue-winged Warblers into the Golden-winged Warbler breeding distribution is followed by a period of hybridization and subsequent replacement of the Golden-winged Warbler phenotype with the Blue-winged Warbler phenotype. The predictable temporal sequence provides a context to examine hybridization in areas with different histories of contact and introgression.

The pattern of advancement and replacement by Blue-winged Warblers has clear conservation implications for Golden-winged Warblers, which have experienced precipitous population declines throughout much of their breeding distribution in the U.S. and Canada over the past 50 years (Gill 1980; Sauer et al. 2012; Chapter 1, this volume). Currently, Golden-winged Warblers are one of the most rapidly declining passerines in North America, with regional declines as high as 25% per year (2001–2011) and an average distribution-wide decline of 2.6% per year (Sauer et al. 2012). Since 2007, Golden-winged Warblers have been listed as a Threatened species under the Canadian Species At Risk Act (SARA 2015) and the species is currently under consideration for federal listing in the U.S. under the Endangered Species Act (U.S. Endangered Species Act 2015). Note that Blue-winged Warbler populations have also declined (as high as 6% per year) in some portions of their breeding distribution (2001–2011; Sauer et al. 2012). However, currently, Blue-winged Warblers are not protected, or under consideration for protection, by federal species-at-risk legislation in either Canada or the U.S.

The mechanisms by which the Blue-winged Warbler phenotype predictably replaces the Golden-winged Warbler phenotype following contact remain unclear. Male Blue-winged

Warblers may be behaviorally dominant over male Golden-winged Warblers (Will 1986), thereby obtaining better quality territories in areas of sympatry, or they may have a higher rate of extra-pair fertilizations with females of both species (Confer and Larkin 1998). In contrast, Ficken and Ficken (1968) suggested that Golden-winged Warblers are generally more aggressive than Blue-winged Warblers. That said, there appears to be no clear pattern of behavioral dominance, suggesting possible genetic dominance of Blue-winged Warblers over Golden-winged Warblers. Moreover, interspecific interactions may vary spatially: in some populations Blue-winged and Golden-winged Warblers do not appear to perceive each other as conspecifics and maintain overlapping territories (Confer and Knapp 1977), whereas in other locations males respond aggressively to individuals singing the other species' song (Ficken and Ficken 1968), albeit often with low frequency (Gill and Murray 1972). Reasons for regional variation in responsiveness to heterospecific song may relate to the history of contact between the species or may be due to variation in levels of genetic introgression.

EXAMINING HYBRIDIZATION USING MITOCHONDRIAL DNA

Molecular tools have been used since the 1990s to better understand the Blue-winged Warbler × Golden-winged Warbler hybridization system. In his pioneering mitochondrial DNA (mtDNA) work, Gill (1997) used restriction fragment length polymorphisms (RFLP) to differentiate Golden-winged Warblers from Blue-winged Warblers. He examined several populations with different histories of contact and hybridization and found that the two parental mitochondrial lineages differed by 3% sequence divergence, a level of separation that is on par with many other pairs of taxa that comprise clear biological species (reviewed by Johnson and Cicero 2004), with all individuals from allopatric populations possessing their species' typical mtDNA lineage. In a sympatric region of recent contact and known hybridization, however, Blue-winged Warbler mtDNA was present in high frequency in phenotypic Golden-winged Warblers and in almost all individuals with hybrid plumage phenotypes. Based on an unexpectedly pervasive and directional mtDNA introgression, Gill (1997) concluded that female Blue-winged

Warblers led the northward advance into the Golden-winged Warbler breeding distribution, and that F$_1$ hybrid females preferentially backcross with male Golden-winged Warblers.

Two studies further explored the pattern of mtDNA introgression at five sites with various histories of contact and documented levels of hybridization (Shapiro et al. 2004, Dabrowski et al. 2005). At none of these sites was there evidence of asymmetric mtDNA introgression, suggesting that Gill's (1997) findings may not be broadly applicable to other populations. Instead, Shapiro et al. (2004) found approximately equal frequencies of mismatches between phenotype and mtDNA haplotypes in a primarily Golden-winged Warbler population from West Virginia. Small samples from mixed populations in Michigan and Ohio also failed to show evidence of asymmetric introgression. Likewise, using similar analyses, Dabrowski et al. (2005) found bidirectional gene flow and long-term persistence of Golden-winged Warbler mtDNA haplotypes in a sympatric population in New York with more than a century of co-existence of these two species (Eaton 1914). Last, in an Ontario population thought to be in the earliest stages of hybridization, Golden-winged Warbler mtDNA haplotypes were found in both Golden-winged Warbler and phenotypic hybrid individuals (Dabrowski et al. 2005). The results indicate that genetic swamping by Blue-winged Warblers does not appear to be occurring in any of these populations.

Most recently, Vallender et al. (2009) surveyed 753 Golden-winged and Blue-winged Warbler samples from nine U.S. states and three Canadian provinces using a PCR-restriction assay modified from earlier mtDNA-sequencing methodologies. The only population in which all phenotypic Golden-winged Warblers had the Golden-winged Warbler mtDNA type was in Manitoba, near the northwestern edge of the Golden-winged Warbler breeding distribution. All other tested populations had some proportion of phenotype-mtDNA haplotype mismatch indicating presence of cryptic hybrids among phenotypically pure individuals of both species. Of 608 Golden-winged Warblers sampled, 49 (8.0%) showed mtDNA-phenotype mismatch, whereas 14 of the 145 (9.6%) Blue-winged Warblers sampled had a mismatch between their phenotype and mtDNA haplotype. Vallender et al. (2009) concluded that there are far fewer genetically pure populations of Golden-winged Warblers than previously believed.

Continued screening of samples from Canada since 2009 has further highlighted a pattern of apparent widespread introgression. Few Golden-winged or Blue-winged Warblers breed within Quebec (45.083°N, 74.217°W) where both species are considered rare (Gauthier and Aubry 1996, Sauer et al. 2012). Application of the PCR-restriction assay developed by Vallender et al. (2009) showed that rates of phenotype-mtDNA mismatch in Quebec are ~1.3% (one of eight Golden-winged Warblers samples screened had Blue-winged Warbler mtDNA). A slightly lower mismatch rate was observed in Golden-winged Warblers sampled from a site in eastern Ontario (44.617°N, 79.483°W; two of 27 Golden-winged Warblers had Blue-winged Warbler mtDNA), and in the westernmost portions of the province (48.833°N, 94.317°W; two of 28 Golden-winged Warblers had Blue-winged Warbler mtDNA; R. Vallender, unpubl. data).

Analyses of samples recently collected in Manitoba are especially informative. The first documented breeding occurrence of Golden-winged Warblers in Manitoba was in 1932 (Manitoba Avian Research Committee, pers. comm.). Consistent with the findings of Vallender et al. (2009), all 191 phenotypically-pure Golden-winged Warblers

sampled (since 2009) from Riding Mountain National Park (50.767°N, 99.500°W) in the midwestern region of the province had Golden-winged Warbler mtDNA (R. Vallender et al., unpubl. data). However, a male "Brewster's Warbler" was found at one of the study sites in 2010 (S. Van Wilgenburg, unpubl. data) and was determined to have Blue-winged Warbler mtDNA (R. Vallender, unpubl. data). Either the male immigrated from elsewhere or there were undetected female Blue-winged Warblers or cryptic hybrid female Golden-winged Warblers breeding in the region.

Given a lack of mtDNA-phenotype mismatch in the western regions of Manitoba, and within Minnesota, it is surprising that the highest rates of mtDNA introgression uncovered in Canada in recent years were in the southeastern most region of Manitoba (49.783°N, 96.467°W). A total of 126 adult Golden-winged Warblers were sampled from this region between 2010 and 2012 and 6.3% (8 of 128) had a phenotype-mtDNA haplotype mismatch (Figure 4.1). Of eight birds with Blue-winged Warbler mtDNA, seven were sampled in 2012 (out of a total sample of 64 collected that year; L. Moulton, unpubl. data). This latest finding suggests that advancement of Blue-winged Warbler-mtDNA is occurring rapidly in this region of the province. However, the number

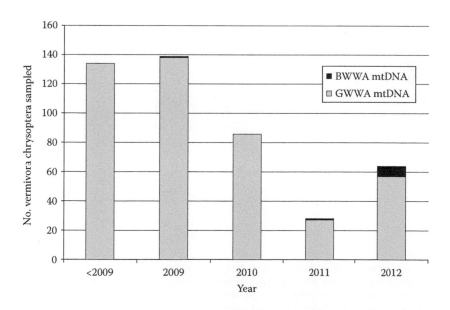

Figure 4.1. Results of mitochondrial DNA analysis of samples collected in Manitoba as of 2013. The light section of each bar represents *Vermivora chrysoptera* with *V. chrysoptera* mtDNA. The dark section of each bar represents *V. chrysoptera* with *V. cyanoptera* mtDNA (i.e., cryptic hybrids). BW mtDNA is *V. cyanoptera* mtDNA type; GW mtDNA is *V. chrysoptera* mtDNA-type. (Based on data from Vallender et al. 2009; R. Vallender, unpubl. data; L. Moulton, unpubl. data.)

of detected hybrids encountered remains low at ~3% of males on territories, and no Blue-winged Warblers were encountered through 2013 (L. Moulton, unpubl. data).

Currently, more Golden-winged Warblers have been sampled in Manitoba than anywhere else in North America (451 Golden-winged Warblers sampled through 2013). Continued sampling in the southeastern and midwestern portions of Manitoba over the next 10–15 years is necessary to monitor introgression. Likewise, additional sampling north to Duck Mountain (Manitoba) and Porcupine Hills (Saskatchewan; Figure 4.2) will be important to determine the genetic status of Golden-winged Warblers at the northwestern extent of their breeding distribution.

Compiling all available mtDNA data reveals a clear picture of widespread introgression in Golden-winged Warblers (Figure 4.2). Of the 12 states and provinces from which Golden-winged Warbler samples have been assayed, all but those from Michigan show some proportion of genetic or cryptic hybrids. In this analysis, Michigan is represented by only two samples. Given a known 30-year history of Golden-winged Warbler × Blue-winged Warbler hybridization (Will 1986), additional sampling is expected to reveal cryptic hybrid Golden-winged Warblers breeding in Michigan.

From a conservation perspective, the mtDNA analysis indicates Golden-winged Warblers in Manitoba appear only minimally affected by hybridization is encouraging. However, putting low levels of mtDNA introgression in Golden-winged Warbler populations breeding in Manitoba into the appropriate context is important. First, despite low rates of mtDNA introgression, there could be substantial introgression in the nuclear (nDNA) genome undetected by the studies completed to date. Second, only 1% of the global Golden-winged Warbler breeding population occurs in Manitoba (Buehler et al. 2007). Surveys provide evidence of increasing numbers of breeding Golden-winged Warblers in Manitoba (Manitoba Avian Research Committee 2003; 22.7%/year increase, n = 2 routes, 1988–2007, P = 0.42; Sauer et al. 2012) and

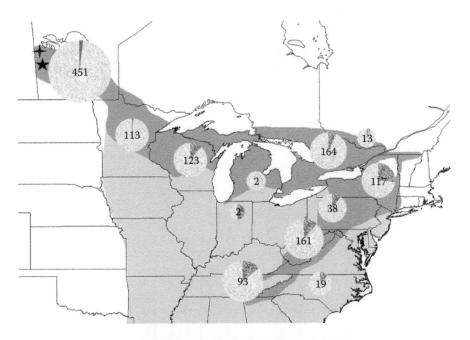

Figure 4.2. Proportion of *Vermivora chrysoptera* individuals assigned to the ancestral *V. chrysoptera* haplotype group across the breeding distribution, shown in dark gray shading on the map (approximate migration areas shown in light gray) based on a compilation of all available mitochondrial DNA results, 1998–2013. Light portions of pie charts indicate the proportion of *V. chrysoptera* samples assigned to the ancestral *V. chrysoptera* haplotype group; dark gray portions represent the proportion of *V. chrysoptera* samples assigned to the *V. cyanoptera* haplotype group as cryptic hybrids. Numbers within pie charts represent the number of *V. chrysoptera* sampled per state or province. Stars indicate sampling sites as the edge of the range at the Duck Mountains (Manitoba) and Porcupine Hills (Saskatchewan). (Based on data from Shapiro et al. 2004; Dabrowski et al. 2005; Vallender et al. 2009; L. Moulton, unpubl. data; R. Vallender, unpubl. data.)

recent, extensive field surveys suggest that earlier estimates of the number of Golden-winged Warblers breeding in Manitoba were underestimates (C. Artuso, unpubl. data). However, the density of Golden-winged Warblers in Manitoba is still low compared to breeding population densities in Minnesota, Wisconsin, or Ontario (which, support ~82% of the estimated global breeding population; Buehler et al. 2007).

Moreover, whereas the use of mtDNA markers has greatly improved understanding of hybridization between Golden-winged and Blue-winged Warblers, it provides only a partial window into patterns of hybridization given that mtDNA is maternally inherited and a non-recombining part of the avian genome. A better understanding of the patterns of introgression of nDNA in Golden-winged and Blue-winged Warblers is likely to be informative (Irwin et al. 2009, Hartman et al. 2012).

EXAMINING HYBRIDIZATION USING NUCLEAR DNA

Nuclear DNA is biparentally inherited and has the potential to improve the understanding of genetic introgression dynamics (Irwin et al. 2009, Vallender et al. 2009, Hartman et al. 2012). In the only nDNA study of Golden-winged and Blue-winged Warblers to date, Vallender et al. (2007b) used a combination of markers (intron sequences, microsatellites, and Amplified Fragment Length Polymorphisms [AFLP]) to differentiate between Golden-winged Warblers and Blue-winged Warblers from allopatric populations with no history of contact with the other species. One of the primary goals of this work was to create a panel of markers that could be applied broadly to quantify the ancestry of hybrid individuals. However, despite screening with a moderate panel of 13 microsatellite markers, analyses of allele frequency patterns suggested that the parental populations were weakly differentiated at these loci. Despite their high variability, microsatellite variation had little utility for distinguishing parental species, and did not prove useful for typing the ancestry of hybrid individuals. Likewise, no evidence of diagnostic allelic variation was found at any of the five intron loci. Sixteen variable nucleotide sites were found in >3,000 nucleotides of intron sequence, but each of the variable character states was found in alleles drawn from both parental populations.

A large panel (>1,000 separate fragments) of AFLP loci was generated using 54 different primer pairs. Despite a large number of variable fragments, Vallender et al. (2007b) found no characters that were fixed within species, although eight characters derived from seven different primer pairs showed significant frequency differences between Golden-winged and Blue-winged Warblers. Accordingly, this panel of AFLP loci was applied to 73 Golden-winged Warbler samples collected in southern Ontario where hybridization was thought to be in the initial stages with a low proportion of phenotypic hybrids and Blue-winged Warblers present. Surprisingly, almost a third of the phenotypic Golden-winged Warblers had introgressed genotypes indicating a substantial history of cryptic hybridization.

AFLP methodologies have been widely employed in plant, fungal, and bacterial research, but they have only recently been explored in animal studies (Bensch and Åkesson 2005, Toews and Irwin 2008, Rush et al. 2009, Sternkopf et al. 2010). Limitations include an inability to detect heterozygotes because characters are scored as either present or absent, but benefits of using AFLPs include the ability to rapidly sift through the genome for recent coalescences, a relatively short start-up time, and generation of data from hundreds of loci (Bensch and Åkesson 2005). In cases where other marker types have revealed weak genetic structure (such as Willow Warbler *Phylloscopus trochilus*; Bensch et al. 2002), the utility of AFLPs has been highlighted and may be useful when examining other animal hybridization systems (Bensch and Åkesson 2005).

However, generating AFLP data is time-consuming and relatively costly and not suitable for population-scale screening, which typically involves hundreds or thousands of samples. To date, attempts to circumvent these limitations by converting diagnostic AFLP fragments into a panel of single-locus, co-dominant markers have not been successful (R. Vallender, unpubl. data). It appears unlikely that AFLPs will be a useful nDNA tool to assess wide-scale patterns and levels of hybridization between Golden-winged and Blue-winged Warblers. Moreover, nDNA results to date indicate that nuclear genome differentiation between Golden-winged and Blue-winged Warblers is low and can only be detected using the most sensitive methods, consistent with the slow mutation rate and longer coalescence times of nDNA relative to mtDNA (Moritz et al. 1987, Zink and Barrowclough 2008).

A lack of effective nDNA markers is currently the factor most limiting improved understanding of the Golden-winged Warbler × Blue-winged Warbler hybridization system. The inclusion of nDNA data from Z-linked loci that coalesce relatively quickly (Carling and Brumfield 2008, Carling et al. 2010) is essential for a better understanding of the current level of hybridization and to track the progression of introgression as Blue-winged Warblers continue to advance northward into the breeding distribution of Golden-winged Warblers.

IMPLICATIONS OF HYBRIDIZATION

Based on extensive distribution-wide molecular research and identifying phenotypic hybrids in many previously allopatric populations, it is likely that a high percentage of breeding Golden-winged Warbler populations are impacted by hybridization with Blue-winged Warblers. This impact is surprisingly far-reaching and can be detected even in Golden-winged Warbler breeding populations that are geographically distant from Blue-winged Warbler populations. Golden-winged Warbler breeding populations in the northwestern portion of their breeding distribution in Minnesota and Manitoba are exhibiting the lowest levels of mtDNA introgression as of 2013, and therefore play a significant role in conservation of Golden-winged Warblers. Continued genetic screening of Golden-winged Warblers from the northwestern portion of their breeding distribution is a conservation priority, coupled with examining related issues such as mate choice, reproductive success, habitat use and selection, parasitic infection, survival and recruitment, and linkages to nonbreeding grounds.

Cumulative genetic evidence suggests that hybrids within the Golden-winged Warbler × Blue-winged Warbler hybridization complex are not affected by post-zygotic selection. To date, post-zygotic selection has been examined in relation to the ability to attract mates of the parental species through use of song by males (Harper et al. 2010), to produce viable offspring (Vallender et al. 2007a), to produce clutches of balanced sex ratio not biased towards the homogametic sex in accordance with "Haldane's Rule" (Neville et al. 2008), by the ability to successfully fledge young (Reed

et al. 2007, Vallender et al. 2007a), and the probability of parasitic infection (Vallender et al. 2012), although there appears to be regional variation in the vigor of hybrids (Confer and Tupper 2000, Confer 2006) and a general lack of knowledge about possible pre-zygotic selection against hybrid offspring. Moreover, the genetic data suggesting that hybrids are not at a disadvantage compared to genetically pure individuals of either parent species do not align with extensive field observations in two populations (New York and West Virginia) where selection against hybrid males by Golden-winged Warbler females has been documented (Confer 2006). Without quantifying reproductive success, together with genetic purity, it is difficult to determine if behavioral avoidance affects reproductive success for either species. However, the hypothesis that hybrid fitness may differ across the Golden-winged Warbler breeding distribution needs further examination.

Gill (1980) predicted that Golden-winged Warblers could be rare, or even extinct, by 2080. The prediction may not be accurate given the now-documented bidirectionality of gene flow between Golden-winged and Blue-winged Warblers, and persistence of the Golden-winged Warbler phenotype in at least one region where these two species co-exist, but the widespread and high frequency of hybridization suggests that it is unlikely that stable coexistence will occur over any large geographic region.

FUTURE APPLICATIONS OF GENETIC TECHNIQUES

The Golden-winged Warbler × Blue-winged Warbler hybridization system is complicated, variable, and unlikely to be halted by directed management actions. However, pursuit of the following five key research directions will increase understanding of the impact of hybridization on both Golden-winged and Blue-winged Warblers.

Identify Plumage Genes

Given their genetic similarity but strikingly dissimilar plumage characteristics, Golden-winged and Blue-winged Warblers are likely well-differentiated at ≥ 1 loci encoding plumage patterns. Plumage genes may be the most appropriate and useful regions of the genome to target for further research examining the genetic differentiation of these species. Moreover, in other avian species the differences in the number of alleles at the *melanocortin-1 receptor* gene influence the degree of melanism, and genes associated with plumage polymorphism influence mate preference (Uy and Borgia 2000, Mundy et al. 2004, Mundy 2005).

High-Throughput Sequencing

With the continuing development of high-throughput sequencing (HTS) technologies, generating genome-scale sequence data for nonmodel organisms continues to become more attainable. Therefore, HTS is being used more frequently in ecological and evolutionary studies (Lerner and Fleischer 2010, Ekblom and Galindo 2011, Steiner et al. 2013, Toews et al. 2016) and is proving effective even when applied to historical specimens, such as museum skins, despite the challenges of DNA degradation (Bi et al. 2013, Gansauge and Meyer 2013, Guschanski et al. 2013). Therefore, with the necessary analytical resources, HTS has great potential to reveal a much more complete picture of the genetics of the Golden-winged Warbler × Blue-winged Warbler hybridization system.

Specifically, applying HTS and comparative genomic analysis to the Golden-winged Warbler × Blue-winged Warbler hybridization system may better reveal the evolutionary history of these two species as it has recently for two other hybridizing species, grizzly bears (*Ursus arctos*) and polar bears (*U. maritimus*; Miller et al. 2012). Genomic analyses may give an accurate estimate of the time of divergence for Golden-winged and Blue-winged Warblers and potential shifts since divergence between allopatry and sympatry revealed as genetic signatures of ancient admixture.

Apart from clarifying the evolutionary history of these two species, data generated by HTS could be used to improve understanding of current populations and the extent of introgression across the breeding distributions of both species. In addition to generating genome-scale datasets, HTS techniques can be used to quickly develop thousands of informative molecular markers such as single-nucleotide polymorphisms (SNPs; Twyford and Enos 2012). Alternatively, HTS can be applied directly to population samples through genotyping-by-sequencing methodologies requiring no *a priori* knowledge of the genome of the study species (Elshire et al. 2011). These simple, low-cost methods are suitable for population-level studies and have recently been used to characterize the hybrid zone of Black-capped and Carolina Chickadees (Taylor et al. 2014). Using these approaches, Golden-winged and Blue-winged Warbler populations could be screened to assess interspecific gene flow much more effectively than is currently possible. Such an approach would likely yield a spatially-explicit representation of the extent and progression of introgressive hybridization. Additionally, estimates of the historical rate of advancement of introgression could be made by screening previously-collected blood samples and museum specimens.

Last, through comparisons with avian genome reference markers (Backström et al. 2008), genomic datasets could be used in studies of the genetic basis of adaptive phenotypic traits in Golden-winged and Blue-winged Warblers. Such an approach could lead to predictions of the potential fitness costs or benefits of introgressive gene flow for both species and such information, along with estimates of interspecific gene flow, is critical for conservation planning (Steiner et al. 2013).

Studying Mate Choice

Mate choice is a key factor driving gene flow in the Golden-winged Warbler × Blue-winged Warbler hybridization system and needs to be better understood. To this end, we recommend paternity analyses, combined with genetic-purity analyses, throughout the breeding distributions of Golden-winged Warblers and Blue-winged Warblers. Such analyses, especially in regions of extensive hybridization, where Golden-winged Warblers, Blue-winged Warblers, and their hybrids co-exist, will likely reveal mate-choice dynamics such as how often females select interspecific and hybrid mates, either as their social partners or via extra-pair copulations, and will enable the examination of why heterospecific mate choice decisions occur (Curry 2005, Vallender et al. 2007a, Hartman et al. 2012). Improving measures of realized reproductive success by including survival of fledglings might also help elucidate the adaptive significance of mate choice in Golden-winged Warblers (Streby et al. 2014).

Sampling at the Northern Boundary of Breeding Distribution

Continued sampling of Golden-winged Warblers in the northwestern portion of their breeding distribution (especially Minnesota and Manitoba) will further understanding of hybridization between Golden-winged and Blue-winged Warblers, by tracking the advancement of Blue-winged Warblers or their genes into otherwise genetically pure Golden-winged Warbler populations. Cryptic hybrids are being found with increasing frequency in the southeastern corner of Manitoba, but Golden-winged Warblers from the northwestern portion of Manitoba have not yet been sampled. If Golden-winged Warblers in northwestern Manitoba are determined to be unaffected by hybridization (based on both mtDNA and nDNA analyses), these populations may represent high-priority locations to focus management activities that benefit Golden-winged Warblers. Given the similar ecological niches of Golden-winged and Blue-winged Warblers (Confer 1992, Gill et al. 2001), it will be important to examine habitat × genetic interactions in regions of sympatry. Such analyses have strong potential to inform on-the-ground conservation action that aims to preferentially maintain and improve habitat for Golden-winged Warblers (A. M. Roth et al., unpubl. plan). Indeed, management and conservation efforts are likely to be most effective when focused on areas where a significant proportion of Golden-winged Warblers have been determined to be genetically pure (at least based on mtDNA assays), even though some admixture may be present in all populations. Habitat loss has also been recognized as a significant threat to Golden-winged Warbler persistence and many other shrubland species (A. M. Roth et al., unpubl. plan), and mitigating this threat may be one of the most straightforward conservation actions to undertake.

Treating Golden-winged Warblers and Blue-winged Warblers as a Species Pair

Of these two species, Golden-winged Warblers have received the vast majority of funding and research attention, and considerable information exists about their ecology and conservation. In contrast, only basic ecological information on breeding, mate choice, nonbreeding-ground habitat use, and migration exists for Blue-winged Warblers. We therefore recommend that research be conducted on life-history characteristics of Blue-winged Warblers, particularly in the phenotypically "pure" Blue-winged Warbler breeding distribution. Such work should include genetic analyses of mate choice in allopatry and sympatry, realized reproductive success, genetic purity analyses, and blood parasite load (e.g., Vallender et al. 2007a,b, 2012; Hartman et al. 2012) as these would provide for interesting comparisons with Golden-winged Warblers and would be broadly informative about the impact of hybridization on Blue-winged Warblers. Additional sampling of Blue-winged Warblers is also important, especially from regions where they have been poorly sampled to date such as both peninsulas of Michigan and within Missouri. Given their propensity to hybridize in all areas of contact with Golden-winged Warblers, unraveling individual life histories of Golden-winged and Blue-winged Warblers is unlikely and therefore studying them as a species pair is more likely to elucidate the causes and impacts of their widespread hybridization.

ACKNOWLEDGMENTS

A great many thanks to fellow members of the Golden-winged Warbler Working Group for support, friendship, providing samples to the various genetic projects, and doing some important collaborative work in recent years, and to F. Gill for carrying out the founding genetics work on the Golden-winged Warbler × Blue-winged Warbler system—it was inspirational. We hope that many of the recommended research avenues are pursued by members of the Golden-winged Warbler Working Group or graduate students in the coming years. Appreciation is given to the two anonymous reviewers who provided helpful comments on an earlier version of this chapter.

LITERATURE CITED

Allendorf, F. W., R. F. Leary, P. Spruell, and J. K. Wenburg. 2001. The problems with hybrids: setting conservation guidelines. Trends in Ecology and Evolution 16:613–622.

Avise, J. C. 1994. Speciation and hybridization. Pp. 252–305 in J. C. Avise (editor), Molecular markers, natural history and evolution. Chapman and Hall, London, UK.

Backström, N., S. Fagerberg, and H. Ellegren. 2008. Genomics of natural bird populations: a gene-based set of reference markers evenly spread across the avian genome. Molecular Ecology 17:964–980.

Barton, N. H., and G. M. Hewitt. 1985. Analysis of hybrid zones. Annual Review of Ecology and Systematics 16:113–148.

Bensch, S., and M. Åkesson. 2005. Ten years of AFLP in ecology and evolution: why so few animals? Molecular Ecology 14:2899–2914.

Bensch, S., M. Åkesson, and D. E. Irwin. 2002. The use of AFLP to find an informative SNP: genetic differences across a migratory divide in Willow Warblers. Molecular Ecology 11:2359–2366.

Bi, K., T. Linderoth, D. Vanderpool, J. M. Good, R. Nielsen, and C. Moritz. 2013. Unlocking the vault: next-generation museum population genomics. Molecular Ecology 22:6018–6032.

Brodsky, L. M., C. D. Ankney, and D. G. Dennis. 1988. The influence of male dominance on social interactions in Black Ducks and Mallards. Animal Behaviour 36:1371–1378.

Bronson, C. L., T. C. Grubb Jr., G. D. Sattler, and M. J. Braun. 2003. Mate preference: a possible causal mechanism for a moving hybrid zone. Animal Behaviour 65:489–500.

Bronson, C. L., T. C. Grubb Jr., G. D. Sattler, and M. J. Braun. 2005. Reproductive success across the Black-capped Chickadee (Poecile atricapillus) and Carolina Chickadee (P. carolinensis) hybrid zone in Ohio. Auk 122:759–772.

Brumfield, R. T., R. W. Jernigan, D. B. McDonald, and M. J. Braun. 2001. Evolutionary implications of divergent clines in an avian (Manacus: Aves) hybrid zone. Evolution 55:2070–2087.

Buehler, D. A., A. M. Roth, R. Vallender, T. C. Will, J. L. Confer, R. A. Canterbury, S. B. Swarthout, K. V. Rosenberg, and L. P. Bulluck. 2007. Status and conservation priorities of Golden-winged Warblers (Vermivora chrysoptera) in North America. Auk 124:1439–1445.

Carling, M. D., and R. T. Brumfield. 2008. Haldane's Rule in an avian system: using cline theory and divergence population genetics to test for differential introgression of mitochondrial, autosomal, and sex-linked loci across the Passerina bunting hybrid zone. Evolution 62:2600–2615.

Carling, M. D., I. J. Lovette, and R. T. Brumfield. 2010. Historical divergence and gene flow: coalescent analyses of mitochondrial, autosomal and sex-linked loci in Passerina buntings. Evolution 64:1762–1772.

Confer, J. L. 1992. Golden-winged Warbler. In A. Poole, P. Stettenheim, and F. Gill (editors), The birds of North America, No. 20. The Academy of Natural Sciences, Philadelphia, PA; The American Ornithologists' Union, Washington, DC.

Confer, J. L. 2006. Secondary contact and introgression of Golden-winged Warblers (Vermivora chrysoptera): documenting the mechanism. Auk 123:958–961.

Confer, J. L., and K. Knapp. 1977. Hybridization and interaction between Blue-winged and Golden-winged Warblers. Kingbird 27:181–190.

Confer, J. L., and J. L. Larkin. 1998. Behavioral interactions between Golden-winged and Blue-winged Warblers. Auk 115:209–214.

Confer, J. L., and S. K. Tupper. 2000. A reassessment of the status of Golden-winged and Blue-winged Warblers in the Hudson Highlands of southern New York. Wilson Bulletin 112:544–546.

Curry, R. L. 2005. Hybridization in chickadees: much to learn from familiar birds. Auk 122:747–758.

Dabrowski, A., R. Fraser, J. L. Confer, and I. J. Lovette. 2005. Geographic variability in mitochondrial introgression among hybridizing populations of Golden-winged (Vermivora chrysoptera) and Blue-winged Warblers (V. pinus). Conservation Genetics 6:843–853.

Eaton, E. G. 1914. Birds of New York (vol. 2). University of the State of New York, Albany, NY.

Ehrlich, P. R., and E. O. Wilson. 1991. Biodiversity studies: science and policy. Science 253:758–762.

Ekblom, R., and J. Galindo. 2011. Applications of next generation sequencing in molecular ecology of nonmodel organisms. Heredity 107:1–15.

Elshire R. J., J. C. Glaubitz, Q. Sun, J. A. Poland, K. Kawamoto, E. S. Buckler, and S. E. Mitchell. [online]. 2011. A robust, simple genotyping-by-sequencing (GBS) approach for high diversity species. PLoS One 6:e19379.

Ficken, M. S., and R. W. Ficken. 1968. Territorial relationships of Blue-winged Warblers, Golden-winged Warblers, and their hybrids. Wilson Bulletin 80:442–451.

Gansauge, M. T., and M. Meyer. 2013. Single-stranded DNA library preparation for the sequencing of ancient or damaged DNA. Nature Protocols 8:737–748.

Gauthier, J., and Y. Aubry. 1996. The breeding birds of Québec. Association Québécoise des groupes d'ornithologues. The Province of Québec Society for the Protection of Birds and the Canadian Wildlife Service, Environment Canada (Québec Region), Montreal, QC.

Gill, F. B. 1980. Historical aspects of hybridization between Blue-winged and Golden-winged Warblers. Auk 97:1–18.

Gill, F. B. 1987. Allozymes and genetic similarity of Blue-winged and Golden-winged Warblers. Auk 104:444–449.

Gill, F. B. 1997. Local cytonuclear extinction of the Golden-winged Warbler. Evolution 51:519–525.

Gill, F. B., R.A. Canterbury, and J. L. Confer. 2001. Blue-winged Warbler. In A. Poole and F. Gill (editors), The birds of North America, No. 584. The Academy of Natural Sciences, Philadelphia, PA; The American Ornithologists' Union, Washington, DC.

Gill, F. B., and B. G. Murray. 1972. Discrimination behavior and hybridization of the Blue-winged and Golden-winged Warblers. Evolution 26:282–293.

Grant, P. R., and B. R. Grant. 1992. Hybridization of bird species. Science 256:193–197.

Grava, A., T. Grava, R. Didier, L. A. Lait, J. Dosso, E. Koran, T. M. Burg, and K. A. Otter. 2012. Interspecific dominance relationships and hybridization between Black-capped and Mountain Chickadees. Behavioral Ecology 23:566–572.

Guschanski, K., J. Krause, S. Sawyer, L. M. Valente, S. Bailey, K. Finstermeier, R. Sabin, E. Gilissen, G. Sonet, Z. T. Nagy, G. Lenglet, F. Mayer, and V. Savolainen. 2013. Next-generation museomics disentangles one of the largest primate radiations. Systematic Biology 62:539–554.

Haig, S. M. 1998. Molecular contributions to conservation. Ecology 79:413–425.

Haldane, J. B. S. 1922. Sex-ratio and unisexual sterility in hybrid animals. Journal of Genetics 12:101–109.

Harper, S. L., R. Vallender, and R. J. Robertson. 2010. Male song variation and female mate choice in the Golden-winged Warbler. Condor 112:105–114.

Hartman, P. J., D. P. Wetzel, P. H. Crowley, and D. F. Westneat. 2012. The impact of extra-pair mating behavior on hybridization and genetic introgression. Theoretical Ecology 5:219–229.

Herrick, H. 1874. Description of a new species of *Helminthophaga*. Proceedings of the National Academy of Sciences 26:220.

Hewitt, G. M. 1988. Hybrid zones—natural laboratories for evolutionary studies. Trends in Ecology and Evolution 3:158–167.

Higashi, M., G. Takimoto, and N. Yamamura. 1999. Sympatric speciation by sexual selection. Nature 402:523–526.

Irwin, D. E., A. Brelsford, D. P. L. Toews, C. MacDonald, and M. Phinney. 2009. Extensive hybridization in a contact zone between MacGillivray's Warblers *Oporornis tolmiei* and Mourning Warblers *O. philadelphia* detected using molecular and morphological analyses. Journal of Avian Biology 40:539–552.

Johnson, N. K., and C. Cicero. 2004. New mitochondrial DNA data affirm the importance of Pleistocene speciation in North American birds. Evolution 58:1122–1130.

Lerner, H. R. L., and R. C. Fleischer. 2010. Prospects for the use of next-generation sequencing methods in ornithology. Auk 127:4–15.

MacDonald, C. A., T. Martin, R. Ludkin, D. J. T Hussell, D. Lamble, and O. P. Love. 2012. First report of a Snow Bunting × Lapland Longspur hybrid. Arctic 65:344–348.

Mank, J. E., J. E. Carlson, and M. C. Brittingham. 2004. A century of hybridization: decreasing genetic distance between American Black Ducks and Mallards. Conservation Genetics 5:395–403.

McCarthy, E. M. 2006. Handbook of avian hybrids of the world. Oxford University Press, New York, NY.

Miller, W., S. C. Schuster, A. J. Welch, A. Ratan, O. C. Bedoya-Reina, F. Shao, H. Lim Kim, R. C. Burhans, D. I. Drautz, N. E. Wittekindt, L. P. Tomsho, E. Ibarra-Laclette, L. Herrera-Estrella, E. Peacock, S. Farley, G. K. Sage, K. Rode, M. Obbard, R. Montiel, L. Bachmann, O. Ingólfsson, J. Aars, T. Mailund, Ø. Wiig, S. L. Talbot, and C. Lindqvist. 2012. Polar and brown bear genomes reveal ancient admixture and demographic footprints of past climate change. Proceedings of the National Academy of Sciences USA 36:E2382–E2390.

Møller, A. P., and T. R. Birkhead. 1994. The evolution of plumage brightness in birds is related to extra-pair paternity. Evolution 48:1089–1100.

Moritz, C., T. E. Dowling, and W. M. Brown. 1987. Evolution of animal mitochondrial DNA: relevance for population biology and systematics. Annual Review of Ecology and Systematics 18:269–292.

Morrison, M. L., and J. W. Hardy. 1983. Hybridization between Hermit and Townsend's Warblers. Murrelet 64:65–72.

Mundy, N. I. 2005. A window on the genetics of evolution: mC1R and plumage colouration in birds. Proceedings of the Royal Society of London Series B 276:3809–3838.

Mundy, N. I., N. S. Badcock, T. Hart, K. Scribner, K. Janssen, and N. J. Nadeau. 2004. Conserved genetic basis of a quantitative plumage trait involved in mate choice. Science 303:1870–1873.

Neville, K. J., R. Vallender, and R. J. Robertson. 2008. Nestling sex ratio of Golden-winged Warblers *Vermivora chrysoptera* in an introgressed population. Journal of Avian Biology 39:599–604.

Parkes, K. C. 1951. The genetics of the Golden-winged × Blue-winged Warbler complex. Wilson Bulletin 63:5–15.

Parkes, K. C. 1978. Still another Parulid intergeneric hybrid (Mniotilta × Dendroica) and its taxonomic and evolutionary implications. Auk 95:682–690.

Pearson, S. F. 2000. Behavioral asymmetries in a moving hybrid zone. Behavioral Ecology 11:84–92.

Randler, C. 2002. Avian hybridization and mixed pairing. Animal Behaviour 63:674–685.

Reed, L. P., R. Vallender, and R. J. Robertson. 2007. Provisioning rates by Golden-winged Warblers. Wilson Journal of Ornithology 119:350–355.

Reudink, M. W., S. G. Mech, and R. L. Curry. 2006. Extra-pair paternity and mate choice in a chickadee hybrid zone. Behavioral Ecology 17:56–62.

Reudink, M. W., S. G. Mech, S. P. Mullen, and R. L. Curry. 2007. Structure and dynamics of the hybrid zone between Black-capped Chickadee (Poecile atricapillus) and Carolina Chickadee (P. carolinensis) in southeastern Pennsylvania. Auk 124:463–478.

Rhymer, J. M., and D. Simberloff. 1996. Extinction by hybridization and introgression. Annual Review of Ecology and Systematics 27:83–109.

Rohwer, S., and C. Wood. 1998. Three hybrid zones between Hermit and Townsend's Warblers in Washington and Oregon. Auk 115:284–31.

Rush, A. C., R. J. Canning, and D. E. Irwin. 2009. Analysis of multilocus DNA reveals hybridization in a contact zone between Empidonax flycatchers. Journal of Avian Biology 40:614–624.

Sauer, J. R., J. E. Hines, J. E. Fallon, K. L. Pardieck, D. J. Ziolkowski Jr., and W. A. Link. 2012. The North American Breeding Bird Survey, results and analysis 1966–2011, version 07.03.2013. USGS Patuxent Wildlife Research Center, Laurel, MD.

Shapiro, L. H., R. A. Canterbury, D. M. Stover, and R. C. Fleischer. 2004. Reciprocal introgression between Golden-winged Warblers (Vermivora chrysoptera) and Blue-winged Warblers (V. pinus) in eastern North America. Auk 121:1019–1030.

Sheldon, B. C., and H. Ellegren. 1999. Sexual selection resulting from extrapair paternity in Collared Flycatchers. Animal Behaviour 57:285–298.

Short, L. L. 1963. Hybridization in the wood warblers Vermivora pinus and V. chrysoptera. Pp. 147–160 in Proceedings of the XIII International Ornithological Congress, American Ornithologists' Union, Louisiana State University, Baton Rouge, LA.

Short, L. L. 1969. "Isolating mechanisms" in the Blue-winged Warbler—Golden-winged Warbler complex. Evolution 23:355–356.

Species At Risk Act Registry. [online]. 2015. Golden-winged Warbler account. <http://www.sararegistry.gc.ca/species/speciesDetails_e.cfm?sid=942> (8 March 2013).

Steiner, C. C., A. S. Putnam, and P. E. A. Hoeck. 2013. Conservation genomics of threatened animal species. Annual Review of Animal Biosciences 1:261–281.

Sternkopf, V., D. Liebers-Helbig, M. S. Ritz, J. Zhang, A. J. Helbig, and P. de Knijff. 2010. Introgressive hybridization and the evolutionary history of the Herring Gull complex revealed by mitochondrial and nuclear DNA. BMC Evolutionary Biology 10:348–366.

Streby, H. M., J. M. Refsnider, S. M. Peterson, and D. E. Andersen. 2014. Retirement investment theory explains patterns in songbird nest-site choice. Proceedings of the Royal Society of London B 281:20131834.

Taylor, S. A., T. A. White, W. M. Hochachka, V. Ferretti, R. L. Curry, and I. Lovette. 2014. Climate-mediated movement of an avian hybrid zone. Current Biology 24:1–6.

Toews, D. P. L., L. Campagna, S. A. Taylor, C. N. Balakrishnan, D. T. Baldassarre, P. E. Deane-Coe, M. G. Harvey, D. M. Hooper, D. E. Irwin, C. D. Judy, N. A. Mason, J. E. McCormack, K. G. McCracken, C. H. Oliveros, R. J. Safran, E. S. C. Scordato, K. F. Stryjewski, A. Tigano, J. A. C. Uy, and B. M. Winger. 2016. Genomic approaches to understanding population divergence and speciation in birds. Auk 133:13–30.

Toews, D. P. L., and D. E. Irwin. 2008. Cryptic speciation in a Holarctic passerine revealed by genetic and bioacoustics analyses. Molecular Ecology 17:2691–2705.

Twyford, A. D., and R. A. Ennos. 2012. Next-generation hybridization and introgression. Heredity 108:179–189.

U.S. Endangered Species Act. [online]. 2015. Golden-winged Warbler account. <http://ecos.fws.gov/speciesProfile/profile/speciesProfile.action?spcode=B0G4> (23 September 2013).

Uy, J. A., and G. Borgia. 2000. Sexual selection drives rapid divergence in bowerbird display traits. Evolution 54:273–278.

Vallender, R., R. D. Bull, L. L. Moulton, and R. J. Robertson. 2012. Blood parasite infection and heterozygosity in pure and genetic-hybrid Golden-winged Warblers (Vermivora chrysoptera) across Canada. Auk 129:716–724.

Vallender, R., V. L. Friesen, and R. J. Robertson. 2007a. Paternity and performance of Golden-winged Warblers (Vermivora chrysoptera) and Golden-winged × Blue-winged Warbler (V. pinus) hybrids at the leading edge of a hybrid zone. Behavioral Ecology and Sociobiology 61:1797–1807.

Vallender, R., R. J. Robertson, V. L. Friesen, and I. J. Lovette. 2007b. Complex hybridization dynamics between Golden-winged and Blue-winged Warblers (*Vermivora chrysoptera* and *V. pinus*) revealed by AFLP, microsatellite, intron and mtDNA markers. Molecular Ecology 16:2017–2029.

Vallender, R., S. Van Wilgenburg, L. Bulluck, A. Roth, R. Canterbury, J. Larkin, R. M. Fowlds, and I. J. Lovette. 2009. Extensive rangewide mitochondrial introgression indicates substantial cryptic hybridization in the Golden-winged Warbler. Avian Conservation and Ecology 4:4.

van Dongen, W. F. D., R. A. Vásquez, and H. Winkler. 2012. The use of microsatellite loci for accurate hybrid detection in a recent contact zone between an endangered and recently-arrived hummingbird. Journal of Ornithology 153:585–592.

Veen, T., T. Borge, S. C. Griffith, G-P Sætre, S. Bures, L. Gustafsson, and B. C. Sheldon. 2001. Hybridization and adaptive mate choice in flycatchers. Nature 411:45–50.

Will, T. C. 1986. The behavioral ecology of species replacement: Blue-winged and Golden-winged Warblers in Michigan. Ph.D. dissertation, University of Michigan, Ann Arbor, MI.

Yuri, T., R. W. Jernigan, R. T. Brumfield, N. K. Bhagabati, and M. J. Braun. 2009. The effect of marker choice on estimated levels of introgression across an avian (Pipridae: *Manacus*) hybrid zone. Molecular Ecology 18:4888–4903.

Zink, R. M., and G. F. Barrowclough. 2008. Mitochondrial DNA under siege in avian phylogeography. Molecular Ecology 17:2107–2121.

CHAPTER FIVE

Space and Habitat Use of Breeding Golden-winged Warblers in the Central Appalachian Mountains*

*Mack W. Frantz, Kyle R. Aldinger, Petra B. Wood, Joseph Duchamp,
Timothy Nuttle, Andrew Vitz, and Jeffery L. Larkin*

Abstract. Spot-mapping, or recording locations of observed use by territorial songbirds, is often used to delineate core breeding territories. However, a recent radiotelemetry study in Minnesota found that male Golden-winged Warblers (*Vermivora chrysoptera*) occurring in high-density populations used resources outside their spot-mapped territories. We compared differences in space use and quantified vegetation characteristics in territories and home ranges of individual male Golden-winged Warblers that we monitored using both spot-mapping and radiotelemetry. Our study sites in Pennsylvania and West Virginia had lower population density than in Minnesota. We recorded 524 telemetry locations among 12 male Golden-winged Warblers in Pennsylvania and 488 telemetry locations among seven males in West Virginia. Telemetry-delineated home ranges (100% and 50% minimum convex polygons [MCPs]) were two to four times larger than spot-mapped territories. Spot-mapped territories had minimal overlap among individual males, but home ranges had extensive space-use overlap in both the number and amount of MCP overlap among several males. Forty percent of telemetry locations were outside of spot-mapped territories. Sapling abundance was greater in home ranges (mean 22.5 saplings ± 2.1 SE) than spot-mapped territories in Pennsylvania (11.8 ± 1.9). In managed pastures of West Virginia, tree abundance was greater in home ranges (7.3 trees ± 0.8) than spot-mapped territories (1.9 ± 0.6). More telemetry locations than spot-mapped locations occurred in forest in both states, and telemetry locations were closer to intact forested edges of shrublands than spot-mapped locations in West Virginia. On several occasions, we observed radiomarked individuals >200 m (maximum of 1.5 km) from their MCP spot-mapped territory boundaries. Why Golden-winged Warblers leave their spot-mapped territories is unknown, but our observations suggest foraging, forays for extra-pair mating, and reconnaissance for postbreeding movements as possible motives. Our results from areas with low Golden-winged Warbler territory densities are similar to patterns reported for a high-density population in Minnesota. Ultimately, spot-mapping alone does not accurately reflect space use of Golden-winged Warblers during the breeding season, nor does it characterize all cover types used even in areas

* Frantz, M. W., K. R. Aldinger, P. B. Wood, J. Duchamp, T. Nuttle, A. Vitz, and J. L. Larkin. 2016. Space and habitat use of breeding Golden-winged Warblers in the central Appalachian Mountains. Pp. 81–94 *in* H. M. Streby, D. E. Andersen, and D. A. Buehler (editors). Golden-winged Warbler ecology, conservation, and habitat management. Studies in Avian Biology (no. 49), CRC Press, Boca Raton, FL.

with relatively low territory densities. Current conservation plans for Golden-winged Warblers that are based on habitat characteristics measured within spot-mapped territories or at the landscape scale may not adequately incorporate space use at intermediate spatial scales of clusters of territories or home ranges.

Key Words: extra-territorial movement, foraging, radiotelemetry, spot-mapping, territory.

In the central Appalachian Mountains of Pennsylvania and West Virginia, population numbers of Golden-winged Warblers (*Vermivora chrysoptera*) have declined ~8% per year during 1966–2011 (Sauer et al. 2012), due in part to decline of shrubland cover types (Yahner 2003). Implementation of recently developed management guidelines for the creation of shrubland that supports breeding Golden-winged Warblers is a conservation priority (A. M. Roth et al., unpubl. plan). The primary foundation for these guidelines is avian and vegetation data collected at the scale of the spot-mapped territory where bird use areas are delineated by recording locations of song perches.

Breeding habitat requirements of songbirds are often studied by quantifying resource use within spot-mapped breeding territories (Confer et al. 2003, Roth and Lutz 2004, Bulluck and Buehler 2008), but recent telemetry studies suggest that individuals access important habitat components outside their defended territories (Anich et al. 2009, Streby et al. 2012). In Minnesota, at a study area with one of the highest known breeding densities of Golden-winged Warblers (~1 territory/ha), telemetry-delineated territories were three times larger than territories delineated by spot-mapping (Streby et al. 2012). Given the potential for conservation plans to be less effective if based solely on spot-mapping (Streby et al. 2012), potential differences between techniques need to be examined in other regions of the breeding distribution where territory densities are lower. Ultimately, determining the propensity for Golden-winged Warblers to use areas outside their spot-mapped breeding territories, and the characteristics of those areas, will be important for informing regional habitat management guidelines.

We used radiotelemetry and spot-mapping to examine habitat and space use of territorial male Golden-winged Warblers in the central Appalachian Mountains region of Pennsylvania and West Virginia. Unlike a recent study in Minnesota, our study occurred in areas with low nesting densities of Golden-winged Warblers (1 pair/10 ha in Pennsylvania; J. Larkin, unpubl. data; 1.4/10 ha in West Virginia; K. Aldinger, unpubl. data). By comparing areas used by Golden-winged Warblers in their home ranges to areas within spot-mapped territories, we can address whether spot-mapping is missing important aspects of Golden-winged Warbler habitat use. Our objectives were to (1) compare estimates of Golden-winged Warbler space use based on radiotelemetry versus spot-mapping, and (2) quantify vegetation characteristics of telemetry and spot-mapped-delineated areas at micro- and macrohabitat scales.

METHODS

Study Area

We conducted our study during May–July 2011 at two study areas in north-central Pennsylvania, and during May–July 2012 at two study areas in Randolph and Pocahontas counties, West Virginia. All study areas were within the Appalachian Mountains Bird Conservation Region (BCR 28). Study areas in Pennsylvania were in Bald Eagle State Park and Sproul State Forest. Bald Eagle State Park (41.033°N, 77.600°W) study sites were located at ~200 m elevation within the Ridge and Valley Province (Briggs 1999). We monitored Golden-winged Warblers at six study sites across a 219-ha portion of the park that was comprised of managed and natural shrublands, and remnant closed-canopy forest patches. Nonnative shrubs such as bush honeysuckle (*Lonicera* spp.), autumn-olive (*Elaeagnus umbellata*), and multiflora rose (*Rosa multiflora*) dominated the plant community. The most common native woody species included red osier dogwood (*Cornus sericea*), gray dogwood (*C. racemosa*), arrowwood viburnum (*Viburnum dentatum*), blackberry (*Rubus* spp.), elm (*Ulmus* spp.), black walnut (*Juglans nigra*), hawthorn (*Crataegus* spp.), and ash (*Fraxinus* spp.).

Study sites at Sproul State Forest (41.233°N, 77.700°W) were located at ~610 m in the

Appalachian Mountains High Plateau (Briggs 1999). We monitored Golden-winged Warblers at three study sites across 270 ha of a 1,619-ha area burned by an arson fire in 1990. Most forest stands in Sproul State Forest were mature (80–100 years old), closed-canopied forest that lacked the structural characteristics used by nesting Golden-winged Warblers. Rather, breeding Golden-winged Warblers in our study area were associated with disturbance-generated plant communities adjacent to older forests (80–100 years old) including areas influenced by timber harvests, wildfire, and abandoned natural gas wells (Larkin and Bakermans 2012). Shrubland cover available to breeding Golden-winged Warblers was characterized by a patchy mosaic of saplings, shrubs, herbaceous openings, and scattered trees (approx. basal area: 2.3–9.2 m²/ha). Regenerating forests in the burned areas were dominated by a mosaic of native species including blackberry, blueberry (*Vaccinium* spp.), mountain laurel (*Kalmia latifolia*), sweet fern (*Comptonia peregrina*), hay-scented fern (*Dennstaedtia punctilobula*), sassafras (*Sassafras albidum*), birch (*Betula* spp.), black locust (*Robinia pseudoacacia*), red maple (*Acer rubrum*), and chestnut oak (*Quercus prinus*).

Study sites (n = 6) in Pocahontas (n = 3; 38.317°N, 80.083°W) and Randolph counties (n = 3; 38.917°N, 79.733°W), West Virginia were fenced pastures 7–179 ha and 800–1,200 m in elevation. Five sites were grazed annually by cattle and horses (0.3–1.3 animal units/ha) between 15 May and 1 October, maintaining shrubland vegetation. One site had not been grazed since 1970 (W. A. Tolin, pers. comm.). All sites were characterized by a patchy mosaic of grasses, herbaceous vegetation, blackberry, shrubs, and scattered saplings and canopy trees (>10 cm dbh) surrounded by intact forest, defined as contiguous closed canopy comprised of sawtimber (>25 cm dbh), and generally lacking the dense understory vegetation characteristic of shrubland areas. Grass cover dominated the central portions of all sites, where livestock grazing was most intense, and transitioned into shrub–scrub cover and intact forest toward the fenced perimeter. Dominant species within areas dominated by shrubland vegetation included sweet vernal grass (*Anthoxanthum odoratum*), velvet grass (*Holcus lanatus*), meadow fescue (*Festuca elatior*), goldenrod (*Solidago* spp.), common cinquefoil (*Potentilla simplex*), Virginia strawberry (*Fragaria virginiana*), devil's darning needles (*Clematis*

virginiana), hawthorn (*Crataegus* spp.), autumn-olive, shrubby St. John's wort (*Hypericum prolificum*), multiflora rose, naturalized apple (*Malus* spp.), black locust, white ash (*Fraxinus americana*), black cherry (*Prunus serotina*), and sugar maple (*Acer saccharum*).

Capture and Handling

Between 1 May and 1 July, we captured male Golden-winged Warblers using targeted mist netting with song playback and a warbler decoy (Anich et al. 2009). We fitted captured individuals with a metal U.S. Geological Survey leg band and a unique combination of 1–3 color bands for identification purposes. Additionally, we fit large-bodied males ≥10 g in Pennsylvania (n = 12) and ≥9 g in West Virginia (n = 7) with a VHF radio transmitter (BD-2N [0.43 g] or LB-2X [0.31 g], Holohil Systems Ltd., Carp, ON). We attached radio transmitters to the interscapular region with a glue-on method that required ~5 min of handling time (Frantz 2013) and allowed the transmitter to fall off during the prebasic molt and prior to migration (Pyle 1997). The transmitter units constituted <5% of a Golden-winged Warbler's mass.

Territory and Home-Range Delineation

We refer to areas that we delineated via the spot-mapping technique as *spot-mapped territories*. We refer to areas repeatedly traversed by an individual and all extra-territorial movements (Kenward 2001) detected via radiotelemetry as *home ranges*. We considered movements outside the spot-mapped territory to be extra-territorial, but we recognize that some movements outside the spot-mapped territory might have been within an individuals' true song territory (Streby et al. 2012).

We used spot-mapping to delineate territories for all radiomarked and color-banded males at our study sites for Golden-winged Warblers, Blue-winged Warblers (*V. cyanoptera*), and hybrids (referred to collectively as *Vermivora* spp.). We followed individual color-banded males every other day through visual observations of feeding, perching, and singing. Monitoring all *Vermivora* spp. allowed us to account for space use overlap with males that neighbored our radiomarked Golden-winged Warblers. Golden-winged Warblers, Blue-winged Warblers, and hybrids appeared to behaviorally treat each other as conspecifics on our study sites. All observations were marked with flagging tape and we recorded each location using Garmin eTrex and GPSMAP 60CSx global positioning system units (accurate to <5 m). We spot-mapped territories for each color-banded male during 30–60-min sampling periods between 05:00 and 19:00 Eastern Standard Time (EST; Barg et al. 2005). Like Streby et al. (2012), our methods differed from Barg et al. (2005) in that we recorded each location only once regardless of how long we observed the bird there. We recorded ≥5 spot-mapping locations per visit and we varied time of observation periods among days (mean 11.0 ± 1.0 SE days per territory, range 6–17 days).

We collected radiotelemetry data every other day to alternate with spot-mapping days for the life of each transmitter using the homing method (Mech 1983, White and Garrott 1990). We approached each radiomarked Golden-winged Warbler on foot based on radio signal strength until we observed the bird feeding, perching, or singing, but we did not approach close enough to perceptibly influence their behavior (Vitz and Rodewald 2010). If we were unable to observe a bird without influencing its behavior, we approached the perceived location from several directions to triangulate the individual's location (Anich et al. 2009). We conducted telemetry monitoring between 05:00 and 19:00 EST. We varied the order and time of day that we monitored individual radiomarked Golden-winged Warblers to prevent any time-of-day effects on activity (Shields 1977). To ensure that telemetry locations were biologically independent, we used a sampling interval long enough to allow an individual to move from any point in its territory to any other point (Lair 1987, Holzenbein and Marchinton 1992, McNay et al. 1994, Silva-Opps and Opps 2011). As such, we allowed at least 1 min (although typically 2–10 min) to elapse between consecutive locations.

Vegetation Sampling

As part of a larger study, we placed vegetation sampling points within spot-mapped territories using a standardized protocol developed by the Golden-winged Warbler Working Group (modification of protocol in appendix [AP-16] of A. M. Roth et al., unpubl. plan). The protocol was used throughout the Golden-winged Warbler breeding range for consistent habitat description

(Buehler et al. 2007). Using a systematic random sampling design, we established transects through each spot-mapped territory and sampled vegetation within 1- and 5-m radius plots at ≥30 points per territory (Figure 5.1). We did not sample vegetation directly at the locations where we observed Golden-winged Warblers during spot-mapping surveys to keep protocol for the larger study. Anich et al. (2012) found that habitat features differed between used and random locations within home ranges of Swainson's Warblers (*Limnothlypis swainsonii*) and concluded that sampling at random points may not accurately describe used habitat features (Anich et al. 2012). However, sampling many random points at the small spatial scale of the spot-mapped territory best characterizes the patchy vegetation typically used by Golden-winged Warblers (Bonham 1989, Confer et al. 2003). We also sampled vegetation outside of spot-mapped territories at points where we observed Golden-winged Warblers via telemetry. We sampled habitat characteristics within home ranges at all telemetry locations ≥12 m outside the spot-mapped territory to allow a buffer between spot-mapped territories and home ranges and prevent overlap of measured areas. We did not sample vegetation at telemetry locations that overlapped with spot-mapped territories. We conducted vegetation sampling at each telemetry location within a 1-m radius plot centered at the location, four additional 1-m radius plots 5 m away from the center plot, one in each cardinal direction, and one 5-m radius plot centered at the telemetry location.

Within each 1-m radius plot, we visually estimated percent cover of vegetation (i.e., grass, forbs, ferns, and goldenrod), shrubs <1 m tall, shrubs >1 m tall, saplings (<10 cm dbh), and percent canopy cover. We visually estimated percent cover of woody ground cover, vines, bare ground, and litter in West Virginia only, and nonvegetation (bare ground and litter) in Pennsylvania. Our visual estimates should be considered indices comparing spot-mapped locations and telemetry locations. Visual estimation of vegetation cover can produce both reliable and unreliable results (Kercher et al. 2003, Helm and Mead 2004, Symstad et al. 2008), but is cost effective and valid for managers to apply across large scales (Cook et al. 2010). All visual estimates were made by a single observer in Pennsylvania, and teams were trained in West

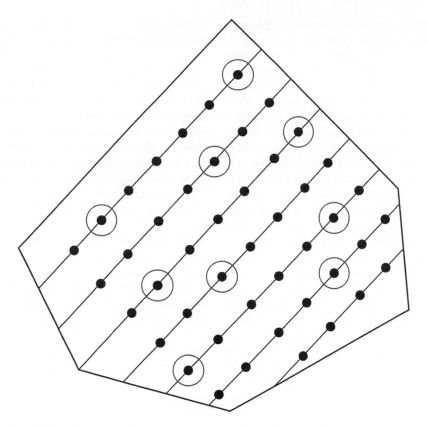

Figure 5.1. Vegetation sampling design for a hypothetical Golden-winged Warbler territory with random transects (lines) and 1-m radius vegetation sampling plots (black dots) and 5-m radius sampling plots (large circles). For each radiomarked male, we averaged habitat variables across all vegetation plots within the spot-mapped territory and across all vegetation plots at home-range telemetry locations that were outside the spot-mapped territory.

Virginia by the same observer using a standard collection protocol to limit bias in visual estimates among observers.

We measured distance from a plot center to a microedge as a change in vegetation height or composition. We also measured the number of snags within 11.3 m and live trees (>10 cm dbh) using a 2.5 m²/ha (10 ft²/ac) prism from each plot center. On the 5-m radius plots, we counted shrubs ≥1 m tall and saplings. We measured distance from plot center to the nearest intact forested edge as defined in the study area description using a measuring tape, rangefinder, or ArcMap Geographic Information System (GIS) (ver. 10; ESRI 2011). We defined forested edge as the edge of the forest canopy forming an interface between intact forest and nonforest (shrubland) cover types. We assigned locations of radiomarked Golden-winged Warblers within forested cover types negative distances and locations on an intact forested edge a distance of 0 m.

To examine macrohabitat characteristics, we assigned a cover type as shrubland or forest to each spot-mapped and telemetry location. We defined shrubland as nonforested, shrub–scrub cover with sparse canopy trees and with an herbaceous understory of forbs and grasses. We classified contiguous, closed-canopy areas with ≥0.25-ha of trees (>10 cm dbh) as forest. If the 0.25 ha circular buffer around a location was not completely forested and contained shrubland cover, then that location was not classified as forest.

Data Analysis

We delineated spot-mapped territories and home ranges using 100% and 50% minimum convex polygons (MCPs; see Chandler 2011) in GIS.

Although MCPs tend to overestimate home-range size (White and Garrot 1990), we used polygons to ensure maximum quantification of area needed to support male Golden-winged Warblers, to be consistent with other Golden-winged Warbler studies that have used spot-mapping and MCPs (Patton et al. 2010; Chapter 7, this volume), and to be consistent with recent radiotelemetry studies conducted on Golden-winged Warblers at other sites (Chandler 2011, Streby et al. 2012). Our spot-mapped territories may have excluded parts of the "real" territory (Streby et al. 2012) but can be considered an approximate estimate of the principal defended areas of the breeding territory used by Golden-winged Warblers (Anich et al. 2009).

We used Selected Cores Analysis in Ranges 7 (Anatrack, Wareham, UK) with the recalculated Ac (RAc) peel center method to determine which points were removed for 50% MCPs because the method reselects the area of densest locations after the farthest radio location is excluded (South et al. 2008). We tested data for normality and applied a $\log_{10}$ transformation to nonnormal data prior to analysis (Zar 2010). We used a paired t-test for normal data or an analogous nonparametric Wilcoxon Signed Rank test when the log transformation did not normalize the data to determine if there were size differences between home ranges and spot-mapped territories.

For each radio-marked male Golden-winged Warbler's spot-mapped territory and home range, we used the intersect tool in GIS to measure the amount of area overlap with all neighboring spot-mapped *Vermivora* territories (Patton et al. 2010). In addition, we counted the number of spot-mapped *Vermivora* territories that overlapped each individual's spot-mapped territory and home range. We used a nonparametric Wilcoxon Signed Rank test to compare the amount of area overlap (in ha) and number of territories that overlapped each individual's spot-mapped territory and home range.

For each radiomarked male, we averaged habitat variables across all vegetation plots within the spot-mapped territory and across all vegetation plots at home-range telemetry locations that were outside the spot-mapped territory. We tested habitat variables for normality and used an appropriate transformation prior to analysis if needed. We used a paired t-test or nonparametric Wilcoxon Signed Rank test to compare averaged habitat variable measurements between spot-mapped territories and home ranges. We used Holm's (1979) correction to control experiment-wise error rate when conducting multiple comparisons [$\alpha/(n-1)$; $P < 0.003$]. We compared distance to intact forested edge between Golden-winged Warbler use locations in spot-mapped territories and home ranges using a paired t-test or nonparametric Wilcoxon Signed Rank test. Forest patches were smaller (typically <2 ha) and more scattered throughout the landscape at Bald Eagle State Park than at Sproul State Forest in Pennsylvania. Therefore, we conducted a Wilcoxon Signed Rank test only on individuals from our study sites on the Sproul State Forest (n = 9) where intact forest surrounded the shrublands and was easy to classify this cover type at a large spatial scale.

We examined macrohabitat characteristics of areas used by Golden-winged Warblers by comparing the number of locations within shrubland or forest at all spot-mapped and telemetry locations using a χ^2-test of independence. We compared only use of each cover type between the two monitoring methods and not use relative to availability (Streby et al. 2012). We used National Land Cover Dataset 2001 Percent Tree Canopy Version 1.0 to assess canopy cover within a 90 m × 90 m window around each location (Homer et al. 2004). We used Focal Statistics in GIS, ground truthing, and review of aerial (years 2011 and 2012) 1- and 0.5-m resolution photographs in ArcMap to classify all locations.

RESULTS

We recorded 524 telemetry and 439 spot-mapped locations for 12 male Golden-winged Warblers in Pennsylvania, and 488 telemetry and 616 spot-mapping locations for seven males in West Virginia. Collectively, we recorded 20–155 locations per Golden-winged Warbler (mean 53.3 ± 6.6 locations). We spot-mapped for 6–17 days (mean 11.0 ± 1.0 days) and conducted radiotelemetry for 3–13 days (mean 8.0 ± 1.0 days) for each individual. For 100% territory MCPs we used 19–49 (37 ± 3) and 33–178 (88 ± 19) spot-mapped locations per individual in Pennsylvania and West Virginia, respectively. For 100% home-range MCPs we used 20–63 (43 ± 4) and 25–155 (70 ± 16) telemetry locations per individual. Out of all telemetry locations collected, 40% of home-range locations fell outside their respective male's spot-mapped territory.

TABLE 5.1

Size comparisons of radiomarked Golden-winged Warbler spot-mapped territories and home ranges using 50% and 100% MCPs, and number and area overlap of neighboring Vermivora spot-mapped territories that overlapped with the 100% MCP spot-mapped territories and 100% MCP home ranges of radiomarked Golden-winged Warblers in Pennsylvania (n = 12) and West Virginia (n = 7) during 2011 and 2012, respectively.

Metric	State	Radiomarked, spot-mapped territory			Radiomarked home range		
		Mean	±SE	Range	Mean	±SE	Range
Territory size (ha):							
100% MCP	PA	1.7	0.2	0.65–3.69	6.3	1.7	1.40–19.76
	WV	2.4	0.5	0.79–4.77	11.8	6.2	2.27–47.99
50% MCP	PA	0.3	0.1	0.12–0.69	0.5	0.1	0.13–1.03
	WV	0.3	0.1	0.13–0.63	0.6	0.1	0.20–1.28
Number of neighboring spot-mapped territories overlapped	PA	0.7	0.3	0–3	2.1	0.6	0–6
	WV	1.0	0.2	0–2	2.4	1.1	0–9
Area of neighboring spot-mapped territories overlapped (ha)	PA	0.2	0.1	0.00–0.54	1.3	0.4	0.00–4.53
	WV	0.4	0.2	0.00–1.18	2.6	1.7	0.00–12.48

Spot-Mapped Territory versus Home-Range Size

Among radiomarked males, home ranges (100% MCPs) were larger than spot-mapped territories in Pennsylvania (1.40–19.76 ha; t-test: t_{11} = 4.16, P = 0.002) and West Virginia (2.27–47.99 ha; Wilcoxon Signed Rank test: Z_7 = −2.37, P = 0.018; Table 5.1). Core home ranges (50% MCP) also were larger than core spot-mapped territories (50% MCP) in Pennsylvania (0.13–1.03 ha; t_{11} = 2.34, P = 0.039) and West Virginia (0.20–1.28 ha; t_6 = −2.75, P = 0.033). Although core areas (50% MCPs) averaged two times larger when delineated by telemetry than spot-mapping (Table 5.1), a majority of telemetry and spot-mapped core areas overlapped (15 of 19 individuals; Figure 5.2).

Spot-Mapped Territory versus Home-Range Overlap

Spot-mapped territories of individual males rarely overlapped territories of neighboring males (0–3 overlapping territories, Table 5.1), but home ranges of individual males overlapped up to nine spot-mapped territories (Figure 5.3). In Pennsylvania and West Virginia, spot-mapped territories were overlapped twice as often by other home ranges than they were by other spot-mapped territories. The difference was statistically significant in Pennsylvania (Z = −2.23, P = 0.026), with a similar nonstatistically significant trend in West Virginia (Z = −1.63,

P = 0.102). In terms of overlap with neighboring spot-mapped territories, home ranges of radiomarked males overlapped six times more area than spot-mapped territories of radiomarked males (Pennsylvania: Z = −2.50, P = 0.013; West Virginia: Z = −2.20, P = 0.028, Table 5.1). Most of the overlapping territories were Golden-winged Warblers (33 of 39 total overlaps), but one Blue-winged Warbler in West Virginia and five hybrids (two in West Virginia and three in Pennsylvania) were included as overlapping radiomarked individuals.

Habitat Characteristics of Spot-Mapped Territories and Home Ranges

We sampled vegetation at 163 and 126 telemetry locations that occurred outside spot-mapped territories in Pennsylvania and West Virginia, respectively. In Pennsylvania, there were fewer saplings in spot-mapped territories than home ranges (t_{11} = −3.81, P = 0.003; Table 5.2). In West Virginia, there were three times as many trees in home ranges as in spot-mapped territories (t_6 = −5.31, P = 0.002; Table 5.2). Distance to intact forested edge in West Virginia was shorter for telemetry locations (14.3 m ± 8.0) than for spot-mapped locations (44.8 m ± 6.7; t_6 = 2.92, P = 0.012). Distance to intact forested edge in Pennsylvania did not differ between telemetry locations (139.5 m ± 22.7) or spot-mapped locations (141.1 m ± 25.3; Z = 0.059, P = 0.952).

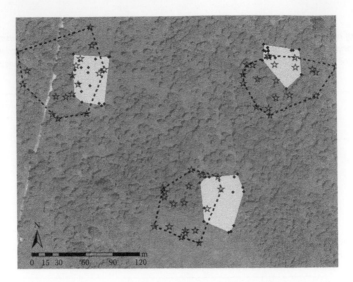

Figure 5.2. An example of overlapping spot-mapped (light gray) and telemetry-based (dashed polygon) core areas (50% MCPs) in Pennsylvania, with spot-mapped locations (circles) and home-range locations (stars) indicated. Core areas were on average two times larger when delineated by telemetry than spot mapping, and the majority of telemetry and spot-mapped core areas overlapped in Pennsylvania and West Virginia (15 of 19 individuals).

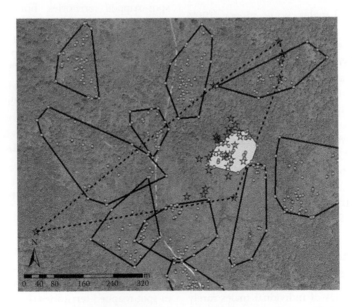

Figure 5.3. A spot-mapped territory (light gray) and home range (dashed polygon) for an individual male Golden-winged Warbler in Pennsylvania. The individual's spot-mapped territory did not overlap with other *Vermivora* spp. spot-mapped territories (hollow polygons), but his home range overlapped with portions of spot-mapped territories of six other individuals. Spot-mapped territory locations are represented as circles and home-range locations as stars.

Locations in Forest versus Shrubland Cover Types

A higher proportion of telemetry locations than spot-mapped locations occurred in forest in Pennsylvania ($\chi^2 = 9.36$, df = 1, P = 0.002) and West Virginia ($\chi^2 = 9.91$, df = 1, P = 0.002). In Pennsylvania, one

spot-mapped location for one individual was in forest, whereas 10 telemetry locations representing four of 12 individuals were in forest. In West Virginia, two spot-mapped locations representing two of seven individuals were located in forest, whereas 12 telemetry locations

TABLE 5.2

Vegetation sampled within Golden-winged Warbler spot-mapped territories and outside territories in home ranges at study sites in Pennsylvania and West Virginia during 2011 and 2012, respectively.

Habitat variable	Pennsylvania				West Virginia			
	Spot-mapped territory (n = 12)		Home range (n = 12)		Spot-mapped territory (n = 7)		Home range (n = 7)	
	Mean	± SE	Mean	± SE	Mean	± SE	Mean	± SE
Grass (%)	5	1	4	1	29	3	21	3
Forbs (%)	7	2	2	<1	27	2	26	3
Ferns (%)	18	2	19	2	<1	<1	1	<1
Rubus spp. (%)	39	4	30	2	7	3	7	1
Goldenrod (*Solidago* spp.: %)	6	1	4	1	17	3	9	2
Woody Cover (%)	—	—	—	—	4	1	7	2
Litter (%)	—	—	—	—	10	2	20	2
Vine (%)	—	—	—	—	2	2	1	<1
Non-vegetation (Bare Ground + Litter) (%)	12	2	17	2	—	—	—	—
Bare Ground (%)	—	—	—	—	5	1	9	2
Shrub <1 m (%)	11	2	12	1	4	<1	7	2
Shrub >1 m (%)	4	1	11	2	7	1	10	2
Sapling (%)	9	2	14	2	2	1	3	1
Canopy cover (1-m scale: %)	15	3	24	3	6	4	31	7
Distance to microedge (m)	1.3	0.1	1.1	0.1	2.2	0.3	2.6	0.6
No. trees	1.8	0.6	3.1	0.5	1.9	0.6	**7.3***	**0.9***
No. Snags	0.1	0.1	0.4	0.1	0.0	0.0	0.3	0.1
No. of shrubs	8.2	3.5	9.2	2.5	11.0	1.7	15.5	1.6
No. of saplings	11.9	2.5	**20.7***	**2.9***	7.3	3.2	6.5	1.6
Distance to intact forested edge (m)	141.1	25.3	139.5	22.7	44.8	6.7	**14.3***	**8.0***

After Holm's correction, we considered P-values to be significant when <0.003. Values in bold and starred (*) were statistically significantly different from each other.

representing four of seven individuals were in forest. Overall, three of 1,055 spot-mapped locations and 22 of 1,012 telemetry locations were classified as forest (Table 5.3). Moderate amounts of high canopy cover at a 90 m × 90 m scale (71% ± 3 Pennsylvania; 68.5% ± 4 West Virginia) occurred at forest locations resulting in presence of shrub, sapling, and herbaceous ground cover at 1- and 5-m plot scales (Table 5.3). Distance to shrubland edge was low (mean 45.9 m ± 4.0; Table 5.3).

DISCUSSION

Golden-winged Warblers in low nesting-density populations in Pennsylvania and West Virginia regularly used areas outside their spot-mapped territories, with 40% of all telemetry locations occurring outside spot-mapped territories. MCP size differences between spot-mapped and telemetry-based techniques were similar to those reported in a high-density population in Minnesota (Streby et al. 2012), although we cannot make direct comparisons between the results of these studies due to differences in field methods and data analyses. However, both our study and Streby et al. (2012) concluded that spot-mapping was limited in its utility to accurately represent space use and movements of Golden-winged Warblers.

Songbird home ranges have been reported to be 1.4–8 times larger than breeding territories

TABLE 5.3

Vegetation characteristics at forested locations where Golden-winged Warblers were observed during spot mapping or via radiotelemetry in Pennsylvania and West Virginia during 2011 and 2012, respectively.

Habitat variable	Pennsylvania Forested locations (n = 10)		West Virginia Forested locations (n = 11)	
	Mean	±SE	Mean	±SE
Grass (%)	1	<1	9	2
Forbs (%)	2	1	34	3
Ferns (%)	6	2	2	1
Rubus spp. (%)	4	2	2	1
Goldenrod (Solidago spp.: %)	1	<1	<1	<1
Woody Cover (%)	—	—	11	3
Litter (%)	—	—	27	3
Vine (%)	—	—	1	<1
Non-vegetation (Bare Ground + Litter) (%)	45	5	—	—
Bare Ground (%)	—	—	14	3
Shrub <1 m (%)	23	4	7	2
Shrub >1 m (%)	18	3	2	1
Sapling (%)	17	3	5	1
Canopy cover (1-m scale: %)	51	6	76	3
Distance to microedge (m)	0.8	0.1	5.4	0.9
No. trees	6.0	1.0	12.6	1.2
No. Snags	0.6	0.2	0.8	0.3
No. of shrubs	12.6	4.5	31.1	8.3
No. of saplings	30.0	10.0	7.6	3.1
Distance to intact forested edge (m)	51.0	7.2	41.3	3.8

Vegetation data represent 21 of 25 forested locations from eight of nine individuals. We did not collect microhabitat vegetation data at the forested spot-mapped locations (n = 3) or at one telemetry location within a spot-mapped territory.

(Hanski and Haila 1988, Leonard et al. 2008, Anich et al. 2009). Home ranges of Golden-winged Warblers based on telemetry locations in our study (6.3–11.8 ha) averaged three-to-four times larger than each individual's spot-mapped territory (1.7–2.4 ha) and were larger than other reported values for Golden-winged Warbler spot-mapped territory size (Confer 1992, Rossell et al. 2003, Patton et al. 2010). For example, one radio-marked Golden-winged Warbler in Pennsylvania traveled 1.6 km outside his spot-mapped territory in late afternoon but was observed back in his spot-mapped territory the following morning (Frantz 2013). Our results are consistent with Streby et al. (2012), who found that Golden-winged Warbler telemetry-delineated home ranges were larger than telemetry-based song territories in Minnesota. Our telemetry-delineated home ranges (Pennsylvania: 6.3 ± 1.7 ha; West Virginia: 11.8 ± 6.2 ha, 100% MCP) were two-to-four times larger than spot-mapped territories (Pennsylvania: 1.7 ± 0.2 ha; West Virginia: 2.4 ± 0.5 ha). However, Streby et al. (2012) also found that telemetry-based song territories (0.45 ± 0.08 ha) were three times larger than spot-mapped song territories (0.16 ± 0.02 ha) in Minnesota. Therefore, comparisons based on spot-mapping between our study and that of Streby et al. (2012) suggest that territories may be considerably larger in low-density populations. Territory size often is inversely related to population density (Maher and Lott 2000), where territories and home ranges are expected to be smaller in areas with more intraspecific competition (Pons et al. 2008, Anich et al. 2010).

Core use areas based on 50% MCPs are generally considered the most important part of an animal's home range (Burt 1943, Samuel and Green 1988) because food and other critical resources often are patchily distributed throughout the landscape (Powell et al. 1997). Avian breeding territories vary in size, but usually represent areas of high use within a larger home range (Harris et al. 1990). Core areas (50% MCPs) for the Golden-winged Warblers we monitored were larger when delineated by telemetry than spot-mapping in almost all cases (16 out of 19 males). However, telemetry and spot-mapped core areas usually overlapped (15 out of 19 cases), suggesting that both methods can reliably delineate the core use areas of territories. Conversely, spot-mapping was less effective at completely characterizing space use of breeding Golden-winged Warblers compared to telemetry. The disparity in estimates of space use resulting from spot-mapped versus telemetry-based methods is important to consider in efforts to create and maintain high quality Golden-winged Warbler breeding habitat across their breeding distribution.

The high overlap among individual home ranges in our study suggests that areas between spot-mapped territories are shared by neighboring Golden-winged Warblers and may provide foraging grounds that do not appear to be defended (Williams 1990). In Pennsylvania, we observed three territorial males foraging in the same area within a 10-m radius circle with no apparent aggressive behavior; two of the males were outside their spot-mapped territories and one was on the periphery of his spot-mapped territory (Frantz 2013). We recognize that MCPs can overestimate the amount of overlap by including areas that a bird may never visit (Barg et al. 2005). However, given the many factors that influence home-range boundaries (Anich et al. 2010, 2012), and that spot-mapping underestimates territory size of Golden-winged Warblers (Streby et al. 2012), it is possible that even more overlap occurred between neighboring Golden-winged Warblers than what we documented.

Spot-mapped territories of neighboring Golden-winged Warblers had little overlap, whereas telemetry-delineated home ranges had extensive overlap with spot-mapped territories (Streby et al.

2012, our study). Movements of male Golden-winged Warblers into other males' breeding territories or home ranges may have been motivated by potential opportunities for extra-pair matings (Frantz 2013). In Ontario, ~30% of Golden-winged Warbler nestlings were the result of extra-pair copulations, occurring in 55% of all nests (Vallender et al. 2007).

Forest cover is important for breeding Golden-winged Warblers at a landscape scale (A. M. Roth et al., unpubl. plan), and the same appears true at the home-range scale. We observed Golden-winged Warblers in forested patches more often using radiotelemetry than through spot-mapping, although visits to forest cover were still relatively rare (25 of 2,067 total spot-map and telemetry locations; 1.2%). The forest cover at locations where we observed radiomarked Golden-winged Warblers were of various stand ages (20–110 years) and typically contained canopy gaps, especially in Pennsylvania, resulting in heterogeneous vegetative structure. Moderate amounts of high canopy cover suggested the presence of canopy disturbance that allowed growth of shrub, sapling, and herbaceous ground cover.

Movements into forest could serve as reconnaissance to identify areas where adults would bring dependent young as fledglings use forest cover in higher proportion than available (Chapter 8, this volume). A radiomarked Golden-winged Warbler male attending fledglings in West Virginia progressively ventured farther into surrounding closed-canopy forest away from its spot-mapped territory. Thus, implementing management activities that result in a diverse array of forest stand conditions will benefit efforts to create and maintain Golden-winged Warbler breeding habitat in the Appalachian region.

Our results inform Golden-winged Warbler monitoring and conservation strategies. Partners in Flight detection distance for population estimates of Golden-winged Warblers was 200 m (Rosenberg and Blancher 2005), which was later determined to be 100–150 m in nonopen cover types (Kubel and Yahner 2007, Aldinger and Wood 2015). Given that many of our telemetry locations were >200 m away from spot-mapped territories, locations where singing males were detected may only be loosely associated with their breeding territories. Furthermore, current conservation plans for Golden-winged Warblers are based on habitat relationships derived solely from spot-mapped territories and lack consideration of an intermediate spatial scale between the territory and landscape scale (Bakermans et al. 2011; A. M. Roth et al., unpubl. plan). Consideration of an intermediate spatial scale may benefit Golden-winged Warbler monitoring and conservation.

ACKNOWLEDGMENTS

The field study was funded by Pennsylvania Department of Conservation and Natural Resources (Wild Resource Conservation Program), West Virginia Division of Natural Resources (WVDNR), U.S. Fish and Wildlife Service, National Fish and Wildlife Foundation, Ruffed Grouse Society, Indiana University of Pennsylvania, Bald Eagle State Park, U.S. Forest Service – Monongahela National Forest, U.S. Natural Resource Conservation Service, and the WV Cooperative Fish and Wildlife Research Unit. We acknowledge field assistance from E. Bellush, D. Clawson, E. Hughes, J. Kreiser, W. Leuenberger, S. McGaughran, V. Olmstead, J. Riffle, L. Smith, and C. Ziegler. We are grateful for additional assistance by T. Simmons, A. Evans, M. Bakermans, H. Streby, J. Benhart, J. Grata, and K. Yoder. J. Ferrara at Bald Eagle State Park and WVDNR provided field season accommodations. The study was conducted under federal permits from the USGS Bird Banding Laboratory. Capture, handling, and radio-tagging procedures were approved by Institutional Animal Care and Use Committees at Indiana University of Pennsylvania (IACUC 03-0708-R) and West Virginia University (IACUC 07-0303 and 10-0201). Use of trade names does not imply endorsement by the Federal Government.

LITERATURE CITED

Aldinger, K., and P. Wood. 2015. Variables associated with detection probability, detection latency, and behavioral responses of Golden-winged Warblers. Condor 117:364–375.

Anich, N. M., T. J. Benson, and J. C. Bednarz. 2009. Estimating territory and home-range sizes: do singing locations alone provide an accurate estimate of space use? Auk 126:626–634.

Anich, N. M., T. J. Benson, and J. C. Bednarz. 2010. Factors influencing home-range size of Swainson's Warblers in eastern Arkansas. Condor 112:149–158.

Anich, N. M., T. J. Benson, and J. C. Bednarz. 2012. What factors explain differential use within Swainson's Warbler (Limnothlypis swainsonii) home ranges? Auk 129:409–418.

Bakermans, M. H., J. L. Larkin, B. W. Smith, T. M. Fearer, and B. C. Jones. 2011. Golden-winged Warbler habitat Best Management Practices for forestlands in Maryland and Pennsylvania. American Bird Conservancy, The Plains, VA, 26pp.

Barg, J. J., J. Jones, and R. J. Robertson. 2005. Describing breeding territories of migratory passerines: suggestions for sampling, choice of estimator, and delineation of core areas. Journal of Animal Ecology 74:139–149.

Bonham, C. D. 1989. Measurements for terrestrial vegetation. John Wiley & Sons, New York, NY.

Briggs, R. P. 1999. Appalachian Plateaus Province and the Eastern Lake Section of the Central Lowland Province. Pp. 362–377 in C. H. Shultz (editor), The geology of Pennsylvania, Chapter 30. Pennsylvania Geological Survey and Pittsburgh Geological Society, Harrisburg, PA.

Buehler, D. A., A. M. Roth, R. Vallender, T. C. Will, J. L. Confer, R. A. Canterbury, S. B. Swarthout, K. V. Rosenberg, and L. P. Bulluck. 2007. Status and conservation priorities of Golden-winged Warbler (Vermivora chrysoptera) in North America. Auk 124:1439–1445.

Bulluck, L. P., and D. A. Buehler. 2008. Factors influencing Golden-winged Warbler (Vermivora chrysoptera) nest-site selection and nest survival in the Cumberland Mountains of Tennessee. Auk 125:551–559.

Burt, W. H. 1943. Territoriality and home range concepts as applied to mammals. Journal of Mammalogy 24:346–352.

Chandler, R. B. 2011. Avian ecology and conservation in tropical agricultural landscapes with emphasis on Vermivora chrysoptera. Dissertation, University of Massachusetts-Amherst, Amherst, MA.

Confer, J. L. 1992. Golden-winged Warbler (Vermivora chrysoptera). In A. Poole, P. Stettenheim, and F. Gill (editors), The birds of North America, No. 20. Academy of Natural Sciences, Philadelphia, PA; American Ornithologists' Union, Washington, DC.

Confer, J. L., J. L. Larkin, and P. E. Allen. 2003. Effects of vegetation, interspecific competition, and brood parasitism on Golden-winged Warbler (Vermivora chrysoptera) nesting success. Auk 120:138–144.

Cook, C. N., G. Wardell-Johnson, M. Keatley, S. A. Gowans, M. S. Gibson, M. E. Westbrooke, and D. J. Marshall. 2010. Is what you see what you get? Visual vs. measured assessments of vegetation condition. Journal of Applied Ecology 47:650–661.

ESRI. 2011. ArcMap GIS Desktop. Release 10. Environmental Systems Research Institute, Redlands, CA.

Frantz, M. 2013. Is spot mapping missing important aspects of Golden-winged Warbler (*Vermivora chrysoptera*) breeding habitat? M.S. thesis, Indiana University of Pennsylvania, Indiana, PA.

Hanski, I. K., and Y. Haila. 1988. Singing territories and home ranges of breeding Chaffinches: visual observation vs. radio-tracking. Ornis Fennica 65:97–103.

Harris, S., W. J. Cresswell, P. G. Forde, W. J. Trewhella, T. Woollard, and S. Wray. 1990. Home-range analysis using radio-tracking data: a review of problems and techniques particularly as applied to the study of mammals. Mammal Review 20:97–123.

Helm, D. J., and B. R. Mead. 2004. Reproducibility of vegetation cover estimates in south-central Alaska forests. Journal of Vegetation Science 15:33–40.

Holm, S. 1979. A simple sequential rejective multiple test procedure. Scandinavian Journal of Statistics 6:65–70.

Holzenbein, S., and R. L. Marchinton. 1992. Spatial integration of maturing-male white-tailed deer into the adult population. Journal of Mammalogy 73:326–334.

Homer, C., C. Huang, L. Yang, B. Wylie, and M. Coan. 2004. Development of a 2001 National Land Cover Database for the United States. Photogrammetric Engineering and Remote Sensing 70:829–840.

Kenward, R. E. 2001. A manual for wildlife radio tagging. Academic Press, San Diego, CA.

Kercher, S. M., C. B. Frieswyk, and J. B. Zedler. 2003. Effects of sampling teams and estimation methods on the assessment of plant cover. Journal of Vegetation Science 19:899–906.

Kubel, J. E., and R. H. Yahner. 2007. Detection probability of Golden-winged Warblers during point counts with and without playback recordings. Journal of Field Ornithology 78:195–205.

Lair, H. 1987. Estimating the location of the focal center in red squirrel home ranges. Ecology 68:1092–1101.

Larkin, J. L. and M. H. Bakermans. 2012. Golden-winged Warbler, *Vermivora chrysoptera*. Pp. 350–351 in A. M. Wilson, D. Brauning, and R. Mulvihill (editors), Second Atlas of Breeding Birds in Pennsylvania. The Penn State University Press, University Park, PA.

Leonard, T. D., P. D. Taylor, and I. G. Warkentin. 2008. Space use by songbirds in naturally patchy and harvested boreal forests. Condor 110:467–481.

Maher, C. R., and D. F. Lott. 2000. A review of ecological determinants of territoriality within vertebrate species. American Midland Naturalist 143:1–29.

McNay, R. S., J. A. Morgan, and F. L. Bunnell. 1994. Characterizing independence of observations in movements of Columbian black-tailed deer. Journal of Wildlife Management 58:422–429.

Mech, L. D. 1983. A handbook of animal radio-tracking. University of Minneapolis Press, Minneapolis, MN.

Patton, L. L., D. S. Maehr, J. E. Duchamp, S. Fei, J. W. Gassett, and J. L. Larkin. 2010. Do the Golden-winged Warbler and Blue-winged Warbler exhibit species-specific differences in their breeding habitat use? Animal Conservation and Ecology 5:2.

Pons, P., J. M. Bas, R. Prodon, N. Roura-Pascual, and M. Clavero. 2008. Territory characteristics and coexistence with heterospecifics in the Dartford Warbler *Sylvia undata* across a habitat gradient. Behavioral Ecology and Sociobiology 62:1217–1228.

Powell, R. A., J. W. Zimmerman, and D. E. Seaman. 1997. Ecology and behavior of North American black bears: home ranges, habitat and social organization. Chapman and Hall, London, UK.

Pyle, P. 1997. Identification guide to North American birds. Part I: Colombidae to Ploceidae. Slate Creek Press, Bolinas, CA.

Rosenberg, K. V., and P. J. Blancher. 2005. Setting numerical population objectives for priority landbird species. Pp. 57–67 in C. J. Ralph and T. D. Rich (editors), Proceedings of the Third International Partners in Flight Conference. Gen. Tech. Rep. PSW-GTR-191. USDA Forest Service, Pacific Southwest Research Station, Albany, CA.

Rossell, C. R. Jr., S. C. Patch, and S. P. Wilds. 2003. Attributes of Golden-winged Warbler territories in a mountain wetland. Wildlife Society Bulletin 31:1099–1104.

Roth, A. M., and S. Lutz. 2004. Relationship between territorial male Golden-winged Warblers in managed aspen stands in northern Wisconsin, USA. Forest Science 50:153–161.

Samuel, M. D., and R. E. Green. 1988. A revised test procedure for identifying core areas within the home range. Journal of Animal Ecology 57:1067–1068.

Sauer, J. R., J. E. Hines, J. Fallon, K. L. Pardieck, D. J. Ziolkowski Jr., and W. A. Link. 2012. The North American Breeding Bird Survey, results and analysis 1966–2009, version 3.23.2011. U.S. Geological Survey Patuxent Wildlife Research Center, Laurel, MD.

Shields, W. M. 1977. The effect of time of day on avian census results. Auk 2:380–383.

Silva-Opps, M., and S. B. Opps. [online]. 2011. Use of telemetry data to investigate home range and habitat selection in mammalian carnivores. Pp. 1–26 in O. Krejcar (editor), Modern telemetry, Chapter 14. InTech Publishing, Rijeka, Croatia. <http://www.intechopen.com/books/modern-telemetry/use-of-telemetry-data-to-investigate-home-range-and-habitat-selection-in-mammalian-carnivores> (10 December 2015).

South, A. B., R. E. Kenward, and S. Walls. [online]. 2008. Ranges 7 demo. Anatrack Ltd., Wareham, UK. <http://www.anatrack.com> (10 December 2015).

Streby, H., J. P. Loegering, and D. E. Andersen. 2012. Spot mapping underestimates song-territory size and use of mature forest by breeding Golden-winged Warblers in Minnesota, USA. Wilson Bulletin 36:40–46.

Symstad, A. J., C. L. Wienk, and A. D. Thorstenson. 2008. Precision, repeatability, and efficiency of two canopy-cover estimate methods in Northern Great Plains vegetation. Rangeland Ecology and Management 61:419–429.

Vallender, R., V. L. Friesen, and R. J. Robertson. 2007. Paternity and performance of Golden-winged Warblers (*Vermivora chrysoptera*) and Golden-winged × Blue-winged Warbler (*V. pinus*) at the leading edge of a hybrid zone. Behavioral Ecology and Sociobiology 61:1797–1807.

Vitz, A.C., and A. D. Rodewald. 2010. Movements of fledgling Ovenbirds (*Seirus aurocapilla*) and Worm-eating Warblers (*Helmitheros vermivorum*) within and beyond the natal home range. Auk 127:364–371.

White, G. C., and R. A. Garrott. 1990. Analysis of radio-tracking data. Academic Press, New York, NY.

Williams, P. L. 1990. Use of radiotracking to study foraging in small terrestrial birds. Pp. 181–186 in M. L. Morrison, C. J. Ralph, J. Verner, and J. R. Jehl Jr. (editors), Avian foraging: theory, methodology, and applications to saline lakes. Studies in Avian Biology (No. 13). Cooper Ornithological Society, Lawrence, KS.

Yahner, R. H. 2003. Terrestrial vertebrates in Pennsylvania: status and conservation in a changing landscape. Northeastern Naturalist 10:343–360.

Zar, J. H. 2010. Biostatistical analysis (5th ed.). Prentice-Hall, Upper Saddle River, NJ.

CHAPTER SIX

Influence of Plant Species Composition on Golden-winged Warbler Foraging Ecology in North-Central Pennsylvania*

Emily C. Bellush, Joseph Duchamp, John L. Confer, and Jeffery L. Larkin

Abstract. Golden-winged Warblers (*Vermivora chrysoptera*) have experienced significant population declines in the Appalachian Mountains for more than 40 years and are currently a focal species for management of young forests throughout their breeding distribution. Avian fitness has been linked to the quality and quantity of insect food supplies, but little information is available on foraging ecology of Golden-winged Warblers. We evaluated shrub and tree species selection by foraging Golden-winged Warblers in north-central Pennsylvania during the 2011 breeding season. Additionally, we compared caterpillar abundance among 13 woody plant species present within breeding territories. Golden-winged Warblers selectively foraged on black locust (*Robinia pseudoacacia*), pin cherry (*Prunus pensylvanica*), white oak (*Quercus alba*), and blackberries (*Rubus* spp.). Tree and shrub species composition differed between Golden-winged Warbler territories and adjacent, unoccupied areas of early successional forest cover, and habitat use was consistent with patterns of caterpillar abundance. Whereas vegetation structure generally dominates management guidelines for breeding habitat of insectivorous songbirds, including Golden-winged Warblers, our research clearly demonstrated the need for land managers to also consider plant species composition. In the case of habitat management for breeding Golden-winged Warblers in north-central Pennsylvania, our assessment suggested favoring the presence of black locust, pin cherry, white oak, and blackberry, over sassafras (*Sassafras albidum*), mountain laurel (*Kalmia latifolia*), and blueberry (*Vaccinium* spp.).

Key Words: black locust, breeding territory, caterpillar, early successional habitat, habitat selection, Neotropical migrant, pin cherry, prey abundance, *Rubus* spp.

Prey availability has a strong effect on reproductive success of breeding songbirds (Martin 1987, Nagy and Holmes 2005). Selection of a breeding territory that provides abundant food resources necessary for rearing young has important consequences for an individual's fitness (Simons and Martin 1990, Rodenhouse and Holmes 1992, Verhulst 1994). Considering the importance of prey availability and provisioning rates to songbird reproductive

* Bellush, E. C., J. Duchamp, J. L. Confer, and J. L. Larkin. 2016. Influence of plant species composition on Golden-winged Warbler foraging ecology in north-central Pennsylvania. Pp. 95–108 in H. M. Streby, D. E. Andersen, and D. A. Buehler (editors). Golden-winged Warbler ecology, conservation, and habitat management. Studies in Avian Biology (no. 49), CRC Press, Boca Raton, FL.

success (Goodbred and Holmes 1996, Moorman et al. 2007), it is not obvious why songbird foraging has received little research attention compared to other aspects of avian ecology. In fact, habitat conservation of imperiled avian species has often focused on vegetative structure (Rotenberry and Wiens 1980, Vale et al. 1982), which herein we define as the physical arrangement of vegetation, irrespective of plant species composition (Wiens 1969). MacArthur and MacArthur (1961) and Diaz et al. (2005) argued that bird community dynamics have little to do with plant species composition. However, several published studies indicate that insectivorous birds selectively search for prey on certain tree species (Holmes and Robinson 1981, Gabbe et al. 2002, George 2009), which may be linked to prey density, an important component of prey availability. Additionally, arthropod densities have been shown to vary among tree species (Marshall and Cooper 2004). In efforts to increase habitat quality to the level where breeding bird populations are self-sustaining, it is equally important to consider both the physical structure of breeding habitat (MacArthur and MacArthur 1961, Gregg et al. 2000, Wood et al. 2006) and foraging ecology of focal species (Weikel and Hayes 1999). Although much work has been done to elucidate breeding habitat structure for many avian species (Recher 1969, Fink et al. 2006, A. M. Roth et al., unpubl. plan), detailed information regarding foraging ecology is often lacking.

Much is known about the structural components of breeding habitat of Golden-winged Warblers (*Vermivora chrysoptera*), but there is little information on foraging ecology. Golden-winged Warblers are an insectivorous, Neotropical migrant that breed in higher elevations of the Appalachian Mountains, the northeast and north-central regions of the U.S., and scattered areas across adjacent southern Canada (Confer et al. 2011; A. M. Roth et al., unpubl. plan; Chapter 1, this volume). Breeding populations of Golden-winged Warblers in the Appalachian Mountains occur in predominantly forested landscapes (A. M. Roth et al., unpubl. plan). Within these forested landscapes, areas used by breeding pairs include patches of young forests, forest and shrub wetlands, managed shrublands, and the edge portions (<55 m) of surrounding unharvested forest (saw timber; >80 years old: Chapter 5, this volume). The Appalachian breeding-distribution segment of Golden-winged Warblers has experienced a population decline

for more than 40 years (Sauer et al. 2011; Chapter 1, this volume) and Golden-winged Warblers are now considered a species of management concern (U.S. Fish and Wildlife Service 2009). Reasons for the species' decline include reduced abundance of breeding habitat, hybridization with Blue-winged Warblers (*V. cyanoptera*), and nest parasitism by Brown-headed Cowbirds (*Molothrus ater*, Buehler et al. 2007). The availability of young forest across the Appalachian landscape has declined during the past century due to ongoing succession of abandoned farmland to closed-canopy forest (Gill 1980).

Many aspects of Golden-winged Warbler biology have been examined but little is known about the species' foraging ecology. There is little documentation of which arthropods are consumed by Golden-winged Warblers but caterpillars and spiders appear to be common prey. Anecdotal evidence suggests that caterpillars, particularly leafrollers (Lepidoptera: Tortricidae), are one of the most important components of the diet of Golden-winged Warblers, and adults provision nestlings with caterpillars (Jacobs 1904, Forbush 1929, Will 1986, Confer et al. 2011). Based on the limited available information on the diet of Golden-winged Warblers (Jacobs 1904, Forbush 1929, Will 1986), we assumed that caterpillars were the dominant prey in our study area.

Forage-site selection, or the plant species on which a bird forages, is a key aspect of avian breeding ecology and has been linked to prey abundance (Willson 1970, Holmes and Robinson 1981, Gabbe et al. 2002, George 2009). Many songbirds are thought to select territories based on plant species composition and selectively search for prey on certain plant species within their territories (Holmes and Robinson 1981, Gabbe et al. 2002, George 2009). If Golden-winged Warblers behave similarly, information regarding the abundance and distribution of prey, foraging-site selection, and vegetative composition and structure associated with territories can improve understanding of the breeding ecology of Golden-winged Warblers. Herein, we describe Golden-winged Warbler foraging ecology, prey abundance, and tree and shrub species use in north-central Pennsylvania. Specifically, our objectives were to (1) examine whether Golden-winged Warblers selectively search for prey on tree and shrub species disproportionate to tree and shrub species availability within their breeding territories, (2) relate Golden-winged Warbler foraging site selection to abundance of caterpillars, and (3) identify plant species composition and vegetation structure associated with Golden-winged Warbler territories compared to unoccupied areas of seemingly appropriate breeding habitat.

METHODS

Study Area

We examined foraging ecology of Golden-winged Warblers at the Sproul State Forest located in north-central Pennsylvania (41.245° N, 77.877° W) from 1 May through 10 July 2011. Sproul State Forest lies in western Clinton and northern Centre counties at an elevation of 600–620 m and encompasses 1,120 km². Sproul State Forest is entirely in the Mountainous High Plateau section of the Appalachian Plateau's physiographic province (Briggs 1999). The region is characterized by high ridges and deep valleys created via headwater erosion of the West Branch of the Susquehanna River (Briggs 1999). There is little human development in the region with the exception of sparsely distributed primitive cabins, natural gas pipelines, a major roadway (State Route 144), several gated and ungated gravel roads, and numerous natural gas wells.

We monitored Golden-winged Warblers in Sproul State Forest at three study sites (140, 60, and 70 ha) within a 1,619-ha area burned by an arson fire in 1990. Sproul State Forest is dominated by northern hardwood and dry oak forests (Johnson et al. 2009). Most forest stands in Sproul State Forest were 80–100-year-old, closed-canopied forest that lacked the structural characteristics used by nesting Golden-winged Warblers. Rather, breeding Golden-winged Warblers in Sproul State Forest were associated with disturbance-generated plant communities adjacent to unharvested forest (saw timber; >80 years old) including areas influenced by timber harvests, wildfire, and abandoned natural gas wells (Larkin and Bakermans 2012). Young forest cover available to breeding Golden-winged Warblers was characterized by a patchy mosaic of saplings, shrubs, herbaceous openings, and scattered trees (approx. basal area: 2.3–9.2 m²/ha). Golden-winged Warbler territory density across the three study sites averaged 1.75 territories/10 ha over a 4-year period (range: 1.08–2.52 territories/10 ha; 2008–2011, J. Larkin, unpubl. data). Common shrub species in descending abundance included blackberry (*Rubus* spp.), mountain laurel (*Kalmia latifolia*), blueberry (*Vaccinium* spp.), and sweet fern (*Comptonia peregrina*). Common tree species in descending abundance included red maple (*Acer rubrum*), black locust (*Robinia pseudoacacia*), pin cherry (*Prunus pensylvanica*), black cherry (*P. serotina*), chokecherry (*P. virginiana*), sweet birch (*Betula lenta*), sassafras (*Sassafras albidum*), white oak (*Quercus alba*), northern red oak (*Q. rubra*), chestnut oak (*Q. montana*), and white pine (*Pinus strobus*).

Field Methods

Golden-winged Warbler Foraging Observations and Territory Mapping

We uniquely banded every territorial male Golden-winged Warbler monitored in our sample to assure individual identification. Our study area was the focus of a larger concurrent study of Golden-winged Warblers beginning in 2008. As such, several males were previously banded with U.S. Geological Survey aluminum leg bands and a unique combination of colored leg bands. To capture and mark unbanded males we used one 6-m mist net, a recording of Golden-winged Warbler type I and II songs (Lang Elliott with Donald and Lillian Stokes: Stokes Field Guide to Bird Songs), and a model of a male Golden-winged Warbler (Ward and Schlossberg 2004). Our study protocol was approved by the Indiana University of Pennsylvania's Animal Care and Use Committee (IACUC #03-0708).

We observed color-banded male Golden-winged Warblers for up to 30 min every other day between the hours of 05:00 and 18:00 Eastern Standard Time (EST) from 9 May 2011 to 20 June 2011. Golden-winged Warblers are rapid fliers and can traverse their territories almost instantaneously; thus, we considered observations on one substrate (shrub or tree) followed by observations on another substrate to be statistically independent (Holmes et al. 1979a). We therefore included each foraging observation recorded from a sequence of foraging observations in our analysis (Holmes et al. 1979a, Morrison 1984, George 2009). When making observations, we randomized the order of individuals observed to account for potential variation in behavior throughout the day (Shields 1977). We recorded behavioral data using field notebooks and digital voice recorders (Olympus VN-5000, Center Valley, PA). During each observation session, we recorded study site, date, time of day, individual identification, individual's sex, geographic coordinates, plant species, observed activity (singing, perching, or foraging), duration of time spent foraging, and forage-site category (shrub or tree). Shrubs were defined as a woody low-growing plant containing multiple stems originating near the base of the plant and trees were defined as >0.5 m tall and >1 cm dbh.

We also conducted territory mapping of male Golden-winged Warblers concurrently with foraging observations using the methods of Wakeley (1987). We delineated territories by following males on ≥8 occasions during the breeding season. Song posts, defined as locations where a bird sang at least once, and foraging sites were marked with flagging tape and geographic coordinates were recorded with a handheld GPS receiver (Garmin eTrex H, North American Datum 1927). We used territory mapping data (geographic coordinates) and the Hawth Tools extension (Beyer 2004) in ArcGIS 9.2 (ESRI 2009) to estimate the area of each male's territory (ha) using Minimum Convex Polygons (Mohr 1947). For the purpose of this manuscript, we delineated territories based on geographic coordinates of observed song posts and foraging sites, which may underestimate the size of home ranges used by breeding Golden-winged Warblers because long-distance foraging is excluded (Streby et al. 2012; Chapter 5, this volume).

Vegetation Sampling

We characterized vegetation from late June through early July 2011, after we had collected Golden-winged Warbler observations. Some vegetation characteristics, such as amount of herbaceous cover, may have changed between when we observed foraging Golden-winged Warblers and when we conducted vegetation surveys but the composition and structure of woody plants remained the same through the period of our study. Within each male's territory we sampled vegetative structure and species composition within 5-m-radius circular plots along transects (Morrison 1981, Canterbury et al. 2000). The number of transects and 5-m-radius sampling plots per territory varied depending on territory size to achieve 10 sampling plots per ha, to ensure

that we sampled all territories proportional to size and that we adequately characterized the patchy nature typical of nesting habitat of Golden-winged Warblers (Confer et al. 2011). To plan appropriate plot spacing we calculated a total transect length by obtaining a random compass bearing (rotation of a handheld compass until a helper indicated "stop") and starting point (north, south, east, or west side of territory, chosen randomly from folded papers). To determine the number of transects and distance between plots along each transect, we then divided total transect length by the number of 5-m-radius plots needed to achieve 10 sampling plots per ha of a warbler territory.

In each 5-m-radius plot, we recorded the number and species of all trees and saplings, which were characterized by the presence of one or a few apical stems or trunks that supported a branching structure. We defined saplings as ≥0.5 m tall and 1.0 cm ≤ dbh < 10.0 cm, whereas trees were ≥10 cm dbh. We did not differentiate between saplings and trees during Golden-winged Warbler foraging observations so we combined those two categories. We also visually estimated to the nearest 1% the percent cover of herbaceous vegetation, tree seedlings, bare ground, and each shrub species within each 5-m-radius plot. To assist with visual estimates, we divided each 5-m-radius circular plot into four quadrants and summed the percent of each cover type in each quadrant to estimate the percent of each cover type in the entire plot. A team of observers surveyed all vegetation plots to avoid variation due to multiple observers. Each team member estimated and the team came to an agreement on the percent cover of each cover type within each quadrant (each quadrant totaling 25% of the area of the plot). We summed the percent cover for each cover type across the four quadrants to create the final percent cover estimates for each cover type within the entire 5-m-radius plot.

We used the same vegetation sampling protocol to characterize adjacent areas of young forest cover that were unoccupied by Golden-winged Warblers. We surveyed these areas for Golden-winged Warblers and detected no birds after listening and visually searching each area two to three times weekly throughout the breeding season; hereafter, we refer to these areas as "unoccupied areas." We used ArcGIS to create random points throughout areas of young forest unoccupied by Golden-winged Warblers. We randomly selected 12 points and used these random points as the center of a 1.7-ha square polygon, as this was the average size of our spot-mapped Golden-winged Warbler territories. Unoccupied areas were located 0.5–3.7 km from known territories of Golden-winged Warblers in our study area.

Arthropod Sampling

We used exclosure netting and branch clippings to estimate arthropod abundance for 13 focal woody plant species present within territories of Golden-winged Warblers. Exclosure netting and branch clipping are effective ways to evaluate arthropod abundance in the absence of bird predation (Holmes et al. 1979a,b; Atlegrim 1989; Johnson 2000), and dietary analyses have been used to verify that this method captures nearly all items eaten by insectivorous birds (Johnson 2000). We constructed exclosure nets using rolled plastic mesh (1-m-wide roll with 2 cm × 2 cm gaps), cut into 2.4-m sections. We folded the mesh and sewed the sides (with cotton string) to create a bag with final exclosure dimensions of 1.2 m × 1 m.

We placed exclosure netting on branches of individual plants of each target species that was common within our study area: blackberry (n = 15), black locust (n = 15), blueberry (n = 10), mountain laurel (n = 10), sweet fern (n = 10), black cherry (n = 10), pin cherry (n = 10), chestnut oak (n = 10), red maple (n = 10), sassafras (n = 10), sweet birch (n = 10), white oak (n = 10), and white pine (n = 10). We placed exclosure nets on branches from 22 May through 24 June 2011 throughout our three study sites. Netting remained on branches for a minimum of two weeks to allow arthropods, including caterpillars, to repopulate vegetation in the absence of bird predation. We clipped all branches from 8 June through 9 July 2011; we timed branch collection to coincide with the nestling stage of Golden-winged Warblers when the demand for food resources was expected to be greatest. To minimize arthropod loss during branch clipping, we gently placed a 147 L, heavy duty plastic bag over each branch, cinched the end of the bag, then clipped the branch and tightly tied the bag shut. To determine if bird predation had an effect on caterpillar densities, we clipped additional, nonnetted branches of black locust (n = 10), pin cherry (n = 5), and blackberry (n = 10), which we compared to additional, paired exclosures placed

on these species. We randomly selected nonnetted branches near the paired exclosure branch using two observers: one observer selected the three closest branches within sight of the branch in the exclosure and assigned them numbers 1–3, then the second observer chose a number from 1 to 3, corresponding to the branch that we subsequently clipped.

Within two days of branch collection, we carefully searched through leaves on all branches to find all arthropods visible to the naked eye, then counted and identified individuals to Order using identification guides (Borror and White 1970, Milne and Milne 1980). We assumed that all caterpillar-like insect larvae acted as one functional group regardless of taxonomic classification and referred to this category as "caterpillars." We collected and stored leaves from each branch clipping in paper bags and hung them to allow air flow and to keep moisture within the bags at a minimum (reducing the risk of rotting and mildew of leaves) until we oven dried (40°C for 48 hr) and weighed samples to determine dry mass (Butler and Strazanac 2000, Marshall and Cooper 2004). We scaled caterpillar abundance to dry leaf mass (number of caterpillars/10 g dry leaf mass) as a measure of relative abundance (Futuyma and Gould 1979).

Statistical Analyses

Plant Species Use versus Availability

Assessment of third-order resource selection compares used and available resources for individual animals within their respective territory (Johnson 1980), which in this study we considered to be the spot-mapped territory. We considered individual Golden-winged Warblers as the sampling units and used compositional analysis based on the function compana in the adehabitat package (Aebischer et al. 1993, Calenge 2006) in Program R (ver. 2.13.0, R Foundation for Statistical Computing, Vienna, Austria). We tested whether Golden-winged Warblers foraged on certain tree and shrub species disproportionate to their availability within a spot-mapped territory. For both tree and shrub compositional analyses, we converted time spent foraging on and availability of each species to a proportional use and availability; using the relative proportional use and availability of resources avoids the problem of the unit-sum constraint, where avoidance of one resource leads to an apparent preference for other resources (Aebischer et al. 1993). To account for nonnormality in our use and availability data, we conducted both tree and shrub analyses based on data randomization with 10,000 repetitions (Aebischer et al. 1993). We maintained an alpha level of 0.05 and controlled for experiment-wise error between two MANOVA tests (Multivariate Analysis of Variance) of shrub and tree composition with the Holm method in Program R (p.adjust; Package: stats).

To reduce the number of resource categories in our analyses, we included only shrubs and trees present in >50% of spot-mapped territories and utilized for foraging on ≥2 occasions. With these criteria, our results reflect relative preference among used plant species within Golden-winged Warbler territories, and not avoidance. Our analyses did not allow us to make inferences about avoidance of trees or shrubs because we only analyzed the subset of plant species used by Golden-winged Warblers.

Caterpillar Abundance

We compared the abundance of caterpillars among woody plant species using a Kruskal–Wallis rank sum test (kruskal.test) in Program R. We used a Kruskal–Wallis test rather than a one-way analysis of variance because our data were not normally distributed (Shapiro–Wilk test) and standard transformations did not result in normally distributed data (Zar 1999). We analyzed caterpillar abundance on trees and shrubs separately because we recorded tree and shrub species use by Golden-winged Warblers separately. If the Kruskal–Wallis test indicated an overall difference in abundance, we conducted multiple comparisons for a Kruskal–Wallis test (kruskalmc) in Program R within the pgirmess package (Siegel and Castellan 1988, Giraudoux 2011) to determine species-specific differences among caterpillar abundances, with an across-comparisons alpha level of 0.05 (Giraudoux 2011). We maintained alpha levels of 0.05 between the two Kruskal–Wallis tests (shrubs and trees) using the Holm method in Program R (p.adjust; Package: stats).

Additionally, we compared caterpillar abundance between netted and unnetted branches of three plant species (blackberry, black locust, and pin cherry) using separate independent-samples t-tests in SPSS (ver. 19.0.0, SPSS Inc., Chicago, IL).

Caterpillar abundance on blackberry was slightly nonnormally distributed (based on a Shapiro–Wilk test) so we $\log_e$ transformed $[\ln(x + 1)]$ the count data to meet assumptions of normality (Zar 1999). We combined caterpillar data from netted and unnetted treatments for further analysis of caterpillar abundance because we found no differences in caterpillar abundance between netted and unnetted branches for any tree or shrub species. However, our sample sizes were small with 10 unnetted pairs per species of black locust and blackberry, and five unnetted pairs of pin cherry, which may have limited our ability to detect differences in caterpillar densities between netted and unnetted treatments.

Analysis of Territories versus Unoccupied Areas

We compared vegetative structure and ground cover of Golden-winged Warbler territories to that in adjacent, unoccupied areas of similar stand age. To evaluate vegetative structure, we compared percent cover of shrubs, number of saplings, and number of trees between territories and unoccupied areas using an independent-sample t-test (Zar 1999). Both tree and sapling counts were nonnormally distributed (based on a Shapiro–Wilk test) and we $\log_e$ transformed $[\ln(x + 1)]$ both datasets prior to analysis (Zar 1999). Data for percent ground cover were also nonnormally distributed (based on a Shapiro–Wilk test) and standard transformations did not result in normally distributed data. We therefore used Mann–Whitney U tests in SPSS to compare differences in ground cover between Golden-winged Warbler territories and adjacent unoccupied areas (Zar 1999). Ground-cover variables included percent cover of bare ground, percent cover of tree seedlings, percent herbaceous cover, and percent cover of each shrub species (blackberry, blueberry, mountain laurel, and sweet fern). We maintained an overall alpha level of 0.05 across all independent sample tests by adjusting the resulting P-values using the Holm method in Program R (p.adjust; Package: stats).

To determine which tree species were associated with territories of Golden-winged Warblers, we conducted an indicator species analysis (Dufrêne and Legendre 1997) in PC-ORD (ver. 6.0, MjM Software, Gleneden Beach, OR). The method is a form of cluster analysis and combines data on the concentration of species abundance in a particular group and faithfulness of species occurrence in a particular group (Dufrêne and Legendre 1997).

The two components are combined to produce indicator values (IV), which range from zero (no group indication) to 100 (perfect group indication). Indicator values are then tested for statistical significance by Monte Carlo randomizations (Dufrêne and Legendre 1997). A statistically significant result indicated that a plant species was characteristic of either territories or unoccupied areas and was present in the majority of the plots in either category (Dufrêne and Legendre 1997). We considered species to be indicative of territories or unoccupied areas when P < 0.05.

RESULTS

Plant Species Use versus Availability

We observed 14 male Golden-winged Warblers searching for prey on trees and shrubs for 253.2 and 51.0 min, respectively. On average, we observed each individual searching for prey on trees for 18.1 min (range: 0.9–76 min) and on shrubs for 4.6 min (range: 0.1–20.7 min). We observed Golden-winged Warbler males searching for prey an average of seven days (range: 4–12 days).

We recorded a total of 25 tree species within Golden-winged Warbler territories. Of 25 tree species, only seven met our criteria for inclusion in compositional analysis (present in ≥50% of territories and at least two observations of foraging visits). Golden-winged Warbler use of these

seven tree species differed from their availability ($\Lambda = 0.052$, P = 0.02; Table 6.1). Black locust and pin cherry were both foraged on more than black cherry, whereas white oak was foraged on more than sassafras and sweet birch, compared to their respective availabilities.

We did not observe three of the 14 Golden-winged Warblers we monitored searching for prey on shrubs; we therefore excluded them from our shrub selection analysis. We recorded four shrub species within territories: blackberry, blueberry, mountain laurel, and sweet fern. We only observed birds foraging on mountain laurel and blackberry. Therefore, these two species were the only shrubs we included in our analysis. Golden-winged Warblers selectively foraged on blackberry compared to mountain laurel ($\Lambda = 0.067$, P = 0.02; Table 6.1).

Structure and Composition of Territories versus Unoccupied Areas

Vegetative structure differed between Golden-winged Warbler territories (n = 14) and unoccupied areas (n = 12; Table 6.2). Territories had a higher percent of herbaceous cover, fewer saplings, and more trees compared to unoccupied areas. Unoccupied areas had a higher percent cover of blueberry and mountain laurel, and a lower percent cover of blackberry and herbaceous plants (Table 6.2). We detected

TABLE 6.1

Simplified ranking matrix from compositional analysis for territories of Golden-winged Warblers based on proportional tree species use (foraging) and proportional tree species availability.

	White oak	Black locust	Pin cherry	Sweet birch	Red maple	Sassafras	Black cherry	Rank
White oak		—	—	—	—	—	—	1
Black locust	ns		—	—	—	—	—	2
Pin cherry	ns	ns		—	—	—	—	3
Sweet birch	P < 0.05	ns	ns		—	—	—	4
Red maple	ns	ns	ns	ns		—	—	5
Sassafras	P < 0.05	ns	ns	ns	ns		—	6
Black cherry	ns	P < 0.05	P < 0.05	ns	ns	ns		7

We observed 14 Golden-winged Warblers foraging at the Sproul State Forest in north-central Pennsylvania during the 2011 breeding season. Cells in the matrix with a significant difference in log ratios (calculated from proportions) of used and available tree species are indicated by P < 0.05. Ranks represent the importance of each tree species in ascending order from most important (1) to least important (7). Significant P-values indicate a difference between the ranking of two tree species.

TABLE 6.2

Differences in vegetation structure and ground cover composition between Golden-winged Warbler territories (n = 14) and unoccupied areas (n = 12) during the 2011 breeding season in Sproul State Forest, Pennsylvania.

Structural variables		Mean	SD	t_{24}	$P \leq$
% Shrub cover[a]	Territory	62	14.9	0.28	1.000
	Unoccupied	64	20.8		
No. saplings[b]	Territory	14.1	6.9	5.42	0.001***
	Unoccupied	58.5	53.3		
No. trees[b]	Territory	2.3	1.2	−3.54	0.010**
	Unoccupied	0.9	0.8		

Percent ground cover composition variables		Median (%)	Interquartile range	Mann–Whitney U	P[c]
Bare ground[a]	Territory	4	0.0–7.5	61.0	0.705
	Unoccupied	9	1.4–14.8		
Tree seedlings[a]	Territory	2	1.1–3.9	40.0	0.095
	Unoccupied	7	1.8–9.3		
Herbaceous cover[a]	Territory	33	20.9–36.1	33.0	0.044*
	Unoccupied	11	5.1–25.6		
Blackberry[a]	Territory	39	31.3–54.8	0	0.001***
	Unoccupied	0	0.0–0.1		
Mountain laurel[a]	Territory	2	0.0–6.7	7.0	0.001***
	Unoccupied	29	14.1–46.3		
Blueberry[a]	Territory	7	3.9–13.8	13.0	0.002**
	Unoccupied	33	23.3–35.2		
Sweet fern[a]	Territory	4	1.4–7.2	71.5	1.000
	Unoccupied	1	0.6–6.3		

Significant differences are indicated with asterisks; * $P \leq 0.05$, ** $P \leq 0.010$, *** $P \leq 0.001$.

[a] Indices based on visual estimates.

[b] No. of saplings and trees were $\log_e$ transformed for analyses; means and standard deviations (SD) are shown for untransformed variables.

[c] P-values are corrected for multiple comparisons using the Holm method.

no difference in percent shrub cover between territories and unoccupied areas.

We detected several floristic differences between Golden-winged Warbler territories and unoccupied areas, based on indicator species analysis. Six tree species were indicative of territories: black locust (IV = 85.7%, P = 0.001), chokecherry (IV = 70.6%, P = 0.001), pin cherry (IV = 64.5%, P = 0.048), black cherry (IV = 63.9%, P = 0.002), white pine (IV = 50.0%, P = 0.005), and red pine (IV = 35.7%, P = 0.041). Three species were indicators of unoccupied areas: sassafras (IV = 99.6%, P < 0.001), chestnut oak (IV = 90.6%, P < 0.001), and red oak (IV = 44.0%, P = 0.044).

Caterpillar Abundance on Woody Plant Species

Caterpillar abundance did not differ between netted and unnetted branches of black locust (t_{23} = 1.731, P = 0.097), pin cherry (t_{13} = 1.353, P = 0.199), or blackberry (t_{23} = 0.185, P = 0.855); therefore, we pooled netted and unnetted treatments for analyses of caterpillar abundance across tree and shrub species. Caterpillar abundance differed among tree species (Kruskal–Wallis test: χ_8^2 = 64.142, P < 0.001), with black locust, pin cherry, and black cherry having the highest caterpillar abundances among trees, and sassafras, sweet birch, and white pine having the lowest (Figure 6.1). Black locust had a higher

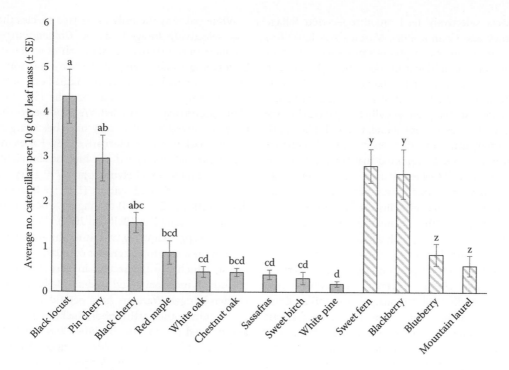

Figure 6.1. Estimated caterpillar abundance on tree and shrub species collected at Sproul State Forest, Pennsylvania, 2011. Means for species with the same letters are not statistically different. Trees (solid fill) and shrubs (line fill) were analyzed separately. n = 10 for all species except black locust (n = 25), blackberry (*Rubus* spp.: n = 25), and pin cherry (n = 15).

abundance of caterpillars than red maple, chestnut oak, sassafras, sweet birch, white pine, and white oak. Pin cherry had a higher abundance of caterpillars than sassafras, sweet birch, white oak, and white pine, whereas black cherry had a higher abundance of caterpillars than white pine. Caterpillar abundance also differed among shrub species (Kruskal–Wallis test: $\chi^2_3 = 20.169$, P < 0.001). Both blackberry and sweet fern had higher caterpillar abundances than mountain laurel and blueberry, whereas blackberry and sweet fern had similar caterpillar abundances (Figure 6.1).

DISCUSSION

Golden-winged Warbler foraging locations and territory placement were influenced by plant species composition and associated prey abundance. Woody plant species that had the highest caterpillar abundances were also selectively foraged on by Golden-winged Warblers and were indicators of Golden-winged Warbler territory locations. Specifically, white oak, blackberry, pin cherry, and black locust were selectively foraged on by the male warblers monitored during this study.

Pin cherry and black locust were also indicators of Golden-winged Warbler territories. Last, black locust and pin cherry had the highest abundance of caterpillars, the primary food source for many breeding insectivorous birds including Golden-winged Warblers (Jacobs 1904). A recent study reported that stomach contents of nestling and young fledgling Golden-winged Warblers in the western Great Lakes region were composed of 89% (based on mass) of leafroller caterpillars (*Archips* spp., Streby et al. 2014). Whereas habitat structure generally dominates management guidelines for Golden-winged Warbler breeding habitat (A. M. Roth et al., unpubl. plan), our research suggests that land managers also need to consider plant species composition, as this habitat characteristic appears to influence prey availability for Golden-winged Warblers.

Black locust is a nitrogen-fixing legume and its leaves have been reported to contain high concentrations of nitrogen (4.0%–5.5% dry leaf mass; Day and Monk 1977, Bassam 1998). Foliar nitrogen content is an important determinant of growth and survival for larval Lepidoptera (Mattson 1980, Schowalter and Crossley 1988). Herbivorous insects selectively feed on nitrogen-rich foliage (Athey and Connor 1989, Marquis and Lill 2010) and grow larger and produce more fecund adults when raised on higher nitrogen diets (Scriber and Slansky 1981). As such, Golden-winged Warblers may have established territories in areas with more black locust, and preferentially foraged on this tree species to access more abundant and higher quality caterpillar prey (Schoener 1971). A previous study in Kentucky reported that black locust was an indicator of Golden-winged Warbler territories on reclaimed surface mines, and that males were often observed gleaning insects from locust leaves (Patton et al. 2010). The importance of black locust to passerine foraging ecology has been reported in Illinois where four of eight passerine species preferred to search for prey on black locust (Hartung and Brawn 2005). However, the Kentucky and Illinois studies did not quantify caterpillar abundance, and thus did not provide a mechanistic explanation for the observed use of black locust by breeding songbirds. Our finding that black locust had the highest abundance of caterpillars among woody species sampled provides the first mechanistic explanation of the Golden-winged Warbler's close association with this early successional tree species in the Appalachian portion of its breeding distribution.

Similar to black locust, pin cherry had high caterpillar abundance, was selectively foraged on by Golden-winged Warblers, and was an indicator of territory use. Interestingly, black cherry was also an indicator of territory use and had a caterpillar abundance similar to that of pin cherry; however, Golden-winged Warblers selectively foraged on pin cherry over black cherry. This observation may be explained by differences in degree of herbivore defense exhibited by these two closely related species. Both black and pin cherry employ chemical defenses against herbivory via alkaloids, sterols, triterpenes, and benzenoids (Ricklefs 2008), but black cherry mounts a higher defense (Burns and Hankala 1990, Eisner and Siegler 2005). Black cherry supported a high caterpillar abundance, but Golden-winged Warblers may have foraged on black cherry less than expected based on its availability because associated prey were less palatable. Further research that examines specific prey species consumed by Golden-winged Warblers would be helpful in explaining observed differences in use of these closely related tree species as foraging substrates.

White oak was the only other tree species that was selectively foraged on by Golden-winged Warblers in our study. The selection of white oak by foraging Golden-winged Warblers was somewhat surprising because white oak had one of the lowest caterpillar densities among the 13 woody plant species that we sampled. We hypothesize that Golden-winged Warblers may have been forging in white oak to access alternative prey. For example, white oak supported the highest abundance of Hymenoptera (3.53 Hymenoptera/10 g dry leaf mass, SE = 2.90) and spiders (1.41 spiders/10 g dry leaf mass, SE = 0.43) among all woody plant species we sampled (Bellush 2012). White oak exhibited approximately two times the abundance of both these arthropod groups compared to black locust, which had the second highest abundance of both Hymenoptera and spiders (Bellush 2012). Golden-winged Warblers have been documented to consume other arthropods, including spiders and winged insects (Jacobs 1904, Forbush 1929) and it is plausible that Golden-winged Warblers were selectively searching for prey on white oak due to the abundance of alternative prey.

Of the four shrub species that dominated our study sites, only two shrubs (blackberry and mountain laurel) were used by foraging Golden-winged Warblers we monitored. Blackberry was selectively foraged on by Golden-winged Warblers and had one of the highest caterpillar abundances (Table 6.1; Figure 6.1). Furthermore, there was considerably more blackberry within Golden-winged Warbler territories compared to unoccupied areas. Previous studies elsewhere have identified positive relationships between songbird communities and blackberry (Kroodsma 1982, 1984; Nur et al. 2008). Our findings support the recommendations within recently published Golden-winged Warbler habitat management guidelines (Bakermans et al. 2011; A. M. Roth et al., unpubl. plan; Chapter 7, this volume) that blackberry provides an important component of Golden-winged Warbler breeding habitat.

Similar to blackberry, sweetfern supported a high abundance of caterpillars. The high number of caterpillars we found on sweetfern is unsurprising given this species' nitrogen fixing properties (Schemnitz 1974). However, we never observed Golden-winged Warblers searching for prey on sweetfern and its abundance did not differ between territories and unoccupied areas. Sweetfern is a low-growing shrub (~0.5 m in height) and was relatively uncommon in both territories and unoccupied areas compared to other shrub species (Table 6.2). Thus, low abundance and short stature may have limited our ability to observe Golden-winged Warblers searching for prey on sweetfern. Determining the degree to which Golden-winged Warblers take advantage of the high caterpillar abundances that occur on sweetfern is worthy of future investigation.

Several woody plant species that we examined appeared to provide limited-to-no benefits for breeding Golden-winged Warblers. Sassafras, mountain laurel, and blueberry were all indicators of areas unoccupied by Golden-winged Warblers, were not selectively foraged on, and had some of the lowest caterpillar densities of all woody species we sampled. It is unsurprising that sassafras was an indicator of areas unoccupied by Golden-winged Warblers as this species is known to have chemical defenses that deter herbivorous insects (Gant and Clebsch 1975). Oils extracted from sassafras leaves contain the monoterpene geraniol (Gant and Clebsch 1975), and various monoterpenes are toxic to herbivorous insects and serve as a feeding deterrent (Barnard and Xue 2004, Gershenzon and Dudareva 2007).

Mountain laurel comprised almost seven times and blueberry about three times the percent ground cover of unoccupied areas compared to Golden-winged Warbler territories. Specifically, these two shrub species alone accounted for over 60% of the ground cover in unoccupied areas (Table 6.2). As such, we speculate that the dominance of mountain laurel and blueberry prevented the development of other vegetation components known to be important to breeding Golden-winged Warblers (i.e., herbaceous plants and blackberry). We propose that low caterpillar abundance we observed on these woody plant species combined with their dominant influence on the presence of other important vegetation components likely explains why they were associated with areas unoccupied by Golden-winged Warblers.

In summary, selective foraging by male Golden-winged Warblers, along with high caterpillar abundances, indicated that black locust, pin cherry, white oak, and blackberry provided important foraging resources to breeding Golden-winged Warblers in north-central Pennsylvania. Thus, we urge land managers intending to create or improve quality of Golden-winged Warbler breeding habitat to consider the potential influence of plant species composition on prey abundance. Vegetation structure generally dominates management guidelines for Golden-winged Warbler breeding habitat, but our findings suggest preferentially maintaining or encouraging the presence of black locust, pin cherry, white oak, and blackberry, while eliminating or discouraging dominance of sassafras, mountain laurel, and blueberry in the High Allegheny Plateau physiographic region. Last, reforestation efforts on drastically disturbed areas in our study region such as reclaimed surface mines should consider including pin cherry, black locust, white oak, and blackberry in revegetation strategies. By doing so, extensive areas of nonforested, reclaimed surface mines in the region could be restored to young forests with high value for breeding Golden-winged Warblers. It is important to recognize that our findings are only applicable to a relatively small portion of the Golden-winged Warbler's breeding distribution. However, they highlight the need for further research that incorporates multiple study sites that represent a diverse group of vegetative communities to determine if plant species composition has a similar influence elsewhere on Golden-winged Warbler foraging ecology and territory placement. Our findings are based on the quantification of all caterpillar types, and that a study concurrent with ours provided the first empirical evidence that tortricid moth larvae (leafrollers, Lepidoptera: Tortricidae) are a major component of Golden-winged Warbler diets in Minnesota (Streby et al. 2014). As such, we strongly encourage future research to specifically quantify tortricid moth larvae to determine the degree to which Golden-winged Warblers depend on this caterpillar group.

ACKNOWLEDGMENTS

Funding for this research was provided by Pennsylvania Department of Conservation and Natural Resources (PA DCNR)-Wild Resource Conservation Fund, Indiana University of Pennsylvania (IUP), and U.S. Fish and Wildlife Service (USFWS)-Division of Migratory Birds.

The authors are grateful to D. D'Amore of the PA DCNR Bureau of Forestry for permission to conduct our study on Sproul State Forest. J. Ferrara at Bald Eagle State Park provided field season lodging and demonstrated much appreciated enthusiasm for our research.

Last, we thank the following individuals for their never-ending dedication in the field: M. Frantz, W. Leuenberger, C. L. Ziegler, S. McGaughran, E. Clawson, D. Clawson, and A. Evans.

LITERATURE CITED

Aebischer, N. J., P. A. Robertson, and R. E. Kenward. 1993. Compositional analysis of habitat use from animal radio-tracking data. Ecology 74:1313–1325.

Athey, L. A., and E. F. Connor. 1989. The relationship between foliar nitrogen content and feeding by *Odonata dorsalis* Thun, on *Robinia pseudoacacia* L. Oecologia 79:390–394.

Atlegrim, O. 1989. Exclusion of birds from bilberry stands: impact on larval density and damage to the bilberry. Oecologia 70:136–139.

Bakermans, M. H., J. L. Larkin, B. W. Smith, T. M. Fearer, and B. C. Jones. 2011. Golden-winged Warbler habitat Best Management Practices for forestlands in Maryland and Pennsylvania. American Bird Conservancy, The Plains, VA.

Barnard, D. R., and R. Xue. 2004. Laboratory evaluation of mosquito repellents against *Aedes albopictus*, *Culex nigripalpus*, and *Ochlerotatus triseriatus* (Diptera: Culicidae). Journal of Medical Entomology 41:726–730.

Bassam, N. E. 1998. Energy plant species: their use and impact on environment and development. James and James (Science), London, UK.

Bellush, E. C. 2012. Foraging ecology of the Golden-winged Warbler (*Vermivora chrysoptera*): does plant species composition matter? M.S. thesis, Indiana University of Pennsylvania, Indiana, PA.

Beyer, H. L. [online]. 2004. Hawth's analysis tools for ArcGIS. <http://www.spatialecology.com/htools> (1 September 2011).

Borror, D. J., and R. E. White. 1970. A field guide to insects of America north of Mexico. Houghton Mifflin, Boston, MA.

Briggs, R. P. 1999. Appalachian Plateaus province and the Eastern Lake section of the Central Lowland province. Pp. 362–377 in C. H. Shultz (editor), The geology of Pennsylvania. Pennsylvania Geologic Survey Special Publication 1. Pennsylvania Geological Survey, Harrisburg, PA.

Buehler, D. A., A. M. Roth, R. Vallender, T. C. Will, J. L. Confer, R. A. Canterbury, S. B. Swarthout, K. V. Rosenberg, and L. P. Bulluck. 2007. Status and conservation priorities of Golden-winged Warbler (*Vermivora chrysoptera*) in North America. Auk 124:1439–1445.

Burns, R. M., and B. H. Hankala. 1990. Silvics of North America. Hardwoods (vol. 2). Agricultural Handbook 654. USDA Forest Service, Washington, DC.

Butler, L., and J. Strazanac. 2000. Occurrence of Lepidoptera on selected host trees in two central Appalachian national forests. Entomological Society of America 93:500–511.

Calenge, C. 2006. The package adehabitat for the R software: a tool for the analysis of space and habitat use by animals. Ecological Modelling 197:516–519.

Canterbury, G. E., T. E. Martin, D. R. Petit, and D. F. Bradford. 2000. Bird communities and habitat as ecological indicators of forest condition in regional monitoring. Conservation Biology 14:544–558.

Confer, J. L., P. Hartman, and A. Roth. 2011. Golden-winged Warbler (*Vermivora chrysoptera*). In A. Poole (editor), The birds of North America online. Cornell Lab of Ornithology, Ithaca, NY.

Day, F. P., and C. D. Monk. 1977. Seasonal nutrient dynamics in the vegetation on a southern Appalachian watershed. American Journal of Botany 64:1126–1139.

Diaz, I. A., J. J. Armesto, S. Feid, K. E. Sieving, and M. F. Willson. 2005. Linking forest structure and composition: avian diversity in successional forests of Chiloe Island, Chile. Biological Conservation 123:91–101.

Dufrêne, M., and P. Legendre. 1997. Species assemblages and indicator species: the need for a flexible asymmetrical approach. Ecological Monographs 67:345–366.

Eisner, M., and M. Siegler. 2005. Secret weapons: defenses of insects, spiders, scorpions, and other many-legged creatures. Harvard University Press, Cambridge, MA.

ESRI (Environmental Systems Resource Institute). 2009. ArcMap 9.2. ESRI, Redlands, CA.

Fink, A. D., F. R. Thompson III, and A. A. Tudor. 2006. Songbird use of regenerating forest, glade, and edge habitat types. Journal of Wildlife Management 70:180–188.

Forbush, E. H. 1929. Birds of Massachusetts and other New England states: land birds from sparrows to thrushes. Commonwealth of Massachusetts, Norwood, MA.

Futuyma, D. J., and F. Gould. 1979. Associations of plants and insects in a deciduous forest. Ecological Monographs 49:33–50.

Gabbe, A. P., S. K. Robinson, and J. D. Brawn. 2002. Tree-species preferences of foraging insectivorous birds: implications of floodplain forest restoration. Conservation Biology 16:462–470.

Gant, R. E., and E. E. C. Clebsch. 1975. The allelopathic influences of *Sassafras albidum* in old-field succession in Tennessee. Ecology 56:604–615.

George, G. 2009. Foraging ecology of male Cerulean Warblers and other Neotropical migrants. Ph.D. dissertation, West Virginia University, Morgantown, WV.

Gershenzon, J., and N. Dudareva. 2007. The function of terpene natural products in the natural world. Nature 3:408–414.

Gill, F. B. 1980. Historical aspects of hybridization between Blue-winged and Golden-winged Warblers. Auk 97: 1–18.

Giraudoux, P. [online]. 2011. PGIRMESS: data analysis in ecology. R package version 1.5.2. <http://CRAN.R-project.org/package=pgirmess> (26 April 2012).

Goodbred, C. O., and R. T. Holmes. 1996. Factors affecting food provisioning of nestling Black-throated Blue Warblers. Wilson Bulletin 108:467–479.

Gregg, I. D., P. B. Wood, and D. E. Samuel. 2000. American Woodcock use of reclaimed surface mines in West Virginia. Pp. 9–22 in D. G. McAuley, J. G. Bruggink, and G. F. Sepik (editors). Proceedings of the Ninth American Woodcock Symposium. U.S. Geological Survey, Biological Resources Division Information and Technology Report USGS/BRD/ITR-2000-0009. Patuxent Wildlife Research Center, Laurel, MD.

Hartung, S. C., and J. D. Brawn. 2005. Effects of savanna restoration on the foraging ecology of insectivorous songbirds. Condor 107:879–888.

Holmes, R. T., R. E. Bonney Jr., and S. W. Pacala. 1979a. Guild structure of the Hubbard Brook bird community: a multivariate approach. Ecology 60:512–520.

Holmes, R . T., and S. K. Robinson. 1981. Tree species preferences of foraging insectivorous birds in a northern hardwoods forest. Oecologia 48:31–35.

Holmes, R. T., J. C. Schultz, and P. Nothnagle. 1979b. Bird predation on forest insects: an exclosure experiment. Science 206:462–463.

Jacobs, J. W. 1904. Gleanings No. III. The haunts of the Golden-winged Warbler (*Helminthophila chrysoptera*) with notes on migration, nest-building, song, food, young, eggs, etc. Independent Printing Company, Waynesburg, PA.

Johnson, D. H. 1980. The comparison of usage and availability measurement for evaluating resource preference. Ecology 61:65–71.

Johnson, M. D. 2000. Evaluation of an arthropod sampling technique for measuring food availability for forest insectivorous birds. Journal of Field Ornithology 71:88–109.

Johnson, P. S., S. R. Shifley, and R. Rogers. 2009. The ecology and silviculture of oaks (2nd ed.). CABI, Cambridge, MA.

Kroodsma, R. L. 1982. Bird community ecology on power-line corridors in east Tennessee. Biological Conservation 23:79–94.

Kroodma, R. L. 1984. Ecological factors associated with degree of edge effect in breeding birds. Journal of Wildlife Management 28:418–425.

Larkin, J. L., and M. Bakermans. 2012. Golden-winged Warbler. Pp. 350–351 in A. M. Wilson, D. W. Brauning, and R. S. Mulvihill (editors), Second atlas of breeding birds in Pennsylvania. The Pennsylvania State University Press, University Park, PA.

MacArthur, R. H., and J. W. MacArthur. 1961. On bird species diversity. Ecology 42:594–598.

Marquis, R. J., and J. T. Lill. 2010. Impact of plant architecture versus leaf quality on attack by leaf-tying caterpillars on five oak species. Oecologia 163:203–213.

Marshall, M. R., and R. J. Cooper. 2004. Territory size of a migratory songbird in response to caterpillar density and foliage structure. Ecology 85:432–445.

Martin, T. E. 1987. Food as a limit on breeding birds: a life history perspective. Annual Review of Ecology and Systematics 18:453–487.

Mattson, W. J. 1980. Herbivory in relation to plant nitrogen content. Annual Review of Ecological Systems 11:119–161.

Milne, L., and M. Milne. 1980. National Audubon Society: field guide to insects and spiders. Alfred A. Knopf, Inc., New York, NY.

Mohr, C. O. 1947. Table of equivalent populations of North American small mammals. American Midland Naturalist 37:223–249.

Moorman, C. E., L. T. Bowen, J. C. Kilgo, C. E. Sorenson, J. L Hanula, S. Horn, and M. D. Ulyshen. 2007. Seasonal diets of insectivorous birds using canopy gaps in a bottomland forest. Journal of Field Ornithology 78:11–20.

Morrison, M. L. 1981. The structure of western warbler assemblages: analysis of foraging behavior and habitat selection in Oregon. Auk 98:578–588.

Morrison, M. L. 1984. Influence of sample size and sampling design on analysis of avian foraging behavior. Condor 86:146–150.

Nagy, L. R., and R. T. Holmes. 2005. Food limits annual fecundity of a migratory songbird: an experimental study. Ecology 86:675–681.

Nur, N., G. Ballard, and G. R. Geupel. 2008. Regional analysis of riparian bird species response to vegetation and local habitat features. Wilson Journal of Ornithology 120:840–855.

Patton, L. L., D. S. Maehr, J. E. Duchamp, S. Fei, J. W. Gassett, and J. L. Larkin. 2010. Do the Golden-winged Warbler and Blue-winged Warbler exhibit species-specific differences in their breeding habitat use? Avian Conservation and Ecology 5:2.

Recher, H. F. 1969. Bird species diversity and habitat diversity in Australia and North America. American Naturalist 103:75–80.

Ricklefs, R. E. 2008. Foliage chemistry and the distribution of Lepidoptera larvae on broad-leaved trees in southern Ontario. Oecologia 157:53–67.

Rodenhouse, N. L., and R. T. Holmes. 1992. Results of experimental and natural food reduction for breeding Black-throated Blue Warblers. Ecology 73:357–372.

Rotenberry, J. T., and J. A. Wiens. 1980. Habitat structure, patchiness, and avian communities in North American steppe vegetation: a multivariate analysis. Ecology 61:1228–1250.

Sauer, J. R., J. E. Hines, J. E. Fallon, K. L. Pardieck, D. J. Ziolkowski Jr., and W. A. Link. 2011. The North American Breeding Bird Survey, results and analysis 1966–2010, version 12.07.2011. U.S. Geological Survey, Patuxent Wildlife Research Center, Laurel, MD.

Schemnitz, S. D. 1974. Sweetfern Comptonia peregrina (L.) Coult. Pp. 138–139 in J. D. Gill and W. M. Healy (editors), Shrubs and vines for northeastern wildlife. General Technical Report NE-9. Department of Agriculture, Forest Service, Northeastern Forest Experiment Station, Upper Darby, PA.

Schoener, T. W. 1971. Theory of feeding strategies. Annual Review of Ecology and Systematics 2:369–404.

Schowalter, T. D., and D. A. Crossley Jr. 1988. Canopy arthropods and their response to forest disturbance. Pp. 183–211 in W. T. Swank and D. A. Crossley Jr. (editors), Forest hydrology and ecology at Coweeta. Springer-Verlag, New York, NY.

Scriber, J. M., and F. Slansky Jr. 1981. The nutritional ecology of immature insects. Annual Review of Entomology 26:183–211.

Shields, W. M. 1977. The effect of time of day on avian census results. Auk 94:380–383.

Siegel, S., and N. J. Castellan. 1988. Nonparametric statistics for the behavioral sciences. McGraw Hill, New York, NY.

Simons, L. S., and T. E. Martin. 1990. Food limitation of avian reproduction: an experiment with the Cactus Wren. Ecology 71:869–876.

Streby, H. M., J. P. Loegering, and D. E. Andersen. 2012. Spot-mapping underestimates song-territory size and use of mature forest by breeding Golden-winged Warblers in Minnesota, USA. Wildlife Society Bulletin 36:40–46.

Streby, H. M., J. M. Refsnider, S. M. Peterson, and D. E. Andersen. 2014. Retirement investment theory explains patterns in songbird nest-site choice. Proceedings of the Royal Society of London B 281:20131834.

U.S. Fish and Wildlife Service. [online]. 2009. Midwest region refuge system: birds of concern—Golden-winged Warbler. <http://www.fws.gov/midwest/midwestbird/birds_golden_winged_warbler.htm> (1 January 2011).

Vale, T. R., A. J. Parker, and K. C. Parker. 1982. Bird communities and vegetation structure in the United States. Annals of the Association of American Geographers 72:120–130.

Verhulst, S. 1994. Supplementary food in the nestling phase affects reproductive success in Pied Flycatchers (Ficedula hypoleuca). Auk 111:714–716.

Wakeley, J. S. 1987. Avian territory mapping. Wildlife Resources Management Manual Technical Report EL-87-7. U.S. Army Corps of Engineers, Department of the Army, Waterways Experiment Station, Vicksburg, MS.

Ward, M. P., and S. Schlossberg. 2004. Conspecific attraction and the conservation of territorial songbirds. Conservation Biology 18:519–525.

Weikel, J. M., and J. P. Hayes. 1999. The foraging ecology of cavity-nesting birds in young forests of the northern coast range of Oregon. Condor 101:58–66.

Wiens, J. A. 1969. An approach to the study of ecological relationships among grassland birds. Ornithological Monographs 8:1–93.

Will, T. C. 1986. The behavioral ecology of species replacement: Blue-winged and Golden-winged Warblers in Michigan. Ph.D. dissertation, University of Michigan, Ann Arbor, MI.

Willson, M. F. 1970. Foraging behavior of some winter birds of deciduous woods. Condor 72:169–174.

Wood, P. B., S. B. Bosworth, and R. Dettmers. 2006. Cerulean Warbler abundance and occurrence relative to large-scale edge and habitat characteristics. Condor 108:154–165.

Zar, J. H. 1999. Biostatistical analysis (4th ed.). Prentice-Hall, Upper Saddle River, NJ.

CHAPTER SEVEN

Golden-winged Warbler Nest-Site Habitat Selection*

Theron M. Terhune II, Kyle R. Aldinger, David A. Buehler,
David J. Flaspohler, Jeffery L. Larkin, John P. Loegering, Katie L. Percy,
Amber M. Roth, Curtis Smalling, and Petra B. Wood

Abstract. Avian habitat selection occurs at multiple spatial scales to incorporate life history requirements. Breeding habitat of Golden-winged Warblers (*Vermivora chrysoptera*) is characterized by largely forested landscapes containing natural or anthropogenic disturbance elements that maintain forest patches in early stages of succession. Breeding habitat occurs in a variety of settings, including shrub and forest swamps, regenerating forests following timber harvest, grazed pastures, and reclaimed mined lands. We identified structural components of nest sites for Golden-winged Warblers by measuring habitat characteristics across five states (North Carolina, New York, Pennsylvania, Tennessee, and West Virginia) in the Appalachian breeding-distribution segment and two states (Minnesota and Wisconsin) in the Great Lakes breeding-distribution segment. We measured habitat characteristics at the nest-site scale with a series of nested plots characterizing herbaceous vegetation (grasses and forbs), woody shrubs and saplings, and overstory trees. We measured similar variables at paired random plots located 25–50 m from the nest within the same territory to evaluate selection. We used conditional logistical regression to identify which parameters were important in habitat selection and Simple Saddlepoint Approximation (SSA) to aid in management interpretation of identified parameters for each study site. Study site was an important determinant for which parameters were significant in nest-site selection, although selection for some parameters was consistent across sites. The amount of woody cover at the nest-site scale was consistently present in the top nest-site selection models across sites, although the direction of the relationship was not the same across all sites. We also identified grass, forb, woody cover, and vegetation density as important components of Golden-winged Warbler nest-site selection. Based on SSA, we identified vegetation thresholds to aid in designing habitat management prescriptions to promote creation or restoration of Golden-winged Warbler nesting habitat across the eastern portion of their breeding distribution.

Key Words: habitat, nest-site selection, saddlepoint approximation.

* Terhune II, T. M., K. R. Aldinger, D. A. Buehler, D. J. Flaspohler, J. L. Larkin, J. P. Loegering, K. L. Percy, A. M. Roth, C. Smalling, and P. B. Wood. 2016. Golden-winged Warbler nest-site habitat selection. Pp. 109–125 in H. M. Streby, D. E. Andersen, and D. A. Buehler (editors). Golden-winged Warbler ecology, conservation, and habitat management. Studies in Avian Biology (no. 49), CRC Press, Boca Raton, FL.

Avian habitat selection is envisioned to occur in a hierarchical fashion from a species' breeding distribution, to the landscape selected by a population, to the territory selected by an individual male, and ultimately to the nest site selected by an individual female (Johnson 1980). Habitat selection by birds is typically assumed to be adaptive such that fitness is higher in selected habitats because of increased probability of successful reproduction, survival, or both (Martin 1998). If nest-site characteristics can be identified and linked to fitness then managers can create or restore sites with these characteristics and a population may respond positively. Population responses may be in the form of colonization of these sites, successful reproduction on these sites, and enhanced survival of adults and fledglings once the initial decision to establish a territory and build a nest has been made.

We documented patterns of nest-site selection in Golden-winged Warblers (*Vermivora chrysoptera*) to help guide management prescriptions intended to ensure availability of nest sites within the larger context of the species' breeding habitat. Golden-winged Warbler populations have declined significantly across most of their breeding range and are considered a species of international conservation concern (Buehler et al. 2007, USFWS 2008, Sauer et al. 2012). In turn, concern about declining populations has led to a major effort to increase the availability of quality breeding habitat on public and private lands (e.g., Bakersman et al. 2011, Golden-winged Warbler Working Group 2013).

Golden-winged Warblers nest across a broad range of conditions (Confer et al. 2011), including young forests and edges of adjacent mature forest stands following timber harvest (Klaus and Buehler 2001, Kubel and Yahner 2008, Roth et al. 2014, Streby et al. 2014), reclaimed surface mines (Bulluck and Buehler 2008, Patton et al. 2010), shrub wetlands (Rossell et al. 2003), swamp forests (Confer et al. 2010), abandoned fields (Confer et al. 2003), power line rights-of-way (Kubel and Yahner 2008), and grazed, high-elevation pastures (Aldinger 2010, Aldinger and Wood 2014). Consistent habitat characteristics used across the breeding distribution include patches of herbaceous cover (grasses and forbs), shrubs and saplings, and mature forest edges. The unique combination of habitat characteristics associated with nest sites has been described for individual studies, but we are not aware of any study comparing habitat characteristics across multiple sites to determine which habitat characteristics are consistent across the breeding distribution. The identification of common characteristics of habitat selection is important for development of distribution-wide conservation strategies. In the absence of such consistent characteristics, habitat management strategies for Golden-winged Warblers may have to be tailored to conditions found on each individual management site or locale where specific habitat characteristics are consistently associated with nest sites.

We designed this study to examine nest-site selection by Golden-winged Warblers across their breeding distribution and across a broad range of ecological conditions in five states in the Appalachian breeding-distribution segment and two states (Minnesota and Wisconsin) in the Great Lakes breeding-distribution segment as outlined in the Golden-winged Warbler Conservation Action Plan (A. M. Roth et al., unpubl. plan; Figure 7.1). The goal of our study was to identify key habitat characteristics associated with nest sites and to compare how nest-site selection varied across the range of study sites we evaluated. Furthermore, we identified threshold values for selection of specific habitat characteristics to aid in development of habitat management prescriptions for Golden-winged Warblers.

METHODS

Study Areas

We selected Golden-winged Warbler breeding sites in five states (North Carolina, New York, Pennsylvania, Tennessee, and West Virginia) in the Appalachian breeding-distribution segment and two states (Minnesota and Wisconsin) in the Great Lakes breeding-distribution segment (Figure 7.1) in vegetation cover types representative of those described in the literature (summarized in Confer et al. 2011). Forest-type descriptions generally follow Braun (2001) unless otherwise noted. Study areas featured recent (<25 years), ongoing, or both recent and ongoing disturbances that created patches of young forest in a predominantly forested landscape. In general, study sites were a mosaic of grasses, forbs, low shrubs, saplings, and few canopy trees surrounded by intact forest.

Study sites at Tamarac National Wildlife Refuge (46.967°N, 95.650°W; 425–500 m in elevation) in Becker County, Minnesota were located at the prairie-forest transition and were comprised

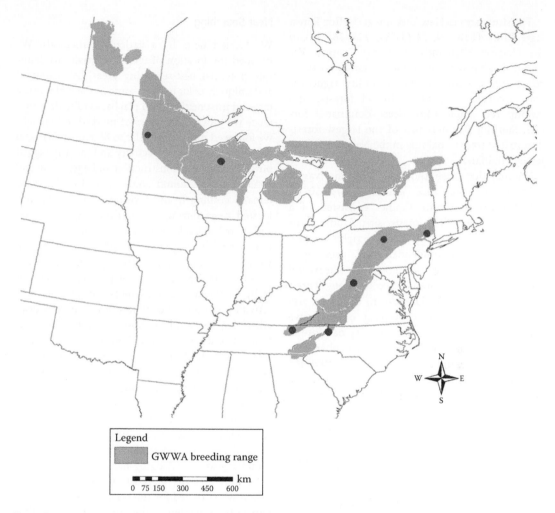

Figure 7.1. Location of individual study sites included in nest-site selection studies of Golden-winged Warblers with distinct conservation management units delineated in gray (Adapted from Roth et al., unpublished report).

of second-growth, moderately closed-canopy aspen (*Populus* spp.) and oak (*Quercus* spp.) forest, interspersed with regenerating forest stands and wetlands. Two upland regenerating forest stands included in this study were created by a combination of gravel removal (one stand), timber harvests, and prescribed burns. Low-quality soils on these stands resulted in a relatively slow rate of succession.

Study areas in Wisconsin were nine aspen-dominated forest stands (45.717°N, 89.533°W; 473–508 m) that had been harvested with varying densities and species of retained canopy trees in Oneida and Vilas counties. Timber harvests involved removal of all aspen and most other tree species except oak (*Quercus* spp.) and pine (*Pinus*

spp.) marked to be retained. The landscape was dominated by northern hardwood forests.

Study areas in North Carolina were located within the Amphibolites (36.400°N, 81.700°W) and Roan Mountain (36.100°N, 82.133°W) Important Bird Areas in Watauga County and extending into Carter County, Tennessee at elevations of 880–1,500 m. A combination of livestock grazing and mechanical brush-removal maintained young forest stands at these sites. Golden-winged Warbler breeding habitat at these study areas was characterized by grass-forb-dominated ground-level vegetation with hawthorn (*Crataegus* spp.) providing the primary woody structure, adjacent to mixed-mesophytic and northern hardwood forests.

The study area in New York was at Sterling Forest State Park (41.183°N, 74.233°W; 245–365 m) in the Hudson Highlands of Orange County. We found Golden-winged Warblers nesting in three cover types: managed upland utility rights-of-way, managed upland hardwood forests, and swamp red maple (*Acer rubrum*)-dominated forests. Sterling Forest is part of the largest forest-dominated tract remaining in the New York–New Jersey Highlands.

We included three study areas in Pennsylvania. Bald Eagle State Park (41.033°N, 77.650°W; 210–245 m) in Centre County consisted of a mosaic of fragmented forests, shrublands, managed grasslands, and powerline rights-of-way. Agricultural and residential land uses were interspersed among the northern hardwood and oak-hickory forests. Sproul State Forest (41.233°N, 77.083°W; 600–620 m) in Centre and Clinton counties was comprised of extensive tracts of northern hardwood and dry oak forests with little human land use. Recent timber harvests since 1990 and a 4,000-ha burn in 1990 led to stands of young forest on the area. Delaware State Forest (41.266°N, 75.100°W; 275–563 m) and adjoining private lands in Monroe and Pike counties contained a variety of plant community types, including scrub oak forest, dry-oak heath, northern hardwood forest and swamps, glacial bogs, and a conifer swamp (Zimmerman et al. 2012).

Study areas in Tennessee consisted of two mountaintops (708–944 m) that were formerly coal surface mines within the North Cumberland Wildlife Management Area (36.217°N, 84.367°W) in Anderson, Campbell, and Scott counties. One of the study areas was managed on a 2-year rotation with prescribed fire. The predominant land cover was a combination of mixed-mesophytic and oak-hickory forests.

Study areas in West Virginia were eight fenced pastures on and around the Monongahela National Forest in Pocahontas (38.317°N, 80.083°W) and Randolph (38.917°N, 79.733°W) counties. Annual low-intensity grazing and periodic mechanical brush removal maintained early stages of succession within the mixed-mesophytic and northern hardwood forest landscape. Overall, pastures were grassland, forb dominated but featured a gradual transition from grassland to shrubland to forest, especially outside the fenced pasture.

Nest Searching

We located nests by a variety of methods. We mapped the location of territorial males to delineate potential nest-searching areas. We captured and uniquely color banded some of the males using targeted mist netting with playbacks of male songs. Once we located territories of marked individuals, we followed methods similar to Martin and Geupel (1993) to locate nesting activity and identify tagged birds. During nest construction and egg laying and incubation, we found some nests by observing females coming from or going to nests. We also located nests during the nestling stage as males and females made repeated trips to nests with food for the young. Rarely (<10% of the sample), we found nests by systematically searching territories where we had previously observed females or by incidentally locating nests during other field activities. We continued to nest search in territories where females had been observed until a nest was found or until we could no longer find the male or female. We checked nests every three days during laying, incubation, and early nestling stages. We checked nests daily as nestlings came close to their potential fledging date (nestling ages 7–10 days). We searched for nestlings around the nest site when we encountered an empty nest with no signs of predation. We deemed nests successful if we found at least one fledging near the nest site or if there were no signs of predation at the nest. If nests failed, we searched for the renest of an individual female by returning to the failed territory within 1–3 days and observing female nesting behavior as above. We limited potential bias in our sample of nests by using behavioral cues of females to lead us to their nest sites.

Vegetation Characterization

We used a nested plot design (1- and 11.3-m-radius plots) to sample vegetation characteristics at each nest site and a single paired random site. Random sites were 25–50 m from nests and located within a defended territory of a male Golden-winged Warbler. Territories were delineated using a standardized spot mapping technique for all sites and that we presume were available for use as nest sites. At the plot center (for both nest and random paired sites), we visually estimated percent cover to the nearest percent in the field (however, to incorporate potential bias associated with visual estimation we rounded to the nearest 5% for presentation purposes) for forbs, grasses and sedges (hereafter, grass cover), blackberry (*Rubus* spp.), woody vegetation (e.g., tree sprouts), litter, vines, and bare ground within the 1-m-radius plot. We measured litter depth (cm) 1 m from plot center in each cardinal direction. We measured nest height defined as the height from the ground to the rim of the cup (cm). We used a standard collection protocol and trained observers participating in vegetation sampling to reduce potential bias associated with visual estimation methods. In addition, we paired nest plots with random plots and had the same observer(s) on each team conduct the visual estimation of cover for those paired plots.

Following Nudds (1977), we measured visual obstruction from vegetation around each nest and random site using a 2-m-tall density board containing two columns of 10, 20 cm × 20 cm cells. One person stood with the board at plot center while an observer viewed it from 10 m away from each cardinal direction. We recorded the number of cells that were >50% covered with vegetation and calculated vegetation density as the number of cells covered/20 × 100 for an overall percent coverage. Within 11.3 m (0.04 ha) of the plot center, we recorded the number of snags (>10 cm dbh), average shrub height (m), average sapling height (m), and measured the distance from plot center to a mature forest edge. We used a 2.5 m²/ha basal area prism to measure basal area. The prism defines a variable-width plot around plot center where trees are counted either "in" or "out" of the plot depending on how close they are to plot center and how large they are. The estimate of basal area is calculated by multiplying the number of "in" trees by 2.5 (Avery 1967).

We classified study sites based on their management history because we predicted site history may be important in nest-site selection of Golden-winged Warblers. Sites came from two distinct origins: a site was originally mature forest but then timber harvest or fire led to secondary succession. We classified these study sites as forest derived. Alternatively, a site lacked mature forest characteristics because of past clearing for agriculture, mining, or other land use but was undergoing succession to forest cover. We classified these sites as grassland derived.

Data Analyses

We evaluated characteristics of nest-site selection using spatially matched sets of plots, with each set consisting of a nest plot and an associated random plot. We used only one set of plots per Golden-winged Warbler territory in our analyses. Our habitat data included two distinct types: continuous (vegetation cover, distance to forest edge, etc.; Table 7.1) and categorical (site, region, and forest-derived or grassland-derived type). We were interested in providing habitat management recommendations across the Golden-winged Warbler breeding distribution and conducted analyses for all sites pooled and based on type (forest derived, grassland derived). Given that our data are primarily from the eastern portion

TABLE 7.1

Descriptions of covariates included in a priori *candidate-model sets for evaluating nest-site selection of Golden-winged Warblers in Minnesota, Wisconsin, New York, Pennsylvania, West Virginia, Tennessee, and North Carolina, 2008–2012.*

Covariates	Notation
Linear and additive effects	
Percent bare ground	BG
Percent forb cover	FC
Percent grass cover	GC
Percent *Rubus* cover	RC
Percent vine cover	VC
Percent shrub cover (shrub + *Rubus* + vines)	Shrub
Percent woody cover	WC
Herbaceous cover (grass cover + forb cover)	GC + FC
Curvilinear effects	
Quadratic relationship with vegetation e.g. (grass cover + grass cover * grass cover)	$GC + GC^2$
Cubic relationship with vegetation e.g. (grass cover + grass cover * grass cover + grass cover * grass cover * grass cover)	$GC + GC^2 + GC^3$
Other structural effects	
Vegetation density	VD
Edge distance	ED

of the distribution, our results and management implications are best suited for the Appalachian breeding-distribution segment.

We used conditional logistic regression to compare continuous covariates (see Table 7.1) of Golden-winged Warbler nest-site plots to random plots (Hosmer et al. 2000). We modeled the differences between matched sets (nest and paired random plots) for all sites and years combined. Given that conditional logistic regression evaluates the difference between matched pairs, categorical variables do not perform as well and can result in spurious results (Hosmer et al. 2000). Therefore, for those covariates that were consistent predictors of nest-site selection, we used Simple Saddlepoint Approximation (SSA) (Renshaw 1998, Matis et al. 2003) to graphically evaluate differences among categorical covariates of interest using program R (R Development Core Team 2013). SSA uses the mean, variance, and skewness of variables to find a general saddlepoint approximation of a probability distribution, which is akin to an approximation of the probability density function (pdf). We used SSA to obtain the respective pdf for nest sites and random points. The SSA approach can provide additional perspective about the data structure of specific covariates identified as important, especially when sample sizes are small and distributions are not approximately normal (Demaso et al. 2011). A selection function value of $\int(x) > 1$ indicates selection, or use greater than availability; a value of $\int(x) < 1$ indicates avoidance, or use less than availability; and a selection value of $\int(x) = 1$ indicates no difference in use compared to availability. As such, we interpret random use with respect to an individual vegetation parameter as $\int(x) = 1$. SSA does not allow, however, for explicitly testing for differences in covariates among sites or among Golden-winged Warbler nest plots and random plots. It is difficult to compute analytic error bounds for Simple Saddlepoint Approximations (Smyth and Podlich 2002, Butler 2007). However, we used a bootstrapping approach following Butler and Bronson (2002) and Bronson (2001) to generate 95% confidence bounds, represented as error clouds, to provide a measure of uncertainty for SSAs.

We developed, a priori, candidate models for comparisons of vegetation characteristics on nest versus random plots based on factors identified in previously published studies (Klaus and Buehler 2001,

Confer et al. 2003, Bullock and Buehler 2008, Aldinger and Wood 2014); these models included additive and factorial combinations of the covariates measured (Table 7.1). We also fit curvilinear models with quadratic and cubic trends to evaluate for possible nonlinear relationships (Table 7.1). We tested hypotheses related to differences in vegetation structure to identify components of vegetation structure unique to forest-derived and grassland-derived sites compared to all sites pooled. We did not include covariates that were highly correlated in analyses (Pearson correlation > 0.7, PROC CORR; SAS Institute 2012). We conducted separate analyses for categorical variables, including site type, forest derived, or grassland derived, to evaluate relative importance of vegetation characteristics among sites with contrasting management origins. In doing so, we could provide habitat management recommendations across the Golden-winged Warbler breeding distribution for all sites pooled and independently based on origin type (forest derived, grassland derived). Furthermore, as noted earlier, categorical variables can result in spurious results in matched-pairs conditional logistic regression (Hosmer et al. 2000). We predicted that nest sites would be associated with a combination of covariates because Golden-winged Warbler nests are typically located in patches of herbaceous vegetation in association with shrubs, saplings, and mature trees (Confer et al. 2011). A primary objective of modeling was to evaluate relative importance of individual predictor variables, and we normalized likelihoods across model subsets within each of our three analyses (sites pooled, forest-derived, and grassland-derived sites) such that Akaike model weights summed to 1 (Burnham and Anderson 2002:167–169, 447–448; Williams et al. 2002:433; Zuur et al. 2009:487).

We conducted conditional logistic regression analyses to model relationships between nest sites and vegetative characteristics using SAS (PROC LOGISTIC; Hosmer et al. 2000, SAS Institute 2012). We ranked competing models using Akaike's Information Criterion corrected for small sample size (AIC$_c$; Hurvich et al. 1998). We identified models of interest as those with $\Delta AIC_c \leq 7$ and reported their associated model weights (w_i; Burnham and Anderson 2002). We further described models with $\Delta AIC_c \leq 2$ as those having "substantial empirical support" and we described models with $2 < \Delta AIC_c \leq 7$ as those having "less empirical support" and models with $\Delta AIC_c > 10$

having "no empirical support" (Burnham and Anderson 2002:70). We also calculated and evaluated profile likelihood 95% confidence intervals (CI) around parameter estimates for those models with substantial empirical support (Venzon and Moolgavkar 1988) and considered CIs not containing zero as evidence for statistical significance. We reported mean values for individual parameter differences between nest and random plots, and 95% CIs around those differences and used log-odds ratios to indicate statistical significance. Descriptors of vegetation are more useful to land managers than mean values for individual parameters differences between nest and random sites and, as such, we also report selection values from SSA for specific nest-site variables based on continuous selection functions (Jensen 1995). We present the SSA analysis for all sites pooled and then stratified the analysis further into site type (forest derived, grassland derived). We report means, precision estimates (CIs), and log-odds ratios to evaluate the selection of individual vegetation characteristics for successful nests compared to failed nests.

RESULTS

We analyzed nest-site characteristics of Golden-winged Warbler nests and associated random sites using 442 paired plots. Sample sizes of nests in landscapes categorized as grassland derived ($n = 219$; North Carolina, West Virginia, and Tennessee) and forest derived ($n = 223$; Minnesota, New York, Pennsylvania, and Wisconsin) were similar. If information for nests was incomplete those vegetation data were excluded from analyses (<1% of all nest data).

Nest-Site Characteristics

Nests were typically located on or near the ground. The average height to nest rim for nests across all sites was 12 cm (±0.3 SE), 11 cm (±0.5 SE) for forest-derived sites, and 13 cm (±0.3 SE) for grassland-derived sites.

We evaluated 44 candidate models (20 linear or additive, 12 curvilinear, and 12 habitat-structure models) at all sites pooled, and forest-derived and grassland-derived sites; however, we only report those models with $\Delta AIC_c < 7.0$. For all sites pooled, two linear or additive models received

substantial empirical support ($\Delta AIC_c < 2.0$; Table 7.2) indicating that bare ground cover, grass cover, forb cover, woody cover, and blackberry may be important vegetation characteristics associated with nest sites of Golden-winged Warblers. The two models with substantial support for all sites pooled carried >93% of the overall model weight (Table 7.2). SSA suggested that Golden-winged Warbler nests were associated with limited bare ground (0%–10% cover; Figure 7.2) and the model including bare ground was 2.3 times more likely than the next best model without bare ground. SSA also revealed that Golden-winged Warbler nests were more commonly associated with grass cover <55% (Figure 7.3) and woody cover within the range of 15%–35% (Figure 7.3). A third model including vine cover received less empirical support ($\Delta AIC_c = 4.53$; Table 7.2) and had a relatively low model weight ($w_i = 0.067$; Table 7.2). Golden-winged Warbler habitat selection for nests located at forest-derived sites was similar to the pooled sites with the exception that vine cover was less important ($w_i = 0.010$; Table 7.3). In contrast, the two models with substantial support for nests at grassland-derived sites both included vine cover and comprised >0.99 of the overall model weight (Table 7.3). The only models receiving substantial empirical support that included curvilinear covariates included bare ground (cubic and quadratic terms) for all sites pooled (Table 7.2) and bare ground (cubic and quadratic terms) for grassland-derived sites (Table 7.4). In addition, models for all sites pooled with grass cover (cubic term), and *Rubus* cover (cubic and quadratic terms) received less empirical support ($\Delta AIC_c < 7.0$; Table 7.2), and models for grassland-derived sites with *Rubus* cover (cubic and quadratic terms) also received less empirical support ($\Delta AIC_c < 7.0$; Table 7.4). In forest-derived sites, the model that included grass cover was overwhelmingly the most empirically supported model and received >0.99 of the overall model weight (Table 7.3). Based on SSA, Golden-winged Warbler nests at forest-derived sites were associated with 10%–65% grass cover (Figure 7.3).

The most important covariates in models for all sites pooled were bare ground cover and grass cover ($\Sigma w_i = 0.69$ and 0.45, respectively; Table 7.5). The most important covariate in models of forest-derived sites was grass cover ($\Sigma w_i = 0.80$; Table 7.5). The most important covariates in models of grassland-derived sites were bare ground cover,

TABLE 7.2

Empirically supported models ($\Delta AIC_c < 7$) of nest-site selection, all sites in all states, included in conditional logistic regression analyses comparing characteristics of Golden-winged Warbler nests and paired randomly sampled sites in Minnesota, Wisconsin, New York, Pennsylvania, West Virginia, Tennessee, and North Carolina, 2008–2012.

Model[a]	K	Deviance	AIC_c	ΔAIC_c	w_i
Linear and additive cover effects					
BC + GC + FC + WC + RB	5	−1063.412	543.706	0.000	0.644
GC + FC + WC + RB	4	−1070.610	545.305	1.599	0.289
GC + FC + WC + RB + VC	5	−1096.478	548.239	4.533	0.067
Curvilinear cover-type effects					
BC + BC^2 + BC^3	3	−1168.354	590.177	0.000	0.433
BC + BC^2	2	−1173.172	590.586	0.409	0.353
GC + GC^2 + GC^3	3	−1173.616	592.808	2.631	0.116
RC + RC^2	2	−1176.634	594.317	4.140	0.055
RC + RC^2 + RC^3	3	−1179.074	595.537	5.360	0.030
Structural cover-type effects					
VD + VD^2 + VD^3	3	−993.370	502.685	0.000	0.973

[a] Covariate notation described in Table 7.1.

Rubus cover, vine cover, and forb cover ($\Sigma w_i = 0.46$, 0.46, 0.40, 0.41, respectively; Table 7.5).

For all sites pooled, nest sites (within the 1-m plot) contained less bare ground compared to random sites within the territory (Table 7.6); this result was similar for nests on forest-derived sites and grassland-derived sites (Table 7.6). Nest sites had greater woody cover than what was present at random sites within the territory for all sites pooled, forest-derived sites, and grassland-derived sites (Table 7.6). In addition, grass cover was much greater at nest sites compared to random sites within the territory for nests located at forest-derived sites (Table 7.6). At grassland-derived sites, however, there was less grass cover at nest sites than that present at random sites within the territory. Blackberry cover, but not forb cover was greater at nest sites compared to that present at random sites within the territory on grassland-derived sites (Table 7.6).

Nest Habitat Structure

Models of habitat structure associated with nests that received substantial empirical support included cubic relationships with vegetation density for all sites pooled ($w_i = 0.973$; Table 7.2) and cubic and quadratic relationships with vegetation density within forest-derived sites ($\Sigma w_i = 0.950$; Table 7.3). At grassland-derived sites, vegetation density appeared to be an important covariate ($w_i = 0.450$ for models containing vegetation density with $\Delta AIC_c < 2.0$; Table 7.4), but shrub cover appeared to be more strongly associated with nest sites (WC + VC + RC; $w_i = 0.550$ with $\Delta AIC_c < 2.0$; Table 7.4). Based on SSA, Golden-winged Warblers selected nest sites with vegetation density ranging from 10% to 35% (Figure 7.3).

Vegetation Characteristics and Nest Success

Successful nests of Golden-winged Warblers at forest-derived sites had greater grass cover compared to unsuccessful nests (Table 7.7), whereas successful nests located at grassland-derived sites had similar amounts of grass cover compared to unsuccessful nests (Table 7.7). No other covariates differed between successful and unsuccessful nests (Table 7.7).

DISCUSSION

Our objective was to identify habitat characteristics associated with nest sites across a broad geographic scale to guide habitat management for breeding Golden-winged Warblers. We hypothesized that

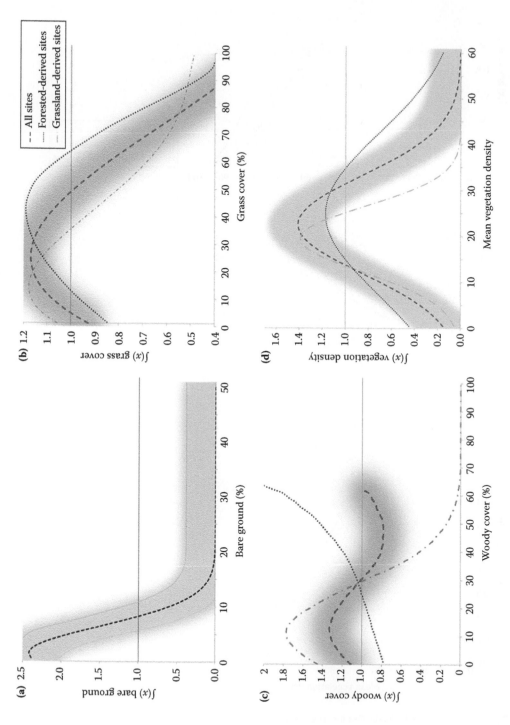

Figure 7.2. Selection function based on SSAs for (a) bare ground cover; (b) grass cover; (c) woody cover; and (d) vegetation density at Golden-winged Warbler nests in Minnesota, New York, North Carolina, Pennsylvania, Tennessee, West Virginia, and Wisconsin for all sites pooled, 2008–2012. Gray-shaded region reflects the 95% bootstrapped confidence limits.

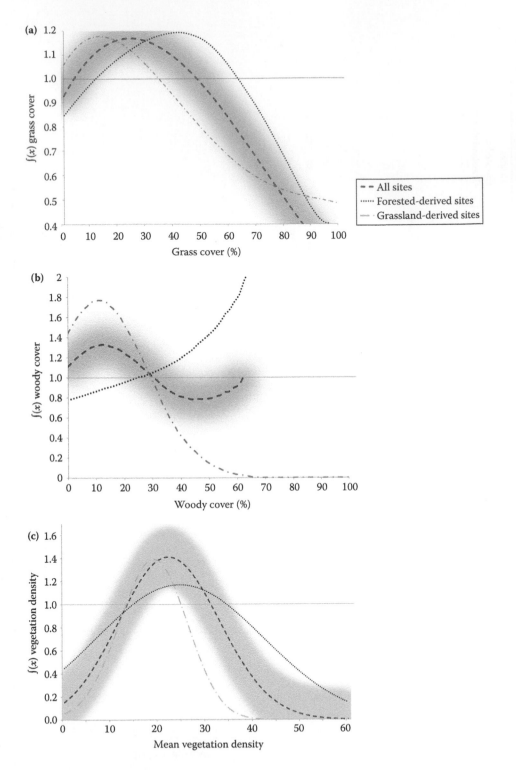

Figure 7.3. Selection function based on SSAs for (a) grass cover; (b) woody cover; and (c) vegetation density at Golden-winged Warbler nests in Minnesota, New York, North Carolina, Pennsylvania, Tennessee, West Virginia, and Wisconsin delineated by all sites, forest-derived sites and grassland-derived sites, 2008–2012. Gray-shaded region reflects the 95% bootstrapped confidence limits.

TABLE 7.3

Empirically supported models (ΔAIC$_c$ < 7) of nest-site selection, in forest-derived sites, in conditional logistic regression analyses comparing characteristics of Golden-winged Warbler nests and paired randomly sampled sites in Minnesota, Wisconsin, New York, Pennsylvania, West Virginia, Tennessee, and North Carolina, 2008–2012.

Model[a]	K	Deviance	AIC$_c$	ΔAIC$_c$	w_i
Linear and additive cover effects					
BC + GC + FC + WC + RB	5	−509.84	266.92	0.00	0.98
Curvilinear cover-type effects					
GC + GC^2 + GC^3	3	−554.41	283.20	0.00	0.61
GC + GC^2	2	−560.13	284.07	0.86	0.39
Structural cover-type effects					
VD + VD^2 + VD^3	3	−450.47	231.23	0.00	0.62
VD + VD^2	2	−456.97	232.48	1.25	0.33
VD	1	−468.51	236.25	5.02	0.05

[a] Covariate notation described in Table 7.1.

TABLE 7.4

Empirically supported models (ΔAIC$_c$ < 7) of nest-site selection, in grassland-derived sites, in conditional logistic regression analyses comparing characteristics of Golden-winged Warbler nests and paired randomly sampled sites in Minnesota, Wisconsin, New York, Pennsylvania, West Virginia, Tennessee, and North Carolina, 2008–2012.

Model[a]	K	Deviance	AIC$_c$	ΔAIC$_c$	w_i
Linear and additive cover effects					
GC + FC + WC + VC + RB	5	−470.92	245.46	0.00	0.72
BC + GC + FC + WC + VC + RB	6	−470.68	247.34	1.88	0.28
Curvilinear cover-type effects					
BG + BG^2 + BG^3	3	−575.52	293.76	0.00	0.48
BG + BG^2	2	−580.34	294.17	0.41	0.39
RC + RC^2	2	−586.00	297.00	3.24	0.09
Structural cover-type effects					
Shrub + shrub^2	2	−541.97	274.99	0.00	0.39
VD + VD^2	2	−544.06	276.03	1.04	0.23
VD + VD^2 + VD^3	3	−540.31	276.16	1.17	0.22
Shrub + shrub^2 + shrub^3	3	−541.56	276.78	1.80	0.16

[a] Covariate notation described in Table 7.1.

in spite of the variety of cover types that Golden-winged Warblers occupied in our study areas, we would be able to identify some characteristics consistently associated with nest sites. Ultimately, we speculated that these characteristics may be linked to increased nest survival (Aldinger et al. 2015). Until this study, no habitat selection studies have been conducted across a broad geographic scale for breeding Golden-winged Warblers. Boves et al. (2013), however, in a recent regional study on Cerulean Warbler (*Setophaga cerulea*) breeding habitat selection, demonstrated that habitat

TABLE 7.5

Cumulative Akaike weights for covariates included in the empirically supported models assessing vegetation characteristics of nest-site selection for Golden-winged Warblers, 2008–2012.

	All sites	Forest-derived sites	Grassland-derived sites
Cover type	Σw_i	Σw_i	Σw_i
Bare ground	0.69	0.39	0.46
Grass cover	0.45	0.80	0.19
Forb cover	0.40	0.40	0.45
Rubus cover	0.43	0.40	0.46
Woody cover	0.40	0.41	0.40
Vine cover	0.03	0.00	0.41

We summed Akaike weights, Σw_i, from individual candidate models with $\Delta AIC_c < 2$ containing each covariate of interest.

selection varied among study sites at the territory scale, although nest-site selection was consistent within Appalachian region sites. If habitat selection in birds is largely driven by local, site-specific conditions, then developing management guidelines for broad geographic areas will be problematic (McNew et al. 2013). Developing management guidelines based on studies of habitat selection without linking the habitat characteristics to nest success and postfledging survival further reduces the likelihood for successful conservation strategies (Jones 2001).

We identified several habitat characteristics that were associated with nest sites across all sites pooled and within subsets of sites based on their origin (forest derived or grassland derived) using visual assessment of vegetation. Several studies have compared methods estimating vegetation cover with varying results among the literature whereby some have suggested that visual estimation produces unreliable results within (Kennedy and Addison 1987) or among differing methods (Symstad et al. 2008). Although other studies suggest that visual estimation is reliable and accurate when conducted by trained observers and when establishing discrete categories (Kercher et al. 2003, Symstad et al. 2008). We recognize that cover estimates and associated precision may be potentially biased in our study due to observer subjectivity. However, the same observers estimated cover on paired sites; hence, comparison of vegetation at random points and nest sites are valid for assessing differences in cover types with respect to nest-site selection.

As such, despite inherent uncertainty around point estimates for individual cover types, general trends about and evaluation of differences in cover type are germane to understanding the biological importance of nest-site selection in the context of site origin.

Nest sites in our study were most strongly associated with low amounts of bare ground (Figure 7.2), and intermediate amounts of grass cover (Figure 7.3a), woody cover (Figure 7.3b), and vegetation density (Figure 7.3c). The nature of the relationships varied depending on whether the site originated from mature forest and was undergoing secondary succession after timber harvest or fire, or whether the site originated from farmland or mining and was undergoing primary succession to forest cover. Furthermore, we observed that Golden-winged Warblers selected for grass cover that was deficient at forest-derived sites, and woody cover is a structural vegetation component that may be linked to nest survival (Aldinger et al. 2015). Selection at the nest-site level may be linked to the landscape context of vegetation at a site and is an ecological phenomenon that has received growing recognition as being of critical importance for understanding the dynamics of natural systems (Borcard and Legendre 2002, Lichstein et al. 2007). Recent studies suggest that incorporating vegetation metrics and spatial habitat context may improve inference from ecological models and help guide on-the-ground management (Thogmartin et al. 2004, Thogmartin and Knutson 2007, van Teeffelen and Ovaskainen 2007). Thus, our results provide

TABLE 7.6

Percent cover of vegetation for Golden-winged Warbler nests (n = 412) compared to paired-random sites in Minnesota (n = 18), New York (n = 78), North Carolina (n = 8), Pennsylvania (n = 101), Tennessee (n = 129), West Virginia (n = 96), and Wisconsin (n = 40) delineated by type (forest derived, grassland derived, and all sites pooled), 2008–2012.

	Nest site			Paired random site			
	Mean	SE	95% CI	Mean	SE	95% CI	Selection
All sites							
Bare ground	1	0.1	(0, 1)	4	0.5	(3, 5)	
Grass cover	21	0.9	(19, 23)	20	1.0	(17, 22)	
Forb cover	38	1.3	(35, 40)	34	1.3	(32, 37)	
Rubus cover	11	0.8	(10, 13)	10	0.9	(8, 12)	
Woody cover	18	1.0	(15, 20)	12	0.9	(10, 14)	+
Vine cover	3	0.4	(2, 3)	3	0.4	(2, 3)	
Forest-derived sites							
Bare ground	1	0.3	(1, 2)	5	0.8	(3, 6)	
Grass cover	21	1.2	(18, 23)	16	1.4	(13, 18)	+
Forb cover	27	1.7	(24, 31)	26	1.7	(23, 29)	
Rubus cover	8	1.1	(5, 10)	8	1.2	(5, 10)	
Woody cover	26	1.7	(23, 29)	19	1.6	(16, 22)	+
Vine cover	2	0.5	(1, 3)	2	0.5	(1, 3)	
Grassland-derived sites							
Bare ground	1	0.2	(0, 1)	3	0.6	(2, 4)	
Grass cover	22	1.1	(19, 24)	24	1.5	(21, 26)	
Forb cover	48	1.5	(45, 51)	42	1.7	(39, 45)	
Rubus cover	15	1.2	(13, 17)	12	1.4	(9, 15)	
Woody cover	9	1.0	(7, 11)	6	0.7	(4, 7)	+
Vine cover	3	0.6	(2, 4)	4	0.7	(2, 5)	

We report standard error (SE) and derived 95% CI; we indicate selection greater than (+) and less than (−) observed at random locations for only those covariates whose CIs are not overlapping.

some of the first empirical evidence that nest-site selection may vary depending on site origin. As such, these selective cues may prove important for understanding how habitat characteristics influence behavioral decisions during nest-site selection, which may depend upon the local environment experienced by breeding females.

Breeding Golden-winged Warblers require diverse vegetation composition comprising herbaceous and woody cover types that create patches of potential nest sites within a larger forested landscape. The strongest patterns of selection among sites we studied were related to amount of herbaceous cover (Figure 7.3) and vegetation density (Figure 7.3). In sites of mature forest origins, woody cover generally dominated the site but herbaceous cover was less abundant. Nest site selection on these sites was associated with greater herbaceous cover (Figure 7.3). In contrast, on sites with grassland-derived origins, herbaceous cover was dominant, with lower availability of woody cover. Nest-site selection on these sites was associated with small amount of herbaceous cover. Selection of woody cover on all sites may be a result of different forms of vegetation providing similar structural elements, including shrubs, saplings, and blackberry cover. Selection for physical visual obstruction at nest sites, in general, was similar among all sites, based on our measures of vegetation density (Figure 7.3), but it is evident

TABLE 7.7

Percent cover (%) of vegetation of Golden-winged Warbler successful (n = 221) and failed nests (n = 171) in Minnesota (n = 18), New York (n = 98), North Carolina (n = 8), Pennsylvania (n = 101), Tennessee (n = 129), West Virginia (n = 96), and Wisconsin (n = 40), 2008–2012.

	Nest successful				Nest failed				
	N	Mean	SE	95% CI	N	Mean	SE	95% CI	Selection
All sites									
Bare ground	221	1	0.2	(1, 2)	171	1	0.3	(1, 2)	
Grass cover	221	23	1.2	(21, 25)	171	19	1.5	(16, 22)	
Forb cover	221	37	1.9	(34, 41)	171	40	2.0	(36, 44)	
Rubus cover	221	11	1.2	(8, 13)	171	13	1.4	(10, 16)	
Woody cover	221	16	1.4	(13, 19)	171	16	1.8	(13, 20)	
Vine cover	221	2	0.4	(1, 3)	171	4	0.8	(2, 5)	
Forest-derived sites									
Bare ground	115	2	0.4	(1, 2)	73	2	0.5	(1, 3)	
Grass cover	115	25	1.9	(21, 29)	73	14	2.2	(10, 18)	+
Forb cover	115	27	2.7	(22, 33)	73	28	3.1	(22, 34)	
Rubus cover	115	8	1.4	(5, 10)	73	8	2.2	(4, 13)	
Woody cover	115	22	2.2	(17, 26)	73	28	3.4	(21, 34)	−
Vine cover	115	1	0.4	(0, 2)	73	4	1.2	(2, 6)	−
Grassland-derived sites									
Bare ground	106	1	0.2	(0, 1)	98	1	0.3	(0, 1)	
Grass cover	106	21	1.5	(17, 24)	98	23	1.9	(18, 26)	
Forb cover	106	48	2.3	(44, 52)	98	48	2.3	(44, 53)	
Rubus cover	106	14	2.0	(10, 18)	98	16	1.7	(13, 19)	
Woody cover	106	10	1.4	(8, 13)	98	8	1.5	(5, 11)	
Vine cover	106	2	0.7	(1, 4)	98	4	1.2	(1, 6)	

We report standard error (SE) and derived 95% CI. We indicate selection greater than (+) and less than (−) observed at random locations for only those covariates whose CIs are not overlapping.

that there is a threshold at which vegetation can be too dense or not dense enough for nesting.

Birds are thought to select breeding habitat in a hierarchical fashion, with first-order selection defining the breeding distribution of a species (Johnson 1980). Second-order selection defines the home range of an individual. Third-order selection by males defines the breeding territory, presumably to attract a female by offering potential suitable nest sites and enabling successful postfledging survival (Streby et al. 2014; Chapter 8, this volume). Females are then constrained within the territories established by males to what is available for nest-site selection and postfledgling habitat. Our study demonstrates that nest-site selection is not random within the territory, but instead is driven by certain vegetation components, presumably because they enhance nest survival (Aldinger et al. 2015) and potentially subsequent fledgling survival (Streby et al. 2014). It is apparent that females are selecting for a variety of conditions to optimize their nest sites to the local environment based on habitat availability or other limitations. Thus, recognizing these habitat limitations in the context of site origin and managing to match habitat selection at the territory level is critical to ensure adequate nesting cover.

CONSERVATION IMPLICATIONS

We documented Golden-winged Warbler nest-site characteristics across a broad geographic scale to help guide management of nesting habitat. As a result of the development of the Golden-winged Warbler Conservation Action Plan (A. M. Roth et al., unpubl. plan) and other associated research, it has become apparent that the Appalachian and Great Lakes regions support disjoint breeding-distribution segments (Chapter 1, this volume) and should be treated as distinct conservation management units. Therefore, the habitat characteristics that we measured and the saddlepoints in the SSA plots can be used as general conservation targets by managers as they improve forest-derived or grassland-derived cover types for Golden-winged Warblers, but our results are likely more relevant to the Appalachian breeding-distribution segment than the Great Lakes breeding-distribution segment. Our results suggest that nest-site selection occurs within the following cover-type ranges: 0%–10% bare ground, 40%–80% herbaceous (forb cover + grass cover), and 15%–35% shrub cover (woody cover + Rubus cover + vine cover). However, depending on the origin of the site as forest-derived or grassland-derived, the upper or lower ranges for these covariates may vary slightly due to vegetation sampling error. In particular among forest-derived sites, it may be necessary to manage for adequate amounts of grass cover. As with management of any ephemeral resource, management of nest habitat for Golden-winged Warblers must address the continual process of forest succession, as sites grow into suitable condition postdisturbance but then mature beyond suitable condition. A continual need for management in some cover types in part explains the challenge to develop successful conservation strategies for Golden-winged Warblers. Incorporating nesting habitat requirements with postfledging habitat requirements within managed landscapes adds another layer of management complexity (Streby et al. 2014; Chapter 8, this volume).

ACKNOWLEDGMENTS

We thank the many agencies and individuals that supported this research, including Audubon North Carolina; Bald Eagle State Park, Pennsylvania; Blue Ridge Conservancy; Cornell Lab of Ornithology; D. and C. Paynter; EcoQuest Travel; Garden Club of America; G. Jones Richardson Trust; Habitat Forever; Indiana University of Pennsylvania; J. Roushdy; Michigan Technological University; National Fish and Wildlife Foundation; National Science Foundation; North Carolina State Parks; North Carolina Wildlife Resources Commission; Oneida County Forests; Pennsylvania Bureau of Forestry; Pennsylvania Department of Conservation and Natural Resources; Potlatch Corporation; Ruffed Grouse Society; Southern Appalachian Highlands Conservancy; Tennessee Wildlife Resources Agency; Tennessee Ornithological Society; The Nature Conservancy; Tomahawk Timberlands; University of Minnesota; University of Tennessee; U.S. Forest Service, Monongahela National Forest; U.S. Forest Service-Northern Research Station; U.S. Geological Survey Cooperative Research Units Program; U. S. Natural Resource Conservation Service; West Virginia Division of Natural Resources; Wisconsin Focus on Energy; Wisconsin Department of Natural Resources; and Wisconsin Society of Ornithology. We also thank the field assistants who worked tirelessly to find and monitor Golden-winged Warbler nests and measure associated vegetation. This study was completed under the auspices of the following IACUC protocols: L0111 and L0200 (Michigan Technological University), 03-0708 (Indiana University of Pennsylvania), 561-1101 (University of Tennessee), 07-0303 and 10-0201 (West Virginia University). Use of trade names does not imply endorsement by the Federal Government.

LITERATURE CITED

Aldinger, K. R. 2010. Playback surveys and breeding habitat characteristics of Golden-winged Warblers (Vermivora chrysoptera) on high-elevation pasturelands on the Monongahela National Forest, West Virginia. M.S. thesis, West Virginia University, Morgantown, WV.

Aldinger, K. R., T. M. Terhune II, P. B. Wood, D. A. Buehler, M. H. Bakermans, J. L. Confer, D. J. Flaspohler, J. L. Larkin, J. P. Loegering, K. L. Percy, and A. M. Roth. 2015. Variables associated with nest survival of Golden-winged Warblers (Vermivora chrysoptera) among vegetation communities commonly used for nesting. Avian Conservation and Ecology 10:6.

Aldinger, K. R., and P. B. Wood. 2014. Reproductive success and habitat characteristics of Golden-winged Warblers in high-elevation pasturelands. Wilson Journal of Ornithology 126:279–287.

Avery, T. E. 1967. Forest measurements. McGraw-Hill, New York, NY.

Bakersman, M. H., J. L. Larkin, B. W. Smith, T. M. Fearer, and B. C Jones. 2011. Golden-winged Warbler habitat Best Management Practices for forested lands in Maryland and Pennsylvania. American Bird Conservancy, The Plains, VA.

Borcard, D., and P. Legendre. 2002. All-scale spatial analysis of ecological data by means of principal coordinates of neighbour matrices. Ecological Modelling 153:51–68.

Boves, T. J., D. A. Buehler, J. Sheehan, P. B. Wood, A. D. Rodewald, J. L. Larkin, P. D. Keyser, F. L. Newell, A. Evans, and G. A. George. 2013. Spatial variation in breeding habitat selection by Cerulean Warblers (Setophaga cerulea) throughout the Appalachian Mountains. Auk 130:46–59.

Braun, E. L. 2001. Deciduous forests of eastern North America. The Blackwell Press, Caldwell, NJ.

Bronson, D. A. 2001. Bootstrapping stochastic systems in survival analysis. Ph.D. dissertation, Department of Statistics, Colorado State University, Fort Collins, CO.

Buehler, D. A., A. M. Roth, R. Vallender, T. C. Will, J. L. Confer, R. A. Canterbury, S. B. Swarthout, K. V. Rosenberg, and L. P. Bulluck. 2007. Status and conservation priorities of Golden-winged Warbler (Vermivora chrysoptera) in North America. Auk 124:1439–1445.

Bulluck, L. P., and D. A. Buehler. 2008. Factors influencing Golden-winged Warbler (Vermivora chrysoptera) nest-site selection and nest survival in the Cumberland Mountains of Tennessee. Auk 125:551–559.

Burnham, K. P., and D. R. Anderson. 2002. Model selection and multimodel inference: a practical information-theoretical approach, 2nd ed. Springer-Verlag, New York, NY.

Butler, R. W. 2007. Saddlepoint approximations with applications. Cambridge University Press, New York, NY.

Butler, R. W., and D. A. Bronson. 2002. Bootstrapping survival times in stochastic systems by using saddlepoint approximations. Journal of the Royal Statistical Society B 64:31–49.

Confer, J. L., K. W. Barnes, and E. C. Alvey. 2010. Golden- and Blue-winged Warblers: distribution, nesting success, and genetic differences in two habitats. Wilson Journal of Ornithology 122:273–278.

Confer, J. L., J. L. Larkin, P. E. Allen, and F. Moore. 2003. Effects of vegetation, interspecific competition, and brood parasites on Golden-winged Warbler (Vermivora chrysoptera) nesting success. Auk 120:138–144.

Confer, J. L., A. M. Roth, and P. J. Hartman. 2011. Golden-winged Warbler (Vermivora chrysoptera). In A. Poole (editor), The birds of North America online. <http://bna.birds.cornell.edu/bna/>. Cornell Lab of Ornithology, Ithaca, NY.

Demaso, S. J., F. Hernández, L. A. Brennan, and R. L. Bingham. 2011. Application of the simple saddlepoint approximation to estimate probability distributions in wildlife research. Journal of Wildlife Management 75:740–746.

Golden-winged Warbler Working Group [online]. 2013. Best Management Practices for Golden-winged Warbler habitats in the Great Lakes region. <www.gwwa.org> (18 June 2014).

Hosmer, D. W., S. Lemeshow, and R. X. Sturdivant. 2000. Applied logistic regression. John Wiley & Sons, New York, NY.

Hurvich, C. M., J. S. Simonoff, and C. L. Tsai. 1998. Smoothing parameter selection in nonparametric regression using an improved Akaike information criterion. Journal of the Royal Statistical Society B 60:271–293.

Jensen, J. L. 1995. Saddlepoint approximations. Oxford University Press, Oxford, UK.

Johnson, D. H. 1980. The comparison of usage and availability measurements for evaluating resource preference. Ecology 61:65–71.

Jones, J. 2001. Habitat selection studies in avian ecology: a critical review. Auk 118:557–562.

Kennedy, K. A., and P. A. Addison. 1987. Some considerations for the use of visual estimates of plant cover in biomonitoring. Journal of Ecology 75:151–157.

Kercher, S. M., C. B. Frieswyk, and J. B. Zedler. 2003. Effects of sampling teams and estimation methods on the assessment of plant cover. Journal of Vegetation Science 14:899–906.

Klaus, N. A., and D. A. Buehler. 2001. Golden-winged Warbler breeding habitat characteristics and nest success in clearcuts in the southern Appalachian Mountains. Wilson Bulletin 113:297–301.

Kubel, J. E., and R. H. Yahner. 2008. Quality of anthropogenic habitats for Golden-winged Warblers in central Pennsylvania. Wilson Journal of Ornithology 120:801–812.

Lichstein, J. W., J. Dushoff, S. A. Levin, and S. W. Pacala. 2007. Intraspecific variation and species coexistence. American Naturalist 170:807–818.

Martin, T. E. 1998. Are microhabitat preferences of coexisting species under selection and adaptive? Ecology 79:656–670.

Martin, T. E., and G. R. Geupel. 1993. Nest-monitoring plots: methods for locating nests and monitoring success. Journal of Field Ornithology 64:507–519.

Matis, J. H., T. R. Kiffe, E. Renshaw, and J. Hassan. 2003. A simple saddlepoint approximation for the equilibrium distribution of the stochastic logistic model of population growth. Ecological Modelling 161:239–248.

McNew, L. B., A. J. Gregory, and B. K. Sandercock. 2013. Spatial heterogeneity in habitat selection: nest site selection by Greater Prairie-Chickens. Journal of Wildlife Management 77:791–801.

Nudds, T. D. 1977. Quantifying the vegetative structure of wildlife cover. Wildlife Society Bulletin 5:113–117.

Patton, L. L., D. S. Maehr, J. E. Duchamp, S. Fei, J. W. Gassett, and J. L. Larkin. 2010. Do the Golden-winged Warbler and Blue-winged Warbler exhibit species-specific differences in their breeding habitat use? Avian Conservation and Ecology 5:2.

R Development Core Team. 2014. R: a language and environment for statistical computing. R Foundation for Statistical Computing, Vienna, Austria.

Renshaw, E. 1998. Saddlepoint approximations for stochastic processes with truncated cumulant generating functions. Mathematical Medicine and Biology 15:41–52.

Rossell, C. R., Jr., S. C. Patch, and S. P. Wilds. 2003. Attributes of Golden-Winged Warbler territories in a mountain wetland. Wildlife Society Bulletin 31:1099–1104.

Roth, A. M., D. J. Flaspohler, and C. R. Webster. 2014. Legacy tree retention in young aspen forest improves nesting habitat quality for Golden-winged Warbler (*Vermivora chrysoptera*). Forest Ecology and Management 321:61–71.

SAS Institute. 2012. Base SAS 9.3 procedures guide. SAS Institute, Cary, NC.

Sauer, J. R., J. E. Hines, J. E. Fallon, K. L. Pardieck, J. D. J. Ziolkowski, and W. A. Link. 2012. The North American Breeding Bird Survey, results and analysis 1966–2011, version 07.03.2013. USGS Patuxent Wildlife Research Center, Laurel, MD.

Smyth, G. K., and Podlich, H. M. 2002. An improved saddlepoint approximation based on the negative binomial distribution for the general birth process. Computational Statistics 17(1):17-28.

Streby, H. M., J. M. Refsnider, S. M. Peterson, and D. E. Andersen. 2014. Retirement investment theory explains patterns in songbird nest-site choice. Proceedings of the Royal Society of London B 281:20131834.

Symstad, A. J., C. L. Wienk, and A. D. Thorstenson. 2008. Repeatability, and efficiency of two canopy-cover estimate methods in Northern Great Plains vegetation. Rangeland Ecology and Management 61:419–429.

Thogmartin, W. E., and M. G. Knutson. 2007. Scaling local species-habitat relations to the larger landscape with a hierarchical spatial count model. Landscape Ecology 22:61–75.

Thogmartin, W. E., J. R. Sauer, and M. G. Knutson. 2004. A hierarchical spatial model of avian abundance with application to Cerulean Warblers. Ecological Applications 14:1766–1779.

USFWS. 2008. Birds of conservation concern. U. S. Department of Interior, Fish and Wildlife Service, Division of Migratory Bird Management, Arlington, VA.

van Teeffelen, A. J., and O. Ovaskainen. 2007. Can the cause of aggregation be inferred from species distributions? Oikos 116:4–16.

Venzon, D., and S. Moolgavkar. 1988. A method for computing profile-likelihood-based confidence intervals. Applied Statistics 37:87–94.

Williams, B. K., J. D. Nichols, and M. J. Conroy. 2002. Analysis and management of animal populations. Academic Press, San Diego, CA.

Zimmerman, E., T. Davis, G. Podniesinski, M. Furedi, J. McPherson, S. Seymour, B. Eichelberger, N. Dewar, J. Wagner, and J. Fike (editors). 2012. Terrestrial and palustrine plant communities of Pennsylvania, 2nd ed. Pennsylvania Natural Heritage Program, Pennsylvania Department of Conservation and Natural Resources, Harrisburg, PA.

Zuur, A., E. N. Ieno, N. Walker, A. A. Saveiliev, and G. M. Smith. 2009. Mixed effects models and extensions in ecology with R. Springer, New York, NY.

CHAPTER EIGHT

Survival and Habitat Use of Fledgling Golden-winged Warblers in the Western Great Lakes Region*

Henry M. Streby, Sean M. Peterson, and David E. Andersen

Abstract. Postfledging habitat use and fledgling survival remain unstudied for most songbirds, but this period is critical for understanding breeding habitat associations and full-season productivity. We used radiotelemetry to study movements, cover-type selection, and survival of fledgling Golden-winged Warblers (*Vermivora chrysoptera*) during the dependent postfledging period in managed forest landscapes of the western Great Lakes region. We used logistic exposure models to determine the relative importance of various habitat characteristics for explaining fledgling survival. In addition, we used compositional analysis, corrected for age-specific fledgling movement capabilities, to test for resource selection, as use versus availability, among cover types. We estimated that 48% of fledglings were depredated before independence from adult care at 25 days after fledging. Fledgling survival was lowest immediately after fledging, and 86% of predation occurred in the first 8 days following fledging. Distance from the nest to forest–shrubland edge was the strongest predictor of young fledgling survival, as survival decreased with nest distance into shrubland cover types

and increased with nest distance into forest cover types. Fledglings from nests in shrubland cover types moved toward the nearest forest–shrubland edge, whereas fledglings from nests in forest cover types did not move toward edge. Fledglings selected mature forest and sapling-dominated clear-cuts over all other cover types during the early postfledging period, and fledgling survival in mature forest and sapling-dominated clear-cuts was greater than in shrub-dominated clear-cuts or wetland shrublands. Fledglings that were 9–25 days postfledging experienced high survival (daily survival >0.99) that was independent of any habitat variables we measured, and birds selected mature forest and shrub-dominated clear-cuts over all other cover types during that period. We conclude that sapling-dominated clear-cuts or mature forest with dense understory and shrub layers, cover types traditionally not associated with breeding, are important for fledgling survival, and therefore full-seasonal productivity in Golden-winged Warblers.

Key Words: forest management, habitat selection, postfledging, songbird.

* Streby, H. M., S. M. Peterson, and D. E. Andersen. 2016. Survival and habitat use of fledgling Golden-winged Warblers in the western Great Lakes region. Pp. 127–140 in H. M. Streby, D. E. Andersen, and D. A. Buehler (editors). Golden-winged Warbler ecology, conservation, and habitat management. Studies in Avian Biology (no. 49), CRC Press, Boca Raton, FL.

full-season productivity in songbirds is defined as the number of young raised to independence from adult care (Streby and Andersen 2011). Fledgling survival is a critical component of full-season productivity but remains poorly understood in most species (Anders et al. 1997, Anders and Marshall 2005, Faaborg et al. 2010, Streby and Andersen 2011). Without information about fledgling survival and habitat associations, estimates of population productivity and descriptions of species habitat requirements can be incomplete or misleading (Streby and Andersen 2011, 2013a). Many songbirds use areas during the postfledging period that differ in vegetation characteristics from areas used for nesting (Pagen et al. 2000, Marshall et al. 2003, Vitz and Rodewald 2007, Streby et al. 2011a). Fledgling survival is affected by habitat use at multiple spatial scales (King et al. 2006, Streby and Andersen 2013b), and fledglings use habitat features associated with greater likelihood of survival disproportionately both before and after independence from adult care (Cohen and Lindell 2004; Streby and Andersen 2012, 2013a; Jackson et al. 2013; Vitz and Rodewald 2013). Predation is the leading cause of fledgling mortality in songbirds, and fledgling predation is greatest during the first few days outside the nest (Anders et al. 1997, Yackel Adams et al. 2006, Moore et al. 2010, Vitz and Rodewald 2011, Streby and Andersen 2013b). Therefore, habitat features associated with the nest site and with early postfledging locations often contribute disproportionately to fledgling survival (Berkeley et al. 2007, Jackson et al. 2013, Streby and Andersen 2013b). However, identifying the features of nest sites and postfledging habitat use associated with fledgling survival requires monitoring fledgling movements and survival.

Songbird population growth rates can be more sensitive to variation in fledgling survival than nest success (Streby and Andersen 2011). Therefore, it is particularly important to investigate postfledging ecology in species of conservation concern for which population growth might be limited by fledgling survival. Nesting habitat associations and nest success of Golden-winged Warblers (*Vermivora chrysoptera*) are relatively well studied (Ficken and Ficken 1968, Will 1986, Klaus and Buehler 2001, Martin et al. 2007, Vallender et al. 2007, Bulluck and Buehler 2008, Confer et al. 2010), especially in recent years (Chapter 7, this volume). Before the current study, information

about Golden-winged Warbler postfledging ecology was limited to one study in which fledglings were banded and resighted in Michigan (Will 1986). Will (1986) provided initial descriptions of brood-division behavior (further explored in Chapter 9, this volume), early postfledging movements, and duration of parental care, although he also described the increasing difficulty of consistently observing fledglings as they aged and developed. Radiotelemetry has been used to monitor fledgling songbirds for two decades starting with large songbirds (Anders et al. 1997), and the technology has advanced to allow repeated and reliable monitoring of birds as small as Golden-winged Warblers (8.5–10 g) for >30 days with no discernible effects on behavior or survival (Streby et al. 2013a).

We used radiotelemetry to study movements, cover-type selection, and survival of fledgling Golden-winged Warblers at two sites in northern Minnesota and one site in southeastern Manitoba. It was particularly important to investigate postfledging habitat associations in Minnesota, which hosts nearly half of the global population of breeding Golden-winged Warblers (Chapter 1, this volume). Population numbers in Minnesota have been stable (Zlonis et al. 2013) or increasing slightly (Chapter 1, this volume), despite a 30% increase in the area of early successional upland forest between 1977 and 2011 (Miles and VanderSchaaf 2012). Our objectives were to determine the relative importance of various habitat characteristics for fledgling survival and to test for cover-type selection based on whether fledglings use cover types associated with high fledgling survival disproportionately relative to availability. As in Streby and Andersen (2013a), we use the term *selection* in a statistical context to refer to comparisons between use and availability of cover types, and not in a behavioral context in which an individual selects or chooses locations, which is probably attributable to the fledglings' parents. Based on previous songbird studies, we expected fledgling survival to increase nonlinearly with age (Anders et al. 1998, Yackel Adams et al. 2006, Berkeley et al. 2007, Moore et al. 2010, Vitz and Rodewald 2011) and to be related to cover-type use and vegetation density at the nest and at early postfledging locations (King et al. 2006, Berkeley et al. 2007, Streby and Andersen 2013a). In addition, we expected fledglings to move into areas of dense vegetation and to experience relatively high survival when fledging from nests near those areas (Cohen and Lindell 2004, Jackson et al. 2013, Streby and Andersen 2013b).

METHODS

Study Area

We studied Golden-winged Warblers at Tamarac National Wildlife Refuge (NWR; 47.049°N, 95.583°W) and Rice Lake NWR (46.529°N, 93.338°W), Minnesota, and at Sandilands Provincial Forest (PF; 49.637°N, 96.247°W), Manitoba, Canada. All three sites were located in the northern hardwood forest transition zone, with boreal forest to the north and east, and tallgrass prairie (mostly converted to agriculture) to the south and west. We collected all data in portions of these NWRs and PF, but each study site had no official boundaries, and animal movements expanded our perceived study sites almost daily. Each study site covered ~50 km^2 by the end of the study. The landscape immediately surrounding each study site (within 5 km) was primarily upland and wetland forest, and shrubland, with limited areas (<10%) of agriculture and other human development. Landscapes within the study sites were characterized by similar cover types containing similar tree and shrub species, but individual cover types comprised different proportions of each site (Table 8.1). At all sites, mature forest stands were dominated by maple (*Acer* spp.), oak (*Quercus* spp.), aspen (*Populus* spp.), paper birch (*Betula papyrifera*), and American basswood (*Tilia americana*), with a few mature stands of jack pine (*Pinus banksiana*) and red pine (*P. resinosa*). The term "mature forest" can be ambiguous in managed forests without a specific age or structure when a stand becomes mature. We describe stands by structure rather than age here because (1) forest structure and age are not reliably correlated in our study area, (2) we assume birds respond to vegetation structure rather than age, and (3) forest structure is readily comparable for others drawing inference from our results. Upland forests on our study sites were primarily even-aged stands; we use "mature forest" here to refer to stands that had canopy >20 m. All mature forest stands on our three study sites averaged >60% canopy closure, which is defined as closed tree canopy forest by the U.S. Forest Service (Brohman and Bryant 2005). Sampled locations within mature forest stands ranged from 50% to 96% canopy cover, and most mature stands contained a patchy and dense shrub layer (vegetation <2 m tall) and understory (vegetation between 2 and ~15 m tall) of maple, aspen, oak, and hazel (*Corylus* spp.). Our estimates of canopy cover percentages are based on analysis of >2,500 digital images and represent percentage of sky obscured by vegetation.

Forested areas at each study site were managed through harvest, prescribed fire, or both, for a combination of timber production and wildlife management resulting in the presence of regenerating forest stands of various seral stages. Age classes provide little useful information and a similar age stand north or south of our study area or on different substrate could have considerably different vegetation structure. Therefore, we include a range of stand ages here, but describe stands primarily by vegetation composition and canopy height. We classified stands dominated by vegetation 1–3 m tall as shrub-dominated clear-cuts. These stands are traditionally described as the vegetative component of Golden-winged Warbler habitat (Chapter 7, this volume) and ranged from five to 15 years postharvest and were composed of shrubs, forbs, grasses, paper birch saplings, and aspen propagule saplings with stems <2 cm in diameter that reached 5 m tall in some areas. Shrub-dominated clear-cuts ranged from one to 30 ha and contained sparse individual or small patches (i.e., <0.25 ha) of trees 10–25 m tall. We classified stands dominated by sapling trees with canopy 5–20 m tall as sapling-dominated clear-cuts. All but two stands classified as sapling-dominated clear-cuts had canopies 10–20 m tall. Sapling-dominated clear-cuts ranged from 15 to 30 years postharvest and were dense stands of aspen, birch, and sometimes green ash (*Fraxinus pennsylvanica*) averaging ~10 cm dbh, but ranging widely in dbh, with sparse individual trees taller than the main canopy, similar to those in the shrub-dominated clear-cuts. We classified stands that were structurally similar to shrub-dominated clear-cuts, but on a wetland substrate and with wetland-associated vegetation, as wetland shrublands. Wetland shrublands were dominated by willow (*Salix* spp.) and alder (*Alnus* spp.) and also contained reeds, grasses, and hazel shrubs. The substrate of wetland shrublands ranged from dry ground to standing or moving water depending on snowmelt and recent rainfall, and in some cases the substrate was sphagnum moss (*Sphagnum papillosum*).

TABLE 8.1

Percentage of each study site comprised by each cover type, and mean percentage of each cover type available to fledgling Golden-winged Warblers the first day outside the nest, based on observed movement distances at Tamarac NWR and Rice Lake NWR, Minnesota, and Sandilands PF, Manitoba, during 2011 and 2012.

Site	Mature forest	Wetland shrubland	Grassland	Forested wetland	Coniferous forest	Shrub-dominated clear-cut	Sapling-dominated clear-cut	Road	Firebreak	Human development
Tamarac NWR	62	12	9	9	<1	2	3	2	<1	<1
Rice Lake NWR	31	30	26	8	<1	3	1	<1	<1	<1
Sandilands PF	47	15	14	<1	14	4	2	2	<1	<1
Available	43	12	10	1	2	28	2	2	<1	0

Percentages for study sites were measured within arbitrary boundaries that encompassed each study site, not geopolitical boundaries, and percentages vary slightly if boundaries are moved. The values demonstrate the necessity to limit availability to accessible space when analyzing cover-type selection by animals with limited movement capabilities. Specifically, cover-type selection by young fledgling Golden-winged Warblers would have been biased toward a common nesting cover type (shrub-dominated clear-cuts) if we did not limit availability to accessible space.

Other, less common cover types at each study site included forested wetlands of tamarack (*Larix laricina*) or black ash (*Fraxinus nigra*), upland and wetland grasslands, firebreaks and powerline rights-of-way (mostly grass with some shrubs), roads ranging from two-track access trails to two-lane paved roads, and small areas of human occupation (houses, outbuildings, and lawns). Each site also contained open water in the form of rivers and lakes, but we excluded open water from our analysis because we assumed it was not available for use by Golden-winged Warbler fledglings. Sandilands PF also included a few small plantations of young jack pine, several areas apparently used as communal trash dumping sites, and at least one marijuana (*Cannabis sativa*) cultivation site. For details on delineation and classification of cover types for analysis, see Cover-type Selection section.

Data Collection

We attached 0.39-g VHF radio transmitters (Blackburn Transmitters Inc., Nacogdoches, TX) to nestling Golden-winged Warblers from nests we found by systematically searching our study areas and by radiomonitoring breeding females. For detailed description of nest searching and monitoring methods, see Chapter 10 (this volume). Briefly, we established 8–16 plots at each study site within which we captured females and searched for nests. Each plot consisted of one shrub-dominated clear-cut stand or one wetland shrubland (described above) and extended ~50 m into the surrounding forest. Plots were 200 m to 10 km from each other to increase independence of nests in the sample, and the 43% of nests we found using radiotelemetry reduced potential bias inherent to standard nest-searching studies (Powell et al. 2005, Peterson et al. 2015). We removed each brood from its nest 0–3 days before they fledged and carried the young in a soft cloth bag >10 m from the nest. We marked each nestling with an aluminum U.S. Geological Survey leg band, and we marked 1–5 (usually 2) randomly selected nestlings with radio transmitters. We placed each brood back in its nest <15 min after removal. In addition, we banded and attached transmitters to fledglings we opportunistically captured and for which we did not know the nest location. We attached transmitters to nestlings and fledglings using an elastic-thread, figure-eight

harness design modified from Rappole and Tipton (1991) and identical to methods used by Streby and Andersen (2013a). Transmitter and harness weighed 0.39 g, which was 4.5% of mean body mass (8.6 g) at the time of fledging.

After transmitter attachment to nestlings, we monitored nests once or twice daily to determine fledging dates. Some nestlings (12%) force fledged or refused to remain in the nest after we handled them, either as individuals or as entire broods. However, force fledging had no effect on fledgling survival (Streby et al. 2013b), so we treated those fledglings as if they had fledged naturally. We monitored each fledgling once per day with some 2–6-day observation intervals because of logistical constraints or inclement weather (<3% of observations), and we recorded fledgling fate as dead or alive at each encounter. Rarely (<3% of observations), when a fledgling was entirely obscured by dense vegetation, we narrowed its location to within ~1 m and recorded the location without visual confirmation of fate. In some cases, the fledgling's parent entered the vegetation with food and we assumed the fledgling was alive. When a transmitter signal emanated from the same obscured location during two consecutive days, we investigated more closely, which typically confirmed fledgling mortality. When an older bird flew away on its own or in response to an approaching observer, we recorded the location where we first observed the fledgling. We recorded each nest and fledgling location with a handheld GPS unit and averaged locations to achieve accuracy <5 m. We measured distance and direction between nest and fledgling locations, and between those locations and the nearest forest–shrubland edge, such as edges between shrub-dominated uplands or wetland shrublands and any taller forest stand, using ArcMap 10.1 (ESRI, Redlands, CA).

At each nest and fledgling location, we estimated canopy cover and lateral cover and recorded observations of fledgling activities as begging, foraging, travelling, or sitting quietly, and adult activities as feeding fledgling, chipping, singing, or not present. We used digital cameras (Vivitar VivCam and Nikon Coolpix, set to 35 mm equivalent focal length) to estimate canopy cover by taking a vertical photograph from 2 m above the ground and analyzing the photograph with ImageJ software (National Institutes of Health) to measure the percentage of sky obscured by vegetation. We estimated lateral cover using a profile-board method modified from MacArthur and MacArthur (1961) and described in Streby and Andersen (2013a). Briefly, we divided a 2 m × 0.25 m sheet of vinyl cloth into eight squares, hung the sheet from a collapsible stand, and then estimated the percentage of each square obscured by vegetation from a distance of 5 m. We then rotated the sheet 90°, repeated the estimates, and averaged all 16 values to produce one estimate of lateral cover for each location.

Fledgling Survival

We used methods of Streby and Andersen (2013b) to investigate the relative effects of various habitat characteristics on fledgling survival from the nest to independence from adult care. Fledgling Golden-winged Warblers can remain with a parent beyond 30 days (Will 1986), but most fledglings we monitored were independent from adult care by 23–25 days post fledge (Streby et al. 2014a). We chose habitat variables that have been found to influence fledgling survival in previous studies (Cohen and Lindell 2004, Jackson et al. 2013, Streby and Andersen 2013b). We considered the following six habitat variables: nest distance to forest–shrubland edge, nest canopy cover, nest lateral cover, fledgling canopy cover, fledgling lateral cover, and road crossings when consecutive fledgling locations were on opposite sides of a road (Streby and Andersen 2013a). We only included fledglings we tracked from nests (n = 246) in these models because nest-site habitat variables were unknown for fledglings captured opportunistically. We used the logistic exposure method to model fledgling survival and we fit models with PROC NLMIXED (Shaffer 2004, SAS Institute 2008). All models included a random effect for brood because preliminary models indicated that survival among broodmates was not independent. Brood identity was a much stronger predictor of survival than age or any habitat variable in single-variable models (H. M. Streby, unpubl. data). In addition, fledgling survival increased nonlinearly with age, as is common in songbirds (Streby et al. 2013a). Therefore, all models except the null model (i.e., constant survival including a random effect for brood) included a quadratic term for fledgling age. We used daily survival measured as whether a fledgling survived a one-day observation interval as the response variable and habitat characteristics associated with the fledgling location on the first day of the interval as explanatory variables. For example, we used habitat variables at the nest as predictors of a fledgling surviving the first day outside the nest, habitat variables from the first daily location outside the nest as predictors of day 2 survival, and so on. The few observation intervals between two and six days were not problematic because the logistic exposure method incorporates variable interval lengths (Shaffer 2004). In those cases, we associated survival of the observation interval with habitat characteristics measured on the first day of the interval, as we did with the 1-day intervals.

Most predation occurred during the first eight days after fledging so we assessed the effects of habitat characteristics on survival separately for days 0–8 and days 9–25. The set of candidate models was identical for each stage. Preliminary analysis indicated no site or year effects on fledgling survival, so we pooled survival data from 2011 and 2012 and from all three sites (Streby et al. 2014a). As in Streby and Andersen (2013b), our objective was to determine the relative importance of each variable for explaining fledgling survival. To that end, we built a set of 23 candidate models that included a null model, a model including only a quadratic term for age, and models including each variable individually and all pairs of two variables. We ranked models using AICc and we used the cumulative Akaike weight (w) of the models including each variable to determine the relative importance of each variable for explaining fledgling survival (Burnham and Anderson 2002, Arnold 2010, Streby and Andersen 2013b). In addition, we modeled cover-type-specific survival for each cover type for which we had >50 fledgling locations. For each of those cover types, we considered a constant survival model and models with linear and quadratic terms for age. Again, all models included a random term for brood. We used AICc to select the best supported model for fledgling survival in each cover type. No individual fledgling spent the entire postfledging period in a single cover type, but each of these cover types was used by enough fledglings of different ages to produce a model that converged for the entire period. Models of cover-type-specific survival can be useful for explaining why some cover types are used more relative to their availability compared to other cover types by fledglings of different ages. We compared daily survival among cover types with posthoc Z-tests (Johnson 1979).

We used locations and condition of fledgling and transmitter remains to identify probable predators for each predation event. We attributed predation to mammals when we found remains buried under leaf litter or in mammal burrows and if we observed tooth marks on transmitters. We attributed predation to avian predators when a transmitter was recovered in a pile of plucked feathers or when a transmitter signal came from a raptor nest. We attributed predation to garter snakes (*Thamnophis radix* and *T. sirtalis*) when a transmitter was ingested by a snake or when we recovered a transmitter with the distinct pungent odor of garter snake feces. Rare causes of fledgling death (<4% total) included exposure during cold and wet nights, drowning during a flood, blowfly (*Trypocalliphora braueri*) infection acquired in the nest, apparent starvation following parental disappearance, and fatal injury from a hailstone. We excluded birds with non-predation-related deaths from analysis because those deaths likely were not related to the habitat characteristics used as explanatory variables in our survival models.

Cover-Type Selection

We used aerial photographs for each study site in ArcMap 10.1 to create a digital cover-type map and added descriptive attributes for each cover-type polygon based on ground-truthing. We visited >2,500 locations throughout our study sites and confirmed or adjusted cover-type classification and delineation. We used 1-m resolution digital orthophoto quadrangles (2009, Minnesota Department of Natural Resources) for the NWRs and we used orthorectified satellite images (2010 imagery) from Google Earth™ (version 6.2) for Sandilands PF. We classified all polygons into 10 classes based on cover types described above: (1) mature forest, (2) sapling-dominated clear-cut, (3) coniferous forest, (4) shrub-dominated clearcut, (5) wetland shrubland, (6) forested wetland, (7) grassland, (8) road, (9) firebreak, and (10) human development.

We analyzed cover-type selection (i.e., use versus availability) with compositional analysis using the adehabitat package in Program R (Aebischer et al. 1993, Calenge 2006). We defined the space or area available to fledglings based on the current location of the fledgling and the fledgling's age-specific movement capability to avoid bias associated with assuming an entire study site is available (Streby and Andersen 2013a). We used the daily estimated availability (DEA) method to determine the space available to fledgling Golden-winged Warblers of each age, days 0–24 postfledge (Streby and Andersen 2013a). We used the greatest distance traveled by any fledgling between consecutive daily locations as an estimate of the maximum straight-line movement capability by a fledgling of each age (Figure 8.1). We used each of those maximum daily distances as the radius of a

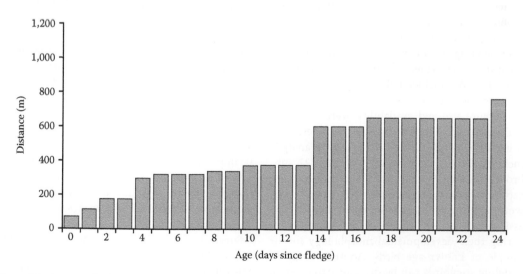

Figure 8.1. Age-specific daily movement capability for fledgling Golden-winged Warblers 0–24 days after fledgling in Minnesota, and Manitoba, Canada during 2011 and 2012. Values were used as radii of circles delineating the DEA of space for cover-type selection analysis.

DEA circle defining available space for fledglings of each age. We then drew DEA circles around each fledgling location in GIS and measured the percentage of each DEA circle comprised by each cover type. By averaging the percentages of each cover type available each day throughout the postfledging period, we standardized cover-type availability by age-specific movement capability for each fledgling Golden-winged Warbler. We defined cover-type use as the cover type occupied by a fledgling within the DEA circle from the previous day. Therefore, we estimated availability for days 0–24 and recorded use for days 1–25.

We analyzed cover-type selection separately for the two stages defined in the survival analysis because cover-type selection might differ between the high-predation period of days 0–8 and the remaining lower predation period of days 9–25. We converted cover-type use and availability to percentages for each bird for compositional analysis. Similar cover-type availability among broodmates does not cause pseudoreplication in compositional analysis because availability is usually identical for entire populations in most such analyses. However, cover-type use was not independent among broodmates during days 0–8, nor among subbroodmates, defined as broodmates with one parent after parents split the brood. During days 9–25, movement patterns but not cover-type use, of male- and female-reared subbroods diverged (Chapter 9, this volume). We therefore averaged use and availability values for broodmates during days 0–8 and for subbroodmates during days 9–25. We excluded fledglings that were depredated in the first 24 h after fledging from cover-type selection analysis because they provided no information for cover-type use. We included fledglings we captured opportunistically in cover-type selection analysis because use and availability were measured daily and did not require knowledge of nest locations after day 1. Cover-type selection analysis included 94 broods for days 0–8 and 127 subbroods for days 9–25. Most of the fledglings we captured opportunistically (n = 25; 11% of total) were captured within a few days of fledging. We assigned these fledglings to an age class by comparing their developmental morphology to our sample of known-age birds. Accurate aging of fledgling songbirds can be difficult (Streby et al. 2013c) and we conducted cover-type selection analyses with and without the opportunistically captured fledglings. Statistical significance and the order of cover-type selection were unaffected if we excluded opportunistically captured fledglings from analysis (H. M. Streby, unpubl. data), so we report results for cover-type selection analysis including all fledglings.

Fledgling Golden-winged Warblers did not consistently move in any cardinal direction during our study (Peterson 2014). However, based on the results of survival and cover-type selection analysis, we examined whether early postfledging movements during first the 24 h after fledging were oriented toward cover types associated with high fledgling survival including sapling-dominated clear-cuts and mature forest. We used Rayleigh V tests for circular uniformity with a specified mean direction (Durand and Greenwood 1958, Zar 2004). We conducted separate tests of oriented movement for fledglings from nests in shrublands (shrub-dominated clear-cuts and wetland shrublands) and for fledglings from nests in forest (mature forest and sampling-dominated clear-cuts). For both groups, we tested whether movements were oriented toward the nearest shrubland–forest edge. We considered all statistical tests to be significant at α = 0.05 level.

RESULTS

Fledgling Survival

We estimated that 48% of fledgling Golden-winged Warblers were depredated before reaching independence from adult care 25 days after fledging. Most (75%) fledgling predation events occurred during the first three days after fledging, and 86% of predation events occurred in the first eight days. Based on location and condition of fledgling and transmitter remains, we attributed predation of day 0–3 fledglings to mammals (86%), snakes (9%), and avian predators (5%). The percentage of fledgling predation attributable to avian predators increased with fledgling age, with 73% of predation during days 9–25 attributed to raptors, whereas 27% and 0% of predation was attributable to mammals and snakes, respectively, during that period.

Survival of fledgling Golden-winged Warblers during the first eight days after fledging was best explained by a model including nest distance to forest–shrubland edge (Figure 8.2; all models included a random term for brood and a quadratic

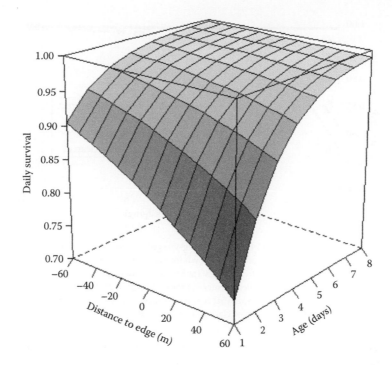

Figure 8.2. Probability of daily survival for 246 fledgling Golden-winged Warblers during the first eight days after fledging from nests in Minnesota, and Manitoba, Canada during 2011 and 2012. The best supported model (lowest AICc) included a quadratic effect of fledgling age and a linear effect of nest distance to edge. Positive values of distance to edge are for nests in shrublands and negative values are for nests in forest. Variance excluded for clarity of presentation.

term for age), and nest distance to edge was the strongest predictor of fledgling survival during that period ($w = 0.73$). Other than fledgling age, which fit the expected pattern for fledgling song-bird survival (Figure 8.2), nest distance to edge was the only variable with a statistically significant relationship with fledgling survival ($\beta = -0.010$; 95% CI $= -0.020$ to -0.001; $t = -2.18$, d.f. $= 93$, $P = 0.03$). Fledgling survival was greatest from nests in forest, lowest from nests in shrublands, and intermediate from nests closest to forest–shrubland edge (Figure 8.2). Nest canopy cover ($w = 0.27$), nest lateral cover ($w = 0.19$), fledgling canopy cover ($w = 0.15$), fledgling lateral cover ($w = 0.22$), and fledgling road crossings ($w = 0.12$) each received less Akaike weight than nest distance to edge during the first eight days after fledging and all of these factors had regression coefficients with confidence intervals overlapping zero. Fledgling survival during days 9–25 was best explained by the null model, suggesting that none of the variables we measured, including fledgling age, explained a meaningful amount of variation

in survival after the first eight days postfledging. For comparison, the null model for the first eight days ranked worse than other models with a ΔAICc of 51.8 from the best supported model.

Cover-type-specific survival models indicated that fledgling survival was least in shrublands and greatest in forests ($Z > 3.00$ and $P < 0.01$ for all comparisons between shrubland types and forest types; Figure 8.3). Daily survival was statistically similar between mature forest and sapling-dominated clear-cuts ($Z = 0.87$, $P = 0.38$) and between the two types of shrublands ($Z = 0.39$, $P = 0.70$). The differences in survival between forested and shrubland cover types were apparent during the first ~8 days, after which fledgling survival was generally high (0.993 ± 0.003 SE) in all cover types. The probability of a fledgling surviving to day 25 and independence from adult care, if it spent the entire dependent postfledging period in one cover type was 0.84 (± 0.09) in sapling-dominated clear-cut, 0.75 (± 0.06) in mature forest, 0.46 (± 0.16) in wetland shrubland, and 0.40 (± 0.10) in a shrub-dominated clear-cut.

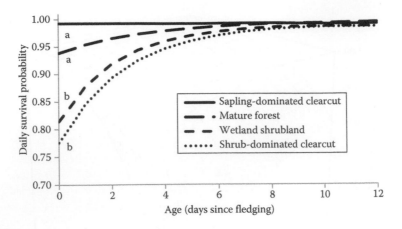

Figure 8.3. Probability of daily survival for 246 fledgling Golden-winged Warblers in each of four cover types in northern Minnesota, and Manitoba, Canada during 2011 and 2012. Shown for each cover type are fitted values for best supported logistic exposure models (lowest AICc) selected from a constant survival model and models with linear and quadratic terms for fledgling age. Values are fitted from models of survival for fledglings 0–24 days after fledging, but days 0–12 are shown to illustrate the differences among cover types that occurred during only the first ~8 days. Letters (a, b) indicate significantly different groups. SE in text, excluded for clarity of presentation.

Cover-Type Selection

Fledglings used cover types disproportionately with respect to availability ($\Lambda = 0.054$, P < 0.01) during the first eight days after fledging. Young fledglings selected for mature forest and sapling-dominated clear-cut over all other cover types, selected against road and grassland, and used shrub-dominated clear-cut, wetland shrubland, firebreak, and forested wetland in accordance with availability. Fledglings also selected ($\Lambda = 0.068$, P < 0.01) among cover types during days 9–25 after fledging. Older fledglings selected for mature forest over all other cover types and selected for shrub-dominated clear-cut over all other cover types except mature forest. During days 9–25, fledglings selected against road and grassland and used wetland shrubland, firebreak, and forested wetland in accordance with availability. First-day movements of fledglings from shrubland nests were oriented toward shrubland–forest edge ($u_{77} = 4.11$, P < 0.01), but first-day movements of fledglings from forest nests were not oriented toward shrubland–forest edge ($u_{20} = 0.24$, P < 0.78).

DISCUSSION

We found that survival of fledgling Golden-winged Warblers was greater in mature forest and sapling-dominated clear-cuts than the shrub-dominated clear-cuts and wetland shrublands with which the species is typically associated (Buehler et al. 2007). In addition, we found that the best predictor of fledgling survival was the location of the nest relative to a forest–shrubland edge. Our results support the contention that fledgling survival is strongly influenced by nest-site choice, which may explain why Golden-winged Warblers rarely nest in areas that maximize nest success (Streby et al. 2014a). The importance of forest stands in later seral stages than stands traditionally associated with breeding Golden-winged Warblers could prove critical for future management plans in the western Great Lakes region. Population trends in Minnesota are stable or growing slightly (Zlonis et al. 2013; Chapter 1, this volume) despite a 30% increase in early successional upland vegetation in the past 35 years (Miles and VanderSchaaf 2012). Refining conservation goals to include postfledging habitat requirements could be key to improving population growth rates in Golden-winged Warblers.

Areas used during the postfledging period differ in vegetation structure from areas used for nesting in many songbirds (Pagen et al. 2000, Marshall et al. 2003, Vitz and Rodewald 2007, Streby et al. 2011a), and postfledging habitat use is associated with fledgling survival, especially in the first few days after fledging (Anders et al. 1998, Yackel Adams et al. 2006, Berkeley et al. 2007, Moore et al. 2010, Vitz and Rodewald 2011). We found that early postfledging movements were oriented toward forest–shrubland edge for fledglings from nests in shrublands but not for fledglings from nests in forested cover types. Our results indicate that adult Golden-winged Warblers direct or lead their fledglings to forested cover types, or keep them in those cover types, regardless of nest location. Similarly, fledgling Ovenbirds (*Seiurus aurocapilla*), Eastern Bluebirds (*Sialia sialis*), and White-throated Robins (*Turdus assimilis*) move toward habitat features associated with high survival shortly after fledging (Cohen and Lindell 2004, Jackson et al. 2013, Streby and Andersen 2013a). Crossing roads had no meaningful influence on survival of fledgling Golden-winged Warblers despite having a considerable impact on survival of fledgling Ovenbirds in the same region (Streby and Andersen 2013b). The difference in mortality associated with road crossings might be related to the manner in which these species move through their environment. Young fledgling ovenbirds spend most of their time on the ground and hop or walk across roads, thereby exposing themselves to predation in the open space of the road (Streby and Andersen 2013b), whereas fledgling Golden-winged Warblers primarily occupy shrubs and understory trees and cross roads relatively quickly by flying (H. M. Streby, pers. obs.).

Fledgling survival differed among cover types for the first eight days, and fledglings selected cover types associated with high fledgling survival (mature forest and sapling-dominated clear-cuts) over all other cover types during that period. Fledgling survival was unaffected by cover-type use or by any other habitat variable we measured during days 9–25, but fledglings still used cover types disproportionately to their availability by selecting mature forest and shrub-dominated clear-cuts over all other cover types. Our results suggest that cover-type selection during the early postfledging period is driven by predator avoidance but is driven by other factors later as fledglings develop and as predation risk declines. Some songbirds choose areas of abundant food resources during the postfledging period (White and Faaborg 2008, Streby et al. 2011b). Leaf-roller caterpillars (*Archips* spp.) made up 89% of stomach contents in fledgling Golden-winged Warbler in our study population (Streby et al. 2014b) and were abundant in broad-leafed trees and shrubs at all of our study sites, possibly explaining the selection of stands abundant in these plants by

fledglings starting to forage. We observed fledglings following foraging adults and starting to attempt foraging at about two weeks after fledging; self-feeding increased until independence (H. M. Streby and S. M. Peterson, unpubl. data).

Many areas within the mature forest stands at our study sites were characterized by dense and patchy understory and shrub layer. Golden-winged Warblers selected mature forest throughout the dependent postfledging period, but young fledglings were most often observed in the shrub layer while their parents foraged elsewhere (Chapter 9, this volume), often >300 m away (H. M. Streby and S. M. Peterson, unpubl. data). Use of shrubs and understory trees within mature forest by young fledglings suggests that early postfledging habitat might be limited for Golden-winged Warblers in landscapes with primarily contiguous closed-canopy forest with limited shrub–sapling vegetation. Ovenbirds, which nest on the ground in mature forest, select sapling-dominated clear-cuts over mature forest during the postfledging period, but only when dense understory vegetation is not available in the mature forest (Streby and Andersen 2013a). Once postfledging habitat associations are considered, it is clear how two species like Golden-winged Warblers and Ovenbirds, with seemingly disparate breeding habitat requirements, can experience strong full-season productivity in the same landscape. Indeed, we regularly observed fledgling Ovenbirds and other mature forest birds while tracking fledgling Golden-winged Warbler in sapling-dominated clear-cuts and mature forest areas with dense understory.

Postfledging survival is important to breeding habitat selection (Streby et al. 2014a) and full-season productivity in songbirds (Streby and Andersen 2011). The presence of cover types associated with high fledgling survival is likely necessary to increase breeding densities and full-season productivity or the number of fledglings raised to independence from adult care. Our results suggest that creating and maintaining young, shrub-dominated stands within an open-understory mature forest matrix is problematic for maximizing density or full-season productivity for Golden-winged Warblers. Rather, the presence of dense forest understory, sapling-dominated clear-cut stands, or both adjacent to shrublands is likely associated with higher density and full-season productivity.

Two potential management options for providing an early postfledging habitat component are thinning of the canopy of mature forest adjacent to shrublands or maintaining a close association between shrublands and sapling-dominated regenerating stands through rotational harvest. In smaller forested areas, or areas where rotational harvest might remove too much mature forest from the landscape to maintain densities of Golden-winged Warblers or other species of conservation concern, thinning the mature forest around maintained shrublands could provide adequate postfledging habitat by increasing density of shrub–sapling layer in the mature forest. Based on distances fledglings move from nests in the early postfledging period (Chapter 9, this volume) when predation is highest, we suggest that a 100-m buffer of thinned mature forest around managed shrublands could provide the vegetation structure associated with high survival of fledgling Golden-winged Warblers.

Our dataset is large and collected from sites spanning a broad geographic range within the western Great Lakes region where a majority of the species breeds, but inference from our results might not apply outside our study region. We described cover types by structure instead of age so the general structure of the vegetative habitat components can be understood and considered elsewhere. However, before applying recommendations based on Golden-winged Warbler-habitat associations in the Great Lakes breeding-distribution segment to the Appalachian breeding-distribution segment, similar studies of postfledging movements, habitat selection, and survival need to be conducted in the Appalachian Mountains portion of the species breeding distribution, where nest productivity is similar to that in our study region (e.g., Confer et al. 2010, Percy 2012), but where many populations of Golden-winged Warblers are declining (Chapter 1, this volume).

ACKNOWLEDGMENTS

Field data were collected during a project funded by the U.S. Fish and Wildlife Service, U.S. Geological Survey, and Minnesota Department of Natural Resources through Research Work Order no. 87 at the Minnesota Cooperative Fish and Wildlife Research Unit. We collected data following Protocol #1004A80575, approved by the University of Minnesota Institutional Animal Care and Use Committee. Use of trade names does not imply endorsement by the U.S. Geological Survey, University of Minnesota, or any other organization supporting this research. We thank P. Blackburn, W. Brininger, W. Faber, W. Ford, J. Loegering, M. McDowell, H. Saloka, and R. Vallender for equipment and logistical support, and J. Kelley for graphics consulting. We thank L. Arnold, S. Barlow, D. Bradshaw, J. Bruggman, R. Carr, M. Dawson, L. Deets, J. Feura, A. Fish, R. Franta, C. Gesmundo, A. Jensen, M. Johnson, G. Kramer, J. Lehman, T. McAllister, D. McNeil, Z. Orbaz, E. Pokrivka, R. Poole, A. Rehmann, J. Refsnider, R. Refsnider, N. Seeger, E. Sinnot, and B. Vernasco for assistance with field data collection, J. Refsnider and S. Loss for constructive comments on early drafts, and two anonymous referees for comments improving the manuscript.

LITERATURE CITED

Aebischer, N. J., P. A. Robertson, and R. E. Kenward. 1993. Compositional analysis of habitat use from animal radio-tracking data. Ecology 74:1313–1325.

Anders, A. D., D. C. Dearborn, J. Faaborg, and F. R. Thompson III. 1997. Juvenile survival in a population of migrant birds. Conservation Biology 11:698–707.

Anders, A. D., J. Faaborg, and F. R. Thompson III. 1998. Postfledging dispersal, habitat use, and home-range size of juvenile Wood Thrushes. Auk 115:349–358.

Anders, A. D., and M. R. Marshall. 2005. Increasing the accuracy of productivity and survival estimates in assessing landbird population status. Conservation Biology 19:66–74.

Arnold, T. W. 2010. Uninformative parameters and model selection using Akaike's information criterion. Journal of Wildlife Management 74:1175–1178.

Berkeley, L. I., J. P. McCarty, and L. L. Wolfenbarger. 2007. Postfledging survival and movement in Dickcissels (*Spiza americana*): implications for habitat management and conservation. Auk 124:396–409.

Brohman, R. J., and L. D. Bryant (editors). 2005. Existing vegetation classification and mapping technical guide version 1.0. USDA Forest Service General Technical Report WO-67. Washington, DC.

Buehler, D. A., A. M. Roth, R. Vallender, T. C. Will, J. L. Confer, R. A. Canterbury, S. B. Swarthout, K. V. Rosenberg, and L. P. Bullock. 2007. Status and conservation priorities of Golden-winged Warbler (*Vermivora chrysoptera*). Auk 124:1439–1445.

Bulluck, L. P., and D. A. Buehler. 2008. Factors influencing Golden-winged Warbler (*Vermivora chrysoptera*) nest-site selection and nest survival in the Cumberland Mountains of Tennessee. Auk 125:551–559.

Burnham, K. P., and D. R. Anderson. 2002. Model selection and multimodel inference: a practical information-theoretic approach, 2nd ed. Springer-Verlag, New York, NY.

Calenge, C. 2006. The package adehabitat for the R software: a tool for the analysis of space and habitat use by animals. Ecological Modelling 197:516–519.

Cohen, E. B., and C. A. Lindell. 2004. Survival, habitat use, and movements of fledgling White-throated Robins (*Turdus assimilis*) in a Costa Rican agricultural landscape. Auk 121:404–414.

Confer, J. L., K. W. Barnes, and E. C. Alvey. 2010. Golden- and Blue-winged Warblers: distribution, nesting success, and genetic differences in two habitats. Wilson Journal of Ornithology 122:273–278.

Durand, D., and J. A. Greenwood. 1958. Modifications of the Rayleigh test in analysis of two-dimensional orientation data. Journal of Geology 66:229–238.

Faaborg, J., R. T. Holmes, A. D. Anders, K. L. Bildstein, K. M. Dugger, S. A. Gauthreaux Jr., P. Heglund, K. A. Hobson, A. E. Jahn, D. H. Johnson, S. C. Latta, D. J. Levey, P. P. Marra, C. L. Mekord, E. Nol, S. I. Rothstein, T. W. Sherry, T. S. Sillett, F. R. Thompson III, and N. Warnock. 2010. Conserving migratory land birds in the New World: do we know enough? Ecological Applications 20:398–418.

Ficken, M. S., and R. W. Ficken. 1968. Territorial relationships of Blue-winged Warblers, Golden-winged Warblers, and their hybrids. Wilson Bulletin 80:442–451.

Jackson, A. K., J. P. Froneberger, and D. A. Cristol. 2013. Habitat near nest boxes correlated with fate of Eastern Bluebird fledglings in an urban landscape. Urban Ecosystems 16:367–376.

Johnson, D. H. 1979. Estimating nest success: the Mayfield method and an alternative. Auk 96:651–661.

King, D. I., R. M. Degraaf, M. L. Smith, and J. P. Buonaccorsi. 2006. Habitat selection and habitat-specific survival of fledgling Ovenbirds (*Seiurus aurocapilla*). Journal of Zoology 269:414–421.

Klaus, N. A., and D. A. Buehler. 2001. Golden-winged Warbler breeding habitat characteristics and nest success in clearcuts in the southern Appalachian Mountains. Wilson Bulletin 113:297–301.

MacArthur, R. H., and J. W. MacArthur. 1961. On bird species diversity. Ecology 42:594–598.

Martin, K. J., R. S. Lutz, and M. Worland. 2007. Golden-winged Warbler habitat use and abundance in northern Wisconsin. Wilson Journal of Ornithology 119:523–532.

Marshall, M. R., J. A. DeCecco, A. B. Williams, G. A. Gale, and R. J. Cooper. 2003. Use of regenerating clearcuts by late-successional bird species and their young during the post-fledging period. Forest Ecology and Management 183:127–135.

Miles, P. D., and C. L. VanderSchaaf. 2012. Minnesota's forest resources, 2011. Northern Research Station Research Note NRS-134. USDA Forest Service, Newtown Square, PA.

Moore, L. C., B. J. M. Stutchbury, D. M. Burke, and K. A. Elliott. 2010. Effects of forest management on postfledging survival of Rose-breasted Grosbeaks (*Pheucticus ludovicianus*). Auk 127:185–194.

Pagen, R. W., F. R. Thompson III, and D. E. Burhans. 2000. Breeding and post-breeding habitat use by forest migrant songbirds in the Missouri Ozarks. Condor 102:738–747.

Percy, K. L. 2012. Effects of prescribed fire and habitat on Golden-winged Warbler (*Vermivora chrysoptera*) abundance and nest survival in the Cumberland Mountains of Tennessee. M.S. thesis, University of Tennessee, Knoxville, TN.

Peterson, S. M. 2014. Landscape productivity and the ecology of brood division in Golden-winged Warblers in the western Great Lakes region. M.S. thesis, University of Minnesota, St. Paul, MN.

Peterson, S. M., H. M. Streby, J. A. Lehman, G. R. Kramer, A. C. Fish, and D. E. Andersen. 2015. High tech or field techs: radio-telemetry is a cost-effective method for reducing bias in songbird nest searching. Condor 117:386–395.

Powell, L. A., J. D. Lang, D. G. Krementz, and M. J. Conroy. 2005. Use of radio telemetry to reduce bias in nest searching. Journal of Field Ornithology 76:274–278.

Rappole, J. H., and A. R. Tipton. 1991. New harness design for attachment of radio transmitters to small passerines. Journal of Field Ornithology 62:335–337.

SAS Institute. 2008. SAS/STAT 9.2 user's guide. SAS Institute, Cary, NC.

Shaffer, T. L. 2004. A unified approach to analyzing nest success. Auk 121:526–540.

Streby, H. M., and D. E. Andersen. 2011. Seasonal productivity in a population of migratory songbirds: why nest data are not enough. Ecosphere 2:78.

Streby, H. M., and D. E. Andersen. 2012. Movement and cover-type selection by fledgling Ovenbirds (*Seiurus aurocapilla*) after independence from adult care. Wilson Journal of Ornithology 124:621–626.

Streby, H. M., and D. E. Andersen. 2013a. Movements, cover-type selection, and survival of fledgling Ovenbirds in managed deciduous and mixed-coniferous forests. Forest Ecology and Management 287:9–16.

Streby, H. M., and D. E. Andersen. 2013b. Survival of fledgling Ovenbirds: influences of habitat characteristics at multiple spatial scales. Condor 115:403–410.

Streby, H. M., S. M. Peterson, and D. E. Andersen. 2011b. Invertebrate availability and vegetation characteristics explain use of non-nesting cover types by mature forest songbirds during the post-fledging period. Journal of Field Ornithology 82:406–414.

Streby, H. M., S. M. Peterson, C. F. Gesmundo, M. K. Johnson, A. C. Fish, J. A. Lehman, and D. E. Andersen. 2013a. Radio-transmitters do not affect seasonal productivity of female Golden-winged Warblers. Journal of Field Ornithology 83:316–321.

Streby, H. M., S. M. Peterson, J. A. Lehman, G. R. Kramer, K. J. Iknayan, and D. E. Andersen. 2013b. The effects of force-fledging and premature fledging on the survival of nestling songbirds. Ibis 155:616–620.

Streby, H. M., S. M. Peterson, J. A. Lehman, G. R. Kramer, B. J. Vernasco, and D. E. Andersen. 2014b. Do digestive contents confound body mass as a measure of relative condition in nestling songbirds? Wildlife Society Bulletin 35:308–310.

Streby, H. M., S. M. Peterson, T. L. McAllister, and D. E. Andersen. 2011a. Use of early-successional managed northern forest by mature-forest species during the post-fledging period. Condor 113:817–824.

Streby, H. M., J. M. Refsnider, S. M. Peterson, and D. E. Andersen. 2014a. Retirement investment theory explains patterns in songbird nest-site choice. Proceedings of the Royal Society of London B 281:20131834.

Streby, H. M., B. Scholtens, A. P. Monroe, S. M. Peterson, and D. E. Andersen. 2013c. The Ovenbird (*Seiurus aurocapilla*) as a model for testing food-value theory. American Midland Naturalist 169:214–220.

Vallender, R., V. L. Friesen, and R. J. Robertson. 2007. Paternity and performance of Golden-winged Warblers (*Vermivora chrysoptera*) and Golden-winged × Blue-winged Warbler (*V. pinus*) hybrids at the leading edge of a hybrid zone. Behavioral Ecology and Sociobiology 61:1797–1807.

Vitz, A. C., and A. D. Rodewald. 2007. Vegetative and fruit resources as determinants of habitat use by mature forest birds during the postbreeding period. Auk 124:494–507.

Vitz, A. C., and A. D. Rodewald. 2011. Influence of condition and habitat use on survival of post-fledging songbirds. Condor 113:400–411.

Vitz, A. C., and A. D. Rodewald. 2013. Behavioral and demographic consequences of access to early-successional habitat in juvenile Ovenbirds (*Seiurus aurocapilla*): an experimental approach. Auk 130:21–29.

White, J. D., and J. Faaborg. 2008. Post-fledging movement and spatial habitat-use patterns of juvenile Swainson's Thrushes. Wilson Journal of Ornithology 120:62–73.

Will, T. C. 1986. The behavioral ecology of species replacement: Blue-winged and Golden-winged Warblers in Michigan. Ph.D. dissertation, University of Michigan, Ann Arbor, MI.

Yackel Adams, A. A., S. K. Skagen, and J. A. Savidge. 2006. Modeling post-fledging survival of Lark Buntings in response to ecological and biological factors. Ecology 87:178–188.

Zar, J. H. 2004. Biostatistical analysis. Pearson, Upper Saddle River, NJ.

Zlonis, E. J., A. Grinde, J. Bednar, and G. J. Niemi. 2013. Summary of breeding bird trends in the Chippewa and Superior National Forests of Minnesota—1995–2013. NRRI technical report NRRI/TR-2013/36. University of Minnesota Duluth, Duluth, MN.

CHAPTER NINE

Spatially Explicit Models of Full-Season Productivity and Implications for Landscape Management of Golden-winged Warblers in the Western Great Lakes Region*

Sean M. Peterson, Henry M. Streby, and David E. Andersen

Abstract. The relationship between landscape structure and composition and full-season productivity (FSP) is poorly understood for most birds. For species of high conservation concern, insight into how productivity is related to landscape structure and composition can be used to develop more effective conservation strategies that increase recruitment. We monitored nest productivity and fledgling survival of Golden-winged Warblers (*Vermivora chrysoptera*), a species of high conservation concern, in managed forest landscapes at two sites in northern Minnesota, and one site in southeastern Manitoba, Canada from 2010 to 2012. We used logistic exposure models to identify the influence of landscape structure and composition on nest productivity and fledgling survival. We used the models to predict spatially explicit, FSP across our study sites to identify areas of low relative productivity that could be targeted for management. We then used our models of spatially explicit, FSP to simulate the impact of potential management actions on our study sites with the goal of increasing total population productivity. Unlike previous studies that suggested wetland cover types provide higher quality breeding habitat for Golden-winged Warblers, our models predicted 14% greater productivity in upland cover types. Simulated succession of a 9-ha grassland patch to a shrubby upland suitable for nesting increased the total number of fledglings produced by that patch and adjacent upland shrublands by 30%, despite decreasing individual productivity by 13%. Further simulated succession of the same patch described above into deciduous forest reduced the total number of fledglings produced to independence on a landscape by 18% because of a decrease in the area available for nesting. Simulated reduction in the cumulative length of shrubby edge within a 50-m radius of any location in our landscapes from 0.6 to 0.3 km increased FSP by 5%. Our models demonstrated that the effects of any single management action depended on the context of the surrounding landscape. We conclude that spatially explicit, FSP models that incorporate data from both the nesting and postfledging periods are useful for informing breeding habitat management plans for Golden-winged Warblers and that similar models can benefit management planning for many other species of conservation concern.

Key Words: fledgling survival, landscape composition, landscape structure, nest success, productivity surface, songbird.

* Peterson, S. M., H. M. Streby, and D. E. Andersen. 2016. Spatially explicit models of full-season productivity and implications for landscape management of Golden-winged Warblers in the western Great Lakes region. Pp. 141–160 in H. M. Streby, D. E. Andersen, and D. A. Buehler (editors). Golden-winged Warbler ecology, conservation, and habitat management. Studies in Avian Biology (no. 49), CRC Press, Boca Raton, FL.

E stimates of productivity are important for modeling population growth and identifying habitat features that affect productivity is important for informing management plans. For example, management directed at identification and elimination of habitat features that comprise ecological traps could increase population growth rate (Battin 2004). Most models of songbird population dynamics include estimates of nest success, but lack consideration of fledgling survival, which can result in estimates of population productivity that are at best incomplete and potentially misleading (Streby and Andersen 2011, Shipley et al. 2013). It is important to include survival of both nests and fledglings in estimates of full-season productivity (FSP) because habitat characteristics can have different effects on different life stages (Streby et al. 2014), and many songbirds appear to have different habitat requirements for nesting than for rearing fledglings (Pagen et al. 2000, Marshall et al. 2003, Vitz and Rodewald 2007, Streby and Andersen 2011).

Previous studies have described the relationships between edge (Askins 1995, Benson et al. 2010), forest fragmentation (Robinson and Wilcove 1994, Faaborg et al. 1995, Bayne and Hobson 1997, Lloyd et al. 2005, Rush and Stutchbury 2008), deforestation (Askins et al. 1987), and urban encroachment (Ausprey and Rodewald 2011) and individual aspects of songbird productivity such as nest success, fledgling survival, or observed population growth. Comparatively few efforts have assessed the influence of landscape structure and composition to model productivity across multiple life stages, or simultaneously assessed both multiple landscape components and multiple life stages (Streby and Andersen 2011). In many landscapes, predation is a primary source of both nest failure (Martin 1993) and fledgling mortality (Chapter 8, this volume). Landscape composition can have substantial impact on the composition of the predator community and thus songbird productivity (Robinson 1992, Porneluzi et al. 1993, Hoover et al. 1995, Brawn and Robinson 1996, Chalfoun et al. 2002). Furthermore, predators may be using a landscape at a different spatial scale than breeding songbirds. As a consequence, some aspects of the landscape may influence productivity more than others (Stephens et al. 2005), and it is only when the entire landscape is considered that productivity can be assessed across a spatial extent relevant for management at the population level.

Golden-winged Warblers (*Vermivora chrysoptera*) are a species of conservation concern that nest in patches of upland shrubland or wetland shrubland within a matrix of mature forest with a dense understory in the Appalachian Mountains, northeastern and north-central United States, and adjacent southern Canada (Confer et al. 2011). Relationships between Golden-winged Warbler breeding and landscape configuration are largely unknown, although Confer et al. (2010) observed significantly higher nest success in swamp forests in a Golden-winged Warbler population in New York, and suggested that populations using those cover types may act as sources for populations using upland cover types. Across much of the breeding distribution, however, declines in populations of Golden-winged Warblers have been attributed to loss of early successional upland forest stands and hybridization with the closely related Blue-winged Warbler (*Vermivora cyanoptera*; Buehler et al. 2007, Confer et al. 2011). Efforts to mitigate or reverse population declines have concentrated on forest management and the creation or maintenance of early successional upland forest stands (Huffman 1997, Roth and Lutz 2004, Kubel and Yahner 2008, Percy 2012) or wetland shrublands (Rossell et al. 2003, Rush and Post 2008, Confer et al. 2010). However, management strategies to date have been developed without a clear understanding of how landscape structure and composition influences FSP and how to best incorporate landscape effects into management plans.

To assess the relationship between landscape structure and composition and productivity, we studied three populations of Golden-winged Warblers in the western Great Lakes region of central North America and derived estimates of FSP at a landscape scale. We constructed spatially explicit models of FSP as a function of landscape structure and composition, and used our models to estimate FSP across our study areas. The resulting estimates of FSP combined estimates of nest success and fledgling survival, each as a function of landscape structure and composition to derive estimates of productivity across our study sites. We used spatially explicit estimates of FSP to evaluate the efficacy of potential management actions.

METHODS

Study Areas

We studied Golden-winged Warblers at Tamarac National Wildlife Refuge (NWR) in Becker County, Minnesota (47.049°N, 95.583°W) from 2010 to 2012, and at Rice Lake NWR in Aitkin

County, Minnesota (46.529°N, 93.338°W) and Sandilands Provincial Forest (PF) in southeastern Manitoba, Canada (49.637°N, 96.247°W) from 2011 to 2012. All three sites were located in the northern hardwood transition zone between boreal forest and tallgrass prairie in the western Great Lakes region of central North America. Each site contained a mix of forest of various seral stages and vegetation structures. The three most abundant cover types at each site were upland forests generally dominated by quaking aspen (*Populus tremuloides*), shrublands dominated by hazel (*Corylus* spp.), and wetland shrublands dominated by alder (*Alnus* spp.). For a detailed description of the study sites, see Chapter 8 (this volume).

Data Collection

We searched for nests of Golden-winged Warblers at all three sites using radiotelemetry to monitor adult females and using standard nest-searching methods (Martin and Geupel 1993). We attached VHF radio transmitters (Blackburn Transmitters, Nacogdoches, TX) to passively mist-netted adult female Golden-winged Warblers using a figure-eight harness design (~4.1% of mean adult mass; Rappole and Tipton 1991, Streby et al. 2015). We used homing on radio signals to locate marked females and find their nests during nest building, egg laying, or early incubation. Radio transmitters had no measurable effect on any aspect of productivity during our study (Streby et al. 2013).

We recorded nest locations using handheld Global Positioning System units (GPSMAP 76 or eTrex Venture HC Global Positioning System, Garmin Ltd., Schaffhausen, Switzerland), averaging 100 points to ensure <5 m accuracy. We monitored nests at 4-day intervals until nestlings fledged (rarely 3-, 5-, or 6-day intervals due to inclement weather or logistical constraints). When possible, we assessed the condition of nests from a distance using binoculars and approached nests from various directions on different visits to minimize nest-site disturbance. We considered nests to be successful if at least one nestling fledged and, to reduce inaccurately assigned nest fates, we considered nests to have failed if we found them empty before a possible fledge date at nestling day 7, if they had cold eggs and were unattended for >2 observation intervals during the incubation stage, or if radio-tagged fledglings were depredated and

no broodmates were detected in the vicinity of the nest (Streby and Andersen 2013).

At 6–9 days after hatching (counting hatch day as day 1), we banded all nestlings with a standard U.S. Geological Survey leg band and attached a radio transmitter to 1–5 randomly selected individuals at each nest (commonly two individuals) using a figure-eight harness (~4.6% of mean nestling mass). Additionally, we attached transmitters to 10 fledglings from known nests captured 1–8 days after fledging. We tracked fledglings daily to assess survival and right-censored 19 individuals (10% of fledglings we monitored) with unknown fates because transmitters were dropped. We focused on the impact of predation in this analysis, so we also censored individuals that died due to exposure (n = 11). We focused analysis on the early postfledging period, days 1–8 after fledging, because the early period included most of the fledgling mortality we observed (86%; Chapter 8, this volume). We divided the early postfledging period into two stages for modeling: days 1–3, characterized by low mobility and high and variable daily mortality; days 4–8, characterized by greater mobility and relatively low mortality (Chapter 8, this volume; Peterson 2014).

Landscape Attributes

To model the impact of cover types on nest success and fledgling survival, we categorized 11 cover types using aerial photographs in Arc 10.1 Geographic Information System (GIS) software (Environmental Systems Research Institute, Redlands, CA). For Tamarac NWR and Rice Lake NWR, we used 1-m resolution digital orthophoto quadrangles (2009; Minnesota Department of Natural Resources). For Sandilands PF, we used georeferenced 1-m resolution satellite images obtained from Google Earth™ 6.2 (2010; Google Inc., Mountain View, CA). We confirmed the cover types derived from aerial photographs and satellite images using >2,500 locations visited at our study sites. Each additional cover type doubled the number of possible unique combinations of cover types present on a landscape, so we collapsed the 11 cover types into six broad categories (deciduous forest, upland shrubland, forested wetland, grassland, wetland shrubland, and coniferous forest), included an additional covariate related to edge density, and used seven categories as potential variables in our FSP model (Table 9.1).

TABLE 9.1

Categorization, definition, and total exposure days used for three logistic exposure models (N = nest success, E = fledgling survival days 1–3, L = fledgling survival days 4–8) of similar cover types present on landscapes used by three Golden-winged Warbler populations studied in the western Great Lakes region of North America, 2010–2012.

Landscape structure and composition cover-type category	Definition	Total exposure days	Cover type[a]
Coniferous forest	Forest dominated by coniferous trees	N = 18	Coniferous forest
		E = 15	
		L = 33	
Deciduous forest	Forest with >60% canopy closure and dominated by deciduous trees >5 m in height	N = 817	Mature forest
		E = 364	Sapling-dominated clear-cut
		L = 463	
Shrubby edge	Edge between shrubland and coniferous forest, forested wetland, or deciduous forest	N = 760	N/A
		E = 442	
		L = 534	
Forested wetland	Perennially wet forest dominated by trees >5 m in height	N = 106	Forested wetland
		E = 57	
		L = 211	
Grassland	Landscape dominated by grass or sedge	N = 573	Grassy wetland
		E = 270	Upland grassland
		L = 279	
Wetland shrubland	Perennially wet shrubland with a canopy <5 m in height	N = 501	Wetland shrubland
		E = 211	
		L = 288	
Upland shrubland	Perennially dry shrubland or sapling-dominated clear-cut with a canopy <5 m in height	N = 716	Firebreak/power-line right-of-way
		E = 359	
		L = 386	Shrub-dominated clear-cut

[a] For detailed description of cover types see Chapter 8 (this volume).

With the exception of coniferous forest, which was adjacent to only one site, and forested wetland, which was an uncommon cover type at each site, we modeled the relationship between each cover-type category and nest success and fledgling survival using ≥200 exposure days for each period (Table 9.1).

In addition to cover-type variables, we included a covariate for edge density (i.e., length of edge within a specified area) in our models to assess how productivity was related to the density of forest–shrubland edge present. We used Arc GIS 10.1 to identify edges between deciduous forest, coniferous forest, or forested wetland with a canopy height >5 m and two shrubland cover types: upland shrubland and wetland shrubland. We limited our measure of edge density to edges between forest and shrubland cover types because those are the cover type edges with which Golden-winged Warblers are most commonly associated (Confer et al. 2011). We excluded less ecologically significant edges such as edges between grassland and shrubland.

For each of the seven model covariates (six cover types and edge density; hereafter, "landscape variables"), we calculated an "impact radius." The impact radius defined the scale at which each landscape variable was most strongly related to survival of nests and fledglings. To calculate the radius, we buffered each nest location with circles with radii in 25-m increments from 25 to 200 m and at 100-m increments from 200 to 500 m. For nest survival and fledgling survival from day 1 to 3, we used a range of 25–200 m for potential impact radii and for fledgling survival from day 4 to 8, we used a range of 25–500 m for potential impact radii, corresponding with the distance that adults moved fledglings (Chapter 10, this volume). We summed the total area (ha) for each cover type and total linear distance of edge (km) for each buffer distance around each nest location. The impact radius of each landscape variable could be at a scale unique to that landscape variable: deciduous forest might be related to nest success at a 50-m radius, whereas wetland shrubland might be related to nest success at a 200-m radius. Thus, we independently estimated survival using each combination of scale and polynomial function (linear, quadratic, or cubic relationships) for each variable by fitting logistic exposure models to survival data from all three sites and years for three different periods (nest survival, fledgling survival day 1–3, and fledgling survival day 4–8) using PROC NLMIXED (SAS Institute, Chicago, IL; Shaffer 2004). For example, to determine the impact radius and polynomial function of deciduous forest in relation to nest success, we compared 24 different deciduous forest models ranging from a linear relationship with a 25-m impact radius to a cubic relationship with a 200-m impact radius.

We treated our models as exploratory and did not attempt to predict what relationships might occur between landscape structure or composition and survival. We used multiple potential polynomial functions to account for the possibility of curvilinear relationships among modeled variables to account for the potential of diminishing returns or exponential increases in the impact of any landscape variable on survival. We included nest or fledgling age as a covariate in models for nest survival and fledgling survival from day 1 to 3. Survival was relatively constant after the first three days (Chapter 8, this volume), and we did not include age as a variable when modeling fledgling survival from day 4 to 8. For models of fledgling survival in both the early and late fledgling periods, we used brood as a random effect. Previous modeling of this study population determined that there were no site or year effects on nest or fledgling depredation (Streby et al. 2014), so we

did not include those variables in our models. We centered all impact radii on the nest because fledgling survival during the first eight days outside the nest was directly related to nest location (Streby et al. 2014; Chapter 8, this volume). Fledglings moved farther from the nest after the first three days (Peterson 2014), so we increased the range of potential impact radii to 25–500 m to model fledgling survival from day 4 to 8, but we still centered radii around the nest because nest location was the strongest predictor of survival during this period in previous models (Chapter 8, this volume). We did not include survival data from day 9 to independence because survival was consistently high and largely unrelated to habitat use or nest location during this period (Chapter 8, this volume). For each landscape variable, we ranked models of nest or fledgling survival using Akaike's Information Criterion corrected for small sample size (AIC$_c$; Burnham and Anderson 2002) and selected the best supported combination of polynomial function and impact radius for use in modeling productivity on a landscape (for complete AIC$_c$ rankings, see Peterson 2014:appendix B). We defined null models with age for nest survival and fledgling survival from day 1 to 3 or constant survival for fledgling survival from day 4 to 8 as null models. If all combinations of polynomial function and impact radius for a variable were less supported than the null model for that survival period, we considered that variable to be noninformative and excluded it from survival models.

Modeling Survival for a Landscape

For each survival period, we used methods similar to techniques used for resource selection functions (Manly et al. 2002) to estimate survival related to landscape structure and composition around any given location at our study sites. We combined the best supported impact radius and polynomial function for each landscape variable into composite survival models that incorporated all landscape variables present at every location (1-m^2 pixel) across our study sites. We used survival models and estimates of renesting rates and brood size to create spatially explicit estimates of the number of fledglings that could be produced to fledgling day 8 at any given location. In contrast to resource selection functions, which estimate the probability of presence or use on a landscape, our models of FSP estimated productivity for a hypothetical breeding pair that may choose to nest at any 1-m² pixel on our study sites.

For each landscape variable, we built a "landscape variable map" that delineated the area over which that variable was related to each component of survival (nest, fledgling days 1–3, fledgling days 4–8). We used the vector cover-type layer delineated using aerial or satellite imagery and isolated each cover type. We individually converted all landscape variable maps to 1 m × 1 m resolution raster layers and then used a neighborhood function in Arc GIS 10.1 to calculate a value at every 1 m × 1 m pixel on the map equal to the quantity (i.e., area or length, "variable quantity map") of each landscape variable within its impact radius for each survival period. For example, the explanatory variable deciduous forest was related to the response variable fledgling survival from day 1 to 3 at a 25-m scale; we therefore created a variable quantity map that for each pixel contained a value equal to the number of ha of deciduous forest within 25 m of that pixel.

We estimated survival separately for each period because Golden-winged Warbler nest and fledgling survival are associated differently with landscape composition around a nest (Streby et al. 2014). For each survival period, we used all variable quantity maps to create a map comprising landscape structure and composition values representing each unique combination of variable quantity values present at every pixel. To do this, we used raster algebra to identify the landscape variables present within their respective impact radii around each pixel (i.e., those with variable quantities >0, including edge density). Our approach created 32 unique combinations of five informative landscape variable compositions (groups; Table 9.2; see Peterson 2014:appendix B for full model results), ranging from simple landscapes with a pixel with only one cover type within its impact radius to more complex landscape areas with a pixel with several cover types and edge density within their respective impact radii. Coniferous forest and forested wetland were not included in the groups used for the results presented here because they were not present at the specific stands analyzed in this manuscript.

For each survival period (nest, fledgling days 1–3, fledgling days 4–8), we used PROC GENMOD in SAS and built logistic exposure survival models corresponding to the landscape variable composition group associated with each pixel (SAS Institute, Chicago, IL; Shaffer 2004). We assigned each of these equations to each pixel based on the landscape structure and composition surface described above. Our approach allowed the effect of each landscape variable to differ depending on landscape structure and composition around each pixel. For example, quantity of deciduous forest might be related to nest survival differently depending upon whether deciduous forest is adjacent to wetland shrubland or grassland. We estimated daily survival (S) within each period for each observed combination of landscape

TABLE 9.2

Scale and polynomial function of top-ranked survival models for each landscape variable and survival period for three populations of Golden-winged Warblers in the western Great Lakes region.

Landscape variable	Nest survival		Day 1–3 fledgling survival		Day 4–8 fledgling survival	
	Scale (m)	Polynomial function	Scale (m)	Polynomial function	Scale (m)	Polynomial function
Coniferous forest	50	Linear	50	Quadratic	N/A	N/A
Deciduous forest	N/A	N/A	25	Linear	25	Linear
Edge	50	Cubic	200	Cubic	400	Cubic
Forested wetland	175	Linear	125	Cubic	400	Cubic
Grassland	200	Quadratic	200	Linear	175	Quadratic
Wetland shrubland	200	Linear	N/A	N/A	300	Cubic
Upland shrubland	N/A	N/A	N/A	N/A	N/A	N/A

Noninformative landscape variables are indicated by "N/A".

structure and composition (l) and survival period (p) on the logit scale as follows:

$$S_{lp} = \frac{\exp(\alpha_{lp} + \beta_{1lp}x_{1lp} + \beta_{2lp}x_{2lp} + \beta_{3lp}x_{3lp} \ldots)}{1 + \exp(\alpha_{lp} + \beta_{1lp}x_{1lp} + \beta_{2lp}x_{2lp} + \beta_{3lp}x_{3lp} \ldots)}$$

where

α is the estimated intercept

β_1 is the estimated coefficient for landscape variable x_1.

To apply the equation defined above to a landscape, we created coefficient maps for each β value derived from each logistic exposure survival equation. We assigned the calculated β values for each survival period (p) to each pixel based on its corresponding landscape structure and composition value (l). For example, if x_1 for an equation represented the amount of wetland shrubland within 200 m (i.e., the impact radius of wetland shrubland associated with nest success), the value at any given pixel on the coefficient map for x_1 was equal to the β_1 value calculated by the logistic exposure survival equation for the landscape structure and composition value at that pixel.

At each pixel on a landscape, we used the previously assigned values of (1) the amount of each landscape variable surrounding that pixel and (2) the β coefficients for the logistic exposure survival equation for the appropriate landscape variable to estimate nest success, fledgling survival from day 1 to 3, and fledgling survival from day 4 to 8. For example, to calculate fledgling survival from day 4 to 8 for a pixel at the center of a circle with 3/4 of the landscape made up of deciduous forest and 1/4 of the landscape made up of shrubby wetland, with a straight shrubby edge separating the cover types at a right angle, the survival equation would be as follows:

$$\text{Daily survival} = \frac{\begin{aligned}\exp(&4.2177 \\ &+ (0.4524 * \text{deciduous forest}) \\ &- (0.0010 * \text{edge}) \\ &+ (2.7450 * 10^{-7} * \text{edge}^2) \\ &+ (1.6789 * \text{shrubby wetland}) \\ &- (0.5959 * \text{shrubby wetland}^2) \\ &+ (0.0454 * \text{shrubby wetland}^3))\end{aligned}}{\begin{aligned}1 + \exp(&4.2177 \\ &+ (0.4524 * \text{deciduous forest}) \\ &- (0.0010 * \text{edge}) \\ &+ (2.7450 * 10^{-7} * \text{edge}^2) \\ &+ (1.6789 * \text{shrubby wetland}) \\ &- (0.5959 * \text{shrubby wetland}^2) \\ &+ (0.0454 * \text{shrubby wetland}^3))\end{aligned}},$$

where each numerical value was the assigned β coefficient for that landscape variable. Each pixel that fell within the deciduous forest and edge landscape would be assigned those β coefficients and the value for the number of ha of deciduous forest within 25 m, the number of ha of shrubby wetland within 300 m, and km of edge within 400 m of that pixel. The hypothetical fledgling's nest described above would exist on a landscape with 0.147 ha of deciduous forest within 25 m, 7.068 ha of shrubby wetland within 300 m, and 0.8 km of edge within 400 m; the fledgling would have a 0.6508 probability of surviving during days 4–8 (daily survival = 0.9177).

We calculated nest productivity as the number of juveniles fledged by a breeding pair (NP), assuming up to two nesting attempts if the first nest failed and a mean fledged brood of four fledglings (H. M. Streby, unpubl. data) as follows:

$$NP = (NS + (1 - NS) * NS) * 4$$

where NS is nest success. We calculated fledgling survival (FS) as follows:

$$FS = ES * LS,$$

where

ES is fledgling survival in the early period (days 1–3)

LS is fledgling survival in the late period (days 4–8).

Assuming negligible mortality until independence (Chapter 10, this volume), we calculated FSP or the number of young raised eight days postfledging as follows:

$$FSP = NP * FS.$$

After applying these equations, each pixel on the map had a value for NP, FS, and FSP, the product of NP and FS that represented the expected productivity for a pair nesting within that pixel. We then used these values to identify areas of high and low productivity on a landscape. A more detailed description of the process we used to estimate spatially explicit productivity is presented in Peterson (2014:appendix A).

No standard method is available for assessment of the robustness of our spatially explicit models of FSP, so we assessed whether our models predicted survival better than null models using k-fold cross-validation (Boyce et al. 2002, Koper and Manseau 2009). We evenly divided the sample for each survival period by randomly assigning nests or broods to eight equal folds. For each fold, we used the remaining seven folds to train a set of spatially explicit models of survival and a null model with either age as a variable (for nests and for fledglings from days 1 to 3) or constant survival (for fledglings from days 4 to 8). We then calculated a Spearman's rank correlation between observed survival and survival predicted by both the null model and the spatially explicit model of survival.

Application of Spatially Explicit Models of Full-Season Productivity

To assess the effects of potential management actions designed to increase FSP of Golden-winged Warblers, we used Arc GIS 10.1 to simulate altered landscapes at our study sites. At each of our study sites, we applied spatially explicit models of FSP to existing and hypothetical landscapes and present estimates of productivity that used all landscape categories in various combinations, except forested wetland and coniferous forest. The scenarios we selected were chosen to illustrate (1) differences between wetland and upland cover types, (2) the effects of grassland succession to upland shrubland and then to mature forest, and (3) the effect of management of shrubby edge density on a landscape. We considered all roads, open water, grassland, or any cover types >100 m from upland shrubland or wetland shrubland to be areas unused for nesting by Golden-winged Warblers and did not include those values in our analyses. We smoothed all graphical representations of spatially explicit productivity estimates using a 25-m mean of productivity in Arc GIS 10.1 to reduce minor, abrupt transitions between landscape structure and composition categories.

In our first assessment, we evaluated the relative FSP of upland and wetland cover types while controlling for the effect of surrounding landscape structure and composition on productivity. Although Golden-winged Warblers use both upland and wetland cover types as nesting habitat, Confer et al. (2010) suggested that productivity in wetland cover types may be greater than productivity in upland cover types. For our study site at Rice Lake NWR (Figure 9.1a) we evaluated the difference in FSP between landscapes dominated by wetland cover types and those dominated by upland cover types. For a wetland-dominated portion of that study site, we used Arc GIS 10.1 to simulate the conversion of the same landscape with all wetland cover types to structurally similar upland cover types. We performed this assessment not to encourage converting wetland to upland cover types on managed landscapes, but to measure the difference in estimated FSP in structurally similar patches. We quantified the difference between wetland and upland cover types by calculating the mean productivity within 100 m of known nest sites for this scenario. We also used logistic exposure to model productivity in the absence of landscape data by dividing nests located in wetland versus upland cover types, as a separate assessment of the difference between cover-type productivity.

In our second assessment, we simulated management to increase productivity at Tamarac NWR, where grassland cover comprised 9 ha of our study area and could be managed for Golden-winged Warblers. We simulated modifying an open grassland (Figure 9.1b) within a forested landscape to evaluate how succession of an open area would affect FSP of Golden-winged Warblers. We simulated converting grassland to upland shrubland and adding shrubby edges where the altered grassland patch abutted deciduous forest to simulate early successional cover. We then simulated upland shrubland continuing to succeed into deciduous forest and merged it with the adjoining deciduous forest patch. The landscape simulation at Tamarac NWR not only altered productivity but also the area available for nesting. We quantified the difference between management scenarios at Tamarac NWR by multiplying the area available for nesting in each scenario by the mean productivity for that area, therefore accounting for both productivity and changes in area available for nesting.

In our final assessment, we identified two areas with low productivity (estimates of <1 fledgling produced per 1 m × 1 m pixel) associated with >0.6 km of edge within 50 m of a pixel at our Sandilands PF site (Figure 9.1c). In these areas, we simulated forest management that would result in lesser edge density, either by allowing upland shrubland to succeed into deciduous forest or by harvesting forest to create a lesser edge density (i.e., <0.3 km of edge length within 50 m of a pixel). We quantified the difference between

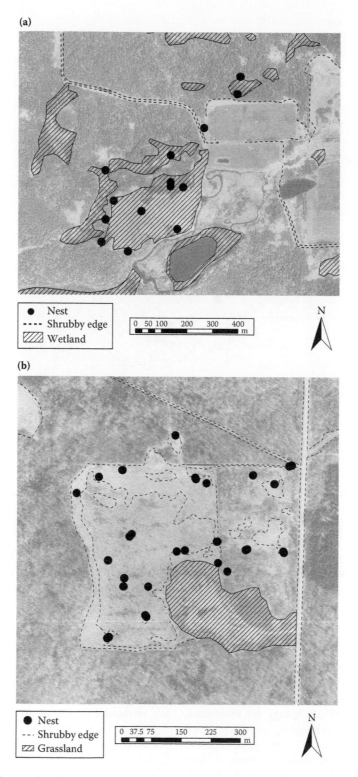

Figure 9.1. Aerial photographs of three Golden-winged Warbler study sites in the western Great Lakes region of North America from 2010 to 2012 with nest locations marked with a circle, soft shrubby edges marked by a dashed line and (a) wetland cover types delineated by thick gray boundary at Rice Lake NWR, (b) grassland delineated by red hatched lines at Tamarac NWR. *(Continued)*

(c)

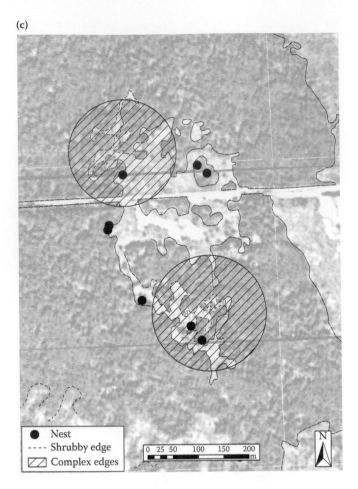

●	Nest
----	Shrubby edge
▨	Complex edges

0 25 50   100   150   200
m

N

Figure 9.1. (*Continued*) Aerial photographs of three Golden-winged Warbler study sites in the western Great Lakes region of North America from 2010 to 2012 with nest locations marked with a circle, soft shrubby edges marked by a dashed line and (c) areas with complex edges indicated by red hatched lines at Sandilands PF.

these edge densities by calculating the mean productivity within the impact radius of altered cover types.

RESULTS

We monitored 29 nests and 49 fledglings in Sandilands PF and 56 nests and 47 fledglings at Rice Lake NWR from 2011 to 2012, and 131 nests and 94 fledglings at Tamarac NWR from 2010 to 2012. Of 216 nests and 190 fledglings we monitored, 127 nests (59%) and 70 fledglings were depredated (37%). We constructed a total of 96 logistic exposure models (for full model results see Peterson 2014: appendix C). For all three survival periods, the spatially explicit models we developed (Nest $r_s = 0.30$, Fledgling days $1 - 3$ $r_s = 0.19$, Fledgling days $4 - 8$ $r_s = 0.11$) explained more variation in survival than

the null model (Nest $r_s = 0.14$, Fledgling days $1 - 3$ $r_s = 0.00$, Fledgling days $4 - 8$ $r_s = -0.14$), indicating that our spatially explicit models were more informative than the null models.

Simulation of Management Options

All three of our simulations of altering landscapes at our study areas led to biologically significant changes in FSP. When we simulated converting wetland cover types to upland cover types at Rice Lake NWR, estimated mean FSP from breeding attempts at a random pixel increased 14% from 1.62 fledglings per pixel (SD = 0.74) to 1.84 fledglings per pixel (SD = 0.65; Figure 9.2a,b). When we modeled productivity in wetland and upland cover types without including landscape variables, we estimated that wetland cover types would produce a mean of 1.05 fledglings per pixel and upland cover types would produce a mean of 1.59 fledglings per pixel, a 51% increase. At Tamarac NWR, when we simulated succession from grassland to upland shrubland, the area available for nesting (i.e., upland shrubland and deciduous forest <100 m from upland shrubland) increased from 18.3 to 27.3 ha (Figure 9.3a,b). However, estimated FSP decreased in this simulation from 1.97 fledglings per pixel (SD = 0.51) to 1.73 fledglings per pixel (SD = 0.40), largely because of decreased fledgling and nest survival in areas that had previously been positively impacted by the presence of nearby grassland cover. Despite estimated mean

productivity decreasing by 13%, the increase in available nesting area caused total landscape productivity to increase by 30%. Simulated further succession from upland shrubland to deciduous forest reduced available nesting area by 22% to 21.2 ha and resulted in estimated landscape productivity 18% lower than what we estimated in the upland shrubland simulation, despite increasing estimated mean FSP from 1.73 fledglings per pixel to 1.86 fledglings per pixel (SD = 0.39; Figure 9.3c). Finally, when we simulated reduced edge density in two small areas with high edge density (i.e., >0.6 km of edge within 50 m of a given pixel; <1 ha of altered area) that had lower estimated FSP than the surrounding landscape at

(a)

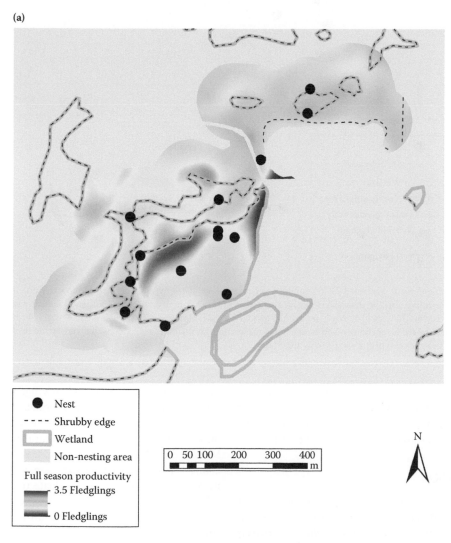

Figure 9.2. Estimated FSP (fledglings per 1-m^2 pixel) modeled from Golden-winged Warbler populations studied from 2010 to 2012 in the western Great Lakes region of North America of (a) a wetland at Rice Lake NWR.   (Continued)

(b)

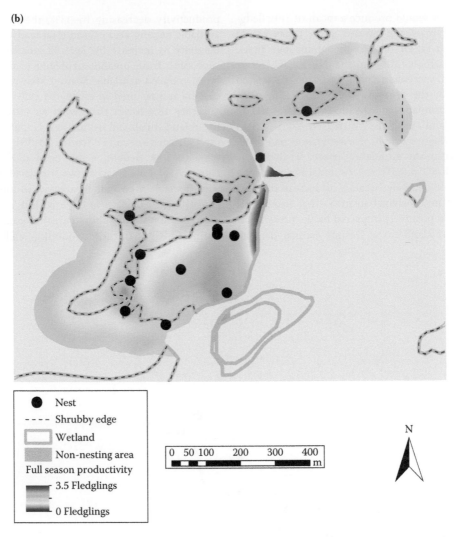

Figure 9.2. (*Continued*) Estimated FSP (fledglings per 1-m² pixel) modeled from Golden-winged Warbler populations studied from 2010 to 2012 in the western Great Lakes region of North America of (b) a hypothetical upland with identical landscape structure. Soft shrubby edges are marked by a dashed line, nests are marked with a circle, wetland cover types are delineated by red hatched lines, and areas unused for nesting (grassland, roads, open water, and deciduous forest >100 m from shrubby cover types) are marked by solid black.

Sandilands PF, estimated mean productivity in 22.5 ha of breeding habitat increased 5% from 1.93 fledglings per pixel (SD = 0.52) to 2.03 fledglings per pixel (SD = 0.43; Figure 9.4a,b).

DISCUSSION

We developed spatially explicit models of FSP across landscapes at three study areas in the western Great Lakes region, where relatively little information about breeding habitat relations exists, but where a significant portion of the global population of Golden-winged Warblers breeds. With spatially explicit models of FSP, we estimated productivity at any given location based on landscape characteristics around that location. Models of FSP allowed us to address questions about low-productivity areas, assess productivity across a landscape, and evaluate management effects on productivity prior to implementation. Perhaps the most important finding from our simulations of potential management options was that any management action can have considerably different effects on Golden-winged Warbler

FSP depending on the context of the surrounding landscape. Therefore, we cannot use the results of the simulations presented here to provide broad, generalizable recommendations with regard to any one-size-fits-all management option. Instead, we provide a tool that can be used to assess the influence of specific management actions on individual landscapes, each with their own intrinsic complexities. Within the western Great Lakes region and, potentially, other regions with similar predator communities and cover types, the models provided here may be used to predict productivity and the impact of management actions.

In contrast to Confer et al. (2010), our results suggested that management of upland cover types on our study sites would increase FSP more than management of wetland cover types, which were generally associated with lesser FSP

(a)

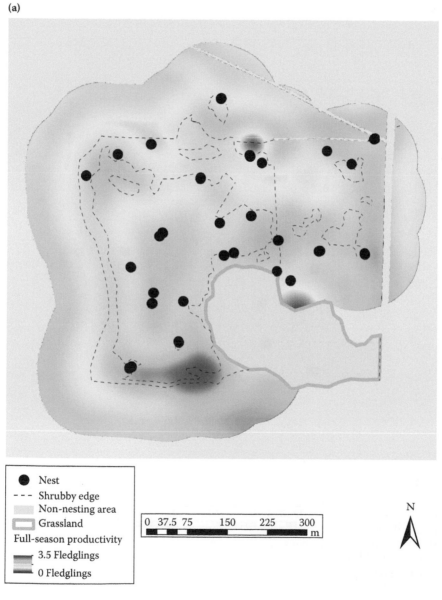

Figure 9.3. Estimated FSP (fledglings per 1-m² pixel) modeled from Golden-winged Warbler populations studied from 2010 to 2012 in the western Great Lakes region of North America of (a) upland shrubland and deciduous forest landscape at Tamarac NWR with an open grassland southeast of the site. (Continued)

in our study area when compared with similar upland cover types. Lesser FSP predicted by our models was supported by similar estimates calculated from nests found in wetland versus upland cover types at our sites. The difference between our findings and those of Confer et al. (2010) with regard to the value of wetlands to Golden-winged Warbler productivity may result from differences in structure and composition between wetland shrub communities in our

study areas and swamp forests studied in New York. The difference may also be due in part to our assessment of FSP. We included fledgling survival, a critical component of productivity that is affected by the wetland cover type differently than nest success, whereas Confer et al.'s (2010) study was based on nest success.

At Tamarac NWR, we evaluated how succession of grassland to upland shrubland cover types influenced Golden-winged Warbler productivity in a

(b)

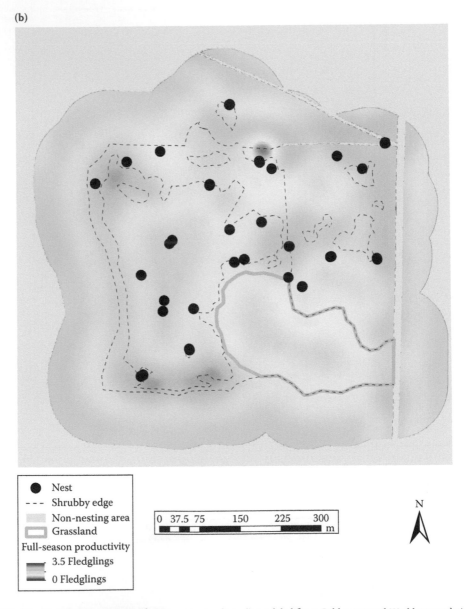

Figure 9.3. (*Continued*) Estimated FSP (fledglings per 1-m² pixel) modeled from Golden-winged Warbler populations studied from 2010 to 2012 in the western Great Lakes region of North America of (b) early stages of succession, with open grassland replaced by upland shrubland. (*Continued*)

(c)

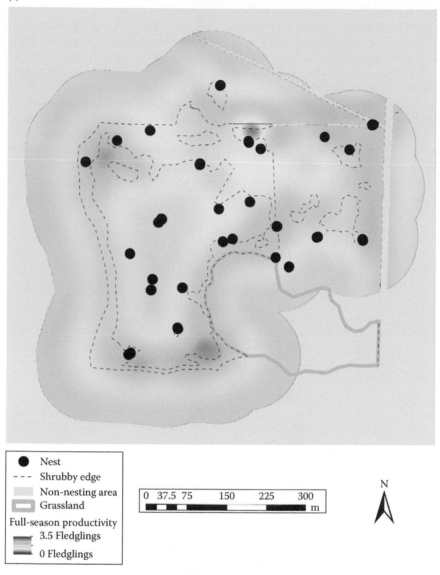

Legend:
- ● Nest
- --- Shrubby edge
- Non-nesting area
- Grassland
- Full-season productivity
  - 3.5 Fledglings
  - 0 Fledglings

0 37.5 75   150   225   300
                                    m

N

Figure 9.3. (*Continued*) Estimated FSP (fledglings per 1-m² pixel) modeled from Golden-winged Warbler populations studied from 2010 to 2012 in the western Great Lakes region of North America of (c) later stages of succession, with open grassland replaced with deciduous forest. Soft shrubby edges are marked by a dashed line, nests are marked with a circle, grassland is delineated by thick gray boundary, and areas unused for nesting (grassland, roads, open water, and deciduous forest >100 m from shrubby cover types) are marked by solid gray.

landscape that already hosted high productivity. Succession of a grassland patch to an upland shrubland patch increased the area available for nesting by 9 ha (49%) and increased the total estimated productivity of the landscape by 30%, in spite of mean estimated FSP decreasing by 0.24 fledglings per 1-m² pixel (12%). Our results demonstrate that there may be scenarios in which increasing the area available for nesting can result in increasing

overall productivity within a landscape, even while decreasing overall nest-site quality. However, our result appears to be management-scale and landscape-context dependent. For example, given the mean estimated FSP presented here for the grassland (1.97 fledglings per pixel) and upland shrubland (1.73 fledglings per pixel) cover types, if we only increased the area available for nesting by 2 ha, total estimated productivity of the landscape would decrease by 3%. If we extended this scenario to include succession of this 2-ha patch to deciduous forest (1.86 fledglings produced per pixel), the deciduous forest scenario would produce 5% more fledglings than the grassland cover type and 8% more fledglings than the upland shrubland cover type. Our results demonstrate how sensitive the models are to the size and landscape context of the proposed management.

Last, our assessment of landscape characteristics related to Golden-winged Warbler nest and fledgling survival indicated that highly complex shrubland-forest edges (e.g., Figure 9.4) were associated with lower rates of FSP of Golden-winged Warblers in the western Great Lakes region. Similar to observations of nest success in Indigo Buntings (*Passerina cyanea*; Weldon and Haddad 2005), we predicted lower FSP near complex forest edges. However, we note that the relationship between edge and productivity can vary substantially depending on the surrounding landscape structure and composition and the amount of edge within the impact radius at any location. For many of the models we developed, the amount of edge was positively related to FSP until an apparent threshold (~0.5 km of shrubby edge within 50 m of any given point), after which increasing the amount of edge led

(a)

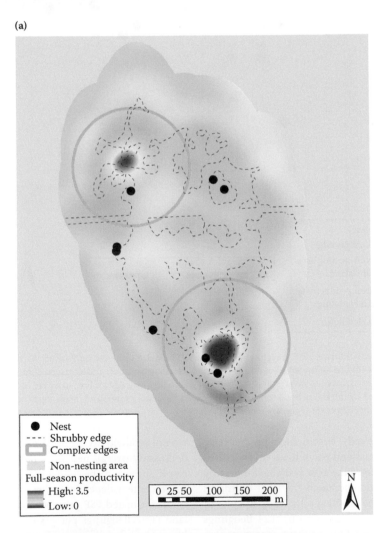

Figure 9.4. Estimated FSP (fledglings per 1-m^2 pixel) modeled from Golden-winged Warbler populations studied from 2010 to 2012 in the western Great Lakes region of North America of (a) an upland shrubland at Sandilands PF with complex edges in the northwest and south portions of the clear-cut. (Continued)

(b)

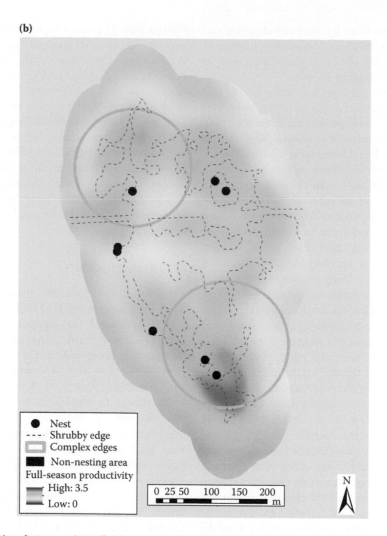

Figure 9.4. *(Continued)* Estimated FSP (fledglings per 1-m² pixel) modeled from Golden-winged Warbler populations studied from 2010 to 2012 in the western Great Lakes region of North America of (b) the same upland with complex edges removed or reduced. Soft shrubby edges are marked by a dashed line, nests are marked with a circle, complex edges are circled with a red hatched lines, and areas unused for nesting (grassland, roads, open water, and deciduous forest >100 m from shrubby cover types) are marked by solid black.

to reduced FSP (Peterson 2014:appendix C). Importantly, the edges we assessed were those between shrubland and forest cover types, and not the smaller scale, micro-edges within shrublands discussed in Chapter 7 (this volume), which may have a different relationship with productivity from the one we observed at a larger spatial scale.

Unlike resource selection functions (Manly et al. 2002), which have been used extensively to assess factors related to species presence (Beerens et al. 2011, Refsnider et al. 2013, Slaght et al. 2013), our spatially explicit models of FSP predict both the amount and quality of breeding habitat, and how the spatial distribution of cover types across a landscape influences productivity. We assessed our models by comparing predicted and estimated values of FSP in a cross-validation framework. The results of that cross-validation indicated that our models of the relationship(s) between FSP and landscape structure and composition explained more variation in both nest success and fledgling survival than models that did not incorporate landscape structure and composition. We suggest, however, that a more thorough validation of our approach is warranted, given the low Spearman's rank correlation from k-fold cross-validation and the general difficulties of validating models of binary and highly stochastic phenomena such as nest success or fledgling survival.

We also note that there are limitations to our analytical approach. As currently presented, our FSP models do not predict the likelihood any location will be selected as a home range or nest site (i.e., second to third order selection, Johnson 1980). Incorporating these probabilities into a composite model may make it possible to identify the most used and the most productive areas on a landscape. Additionally, the renesting rate and brood sizes we used in our models may not apply to populations other than those in the western Great Lakes region. Applying our models to other regions would also likely require region-specific information for demographic parameters.

Our models of FSP predict the potential productivity of a specific location, regardless of whether a breeding pair uses that location. For each pixel across our study sites, our nesting model produced a value that represented the probability that a successful nest could occur in that pixel during the season; that value includes the probability of a first nest succeeding, the probability that a second nest is possible at that location (i.e., did the first nest fail?), and the probability of a second nest succeeding. Our model produces the same estimates of productivity regardless of movements between nesting attempts because the estimates are for any first nest and any second nest at each location, and does not require that those attempts be from the same female or breeding pair. For simplicity, the FSP estimate for each location can be viewed as a modeled estimate of the number of young raised to independence by a breeding pair that nested, and potentially renested, at that specific location. Generally Golden-winged Warblers do not renest at the same location (Streby et al. 2014), although violation of this simplifying assumption would not change our results or conclusions.

For the landscapes we studied, our models provided several insights into Golden-winged Warbler ecology and conservation. First, modeling FSP across landscapes allowed us to identify specific areas where management could be directed to have the largest, positive influence on productivity. Second, simulation of the effects of proposed management can inform decisions about how best to use resources to affect population dynamics. Third, spatially explicit models of FSP can identify areas of high productivity, which may be areas to avoid manipulating or to emulate in other landscapes. Fourth, this modeling process may alter previous management recommendations that wetlands provide better habitat for Golden-winged Warblers than shrubby uplands. Last, assuming comparable demographic data are collected, our approach can be used to simultaneously assess likely impacts on FSP of other species in the same associated with the same landscape successional forests, such as American Woodcock (*Scolopax minor*), to better understand how management for a single species can benefit the avian community.

ACKNOWLEDGMENTS

We thank P. Blackburn, W. Brininger, W. Faber, W. Ford, J. Loegering, M. McDowell, H. Saloka, and R. Vallender for equipment and logistical support. We thank L. Arnold, S. Barlow, D. Bradshaw, J. Bruggman, R. Carr, M. Dawson, L. Deets, J. Feura, A. Fish, R. Franta, C. Gesmundo, A. Jensen, M. Johnson, G. Kramer, J. Lehman, T. McAllister, D. McNeil, E. Pokrivka, R. Poole, A. Rehmann, J. Refsnider, R. Refsnider, N. Seeger, E. Sinnot, and B. Vernasco for assistance with field data collection, J. Fieberg for assistance with model validation, and P. Wood, J. Loegering, L. Schofield, and two anonymous referees for constructive comments on this manuscript. Field data were collected during a project funded by the U.S. Fish and Wildlife Service, U.S. Geological Survey, and Minnesota Department of Natural Resources through Research Work Order no. 87 at the Minnesota Cooperative Fish and Wildlife Research Unit. We collected data following Protocol #1004A80575, approved by the University of Minnesota Institutional Animal Care and Use Committee. Use of trade names does not imply endorsement by the U.S. Geological Survey, University of Minnesota, or any other organization supporting this research.

LITERATURE CITED

Askins, R. A. 1995. Hostile landscapes and the decline of migratory songbirds. Science 267:1956–1957.

Askins, R. A., M. J. Philbrick, and D. S. Sugeno. 1987. Relationship between the regional abundance of forest and the composition of forest bird communities. Biological Conservation 39:129–152.

Ausprey, I. J., and A. D. Rodewald. 2011. Postfledging survivorship and habitat selection across a rural-to-urban landscape gradient. Auk 128:293–302.

Battin, J. 2004. When good animals love bad habitats: ecological traps and the conservation of animal populations. Conservation Biology 18:1482–1491.

Bayne, E. M., and K. A. Hobson. 1997. Comparing the effects of landscape fragmentation by forestry and agriculture on predation of artificial nests. Conservation Biology 11:1418–1429.

Beerens, J. M., D. E. Gawlik, G. Herring, and M. I. Cook. 2011. Dynamic habitat selection by two wading bird species with divergent foraging strategies in a seasonally fluctuating wetland. Auk 218:651–662.

Benson, T. J., N. M. Anich, J. D. Brown, and J. C. Bednarz. 2010. Habitat and landscape effects on brood parasitism, nest survival, and fledgling production in Swainson's Warblers. Journal of Wildlife Management 74:81–93.

Boyce, M. S., P. R. Vernier, S. E. Nielsen, and F. K. A. Schmiegelow. 2002. Evaluating resource selection functions. Ecological Modeling 157:281–300.

Brawn, J. D., and S. K. Robinson. 1996. Source-sink population dynamics may complicate the interpretation of long-term census data. Ecology 77:3–12.

Buehler, D. A., A. M. Roth, R. Vallender, T. C. Will, J. L. Confer, R. A. Canterbury, S. B. Swarthout, K. V. Rosenberg, and L. P. Bullock. 2007. Status and conservation priorities of Golden-winged Warbler (*Verminora chrysoptera*). Auk 124:1439–1445.

Burnham, K. P., and D. R. Anderson. 2002. Model selection and multi-model inference. Springer Publishing Company, New York, NY.

Chalfoun, A. D., F. R. Thompson III, and M. J. Ratnaswamy. 2002. Nest predators and fragmentation: a review and meta-analysis. Conservation Biology 16:306–318.

Confer, J. L., K. W. Barnes, and E. C. Alvey. 2010. Golden- and Blue-winged Warblers: distribution, nesting success, and genetic differences in two habitats. Wilson Journal of Ornithology 122:273–278.

Confer, J. L., P. Hartman, and A. Roth. [online]. 2011. Golden-winged Warbler (*Vermivora chrystoptera*). In A. Poole (editor), The birds of North America online. <http://bna.birds.cornell.edu.ezp1.lib.umn.edu/bna/species/020> (5 October 2013).

Faaborg, J., M. C. Brittingham, T. M. Donovan, and J. Blake. 1995. Habitat fragmentation in the temperate zone. Pp. 357–380 in T. E. Martin and D. M. Finch (editors), Ecology and management of Neotropical migratory birds. Oxford University Press, Oxford, England.

Hoover, J. P., M. C. Brittingham, and L. J. Goodrich. 1995. Effects of forest patch size on nesting success of Wood Thrushes. Auk 112:146–155.

Huffman, R. D. 1997. Effects of residual overstory on bird use and aspen regeneration in aspen harvest sites in Tamarac National Wildlife Refuge, Minnesota. M.S. thesis, West Virginia University, Morgantown, WV.

Johnson, D. H. 1980. The comparison of usage and availability measurements for evaluating resource preference. Ecology 61:65–71.

Koper, N., and M. Manseau. 2009. Generalized estimating equations and generalized linear mixed-effects models for modeling resource selection. Journal of Applied Ecology 46:590–599.

Kubel, J. E., and R. H. Yahner. 2008. Quality of anthropogenic habitats for Golden-winged Warblers in central Pennsylvania. Wilson Journal of Ornithology 120:801–812.

Lloyd, P., T. E. Martin, R. L. Redmond, U. Langner, and M. M. Hart. 2005. Linking demographic effects of habitat fragmentation across landscapes to continental source-sink dynamics. Ecological Applications 15:1504–1514.

Manly, B. F. J., L. L. McDonald, D. L. Thomas, T. L. McDonald, and W. P. Erickson. 2002. Resource selection by animals: statistical design and analysis for field studies. Kluwer Academic Press, Boston, MA.

Marshall, M. R., J. A. DeCecco, A. B. Williams, G. A. Gale, and R. J. Cooper. 2003. Use of regenerating clearcuts by late-successional bird species and their young during the post-fledging period. Forest Ecology and Management 183:127–135.

Martin, T. E. 1993. Nest predation and nest sites. Bioscience 43:523–532.

Martin, T. E., and G. R. Geupel. 1993. Nest monitoring plots: methods for locating nests and monitoring success. Journal of Field Ornithology 64:507–519.

Pagen, R. W., F. R. Thompson, and D. E. Burhans. 2000. Breeding and post-breeding habitat use by forest migrant songbirds in the Missouri Ozarks. Condor 102:738–747.

Percy, K. L. 2012. Effects of prescribed fire and habitat on Golden-winged Warbler (*Vermivora chrysoptera*) abundance and nest survival in the Cumberland Mountains of Tennessee. M.S. thesis, University of Tennessee, Knoxville, TN.

Peterson, S. M. 2014. Landscape productivity and the ecology of brood division in Golden-winged Warblers in the western Great Lakes region. M.S. thesis, University of Minnesota, Minneapolis, MN.

Porneluzi, P., J. C. Bednarz, L. J. Goodrich, N. Zawada, and J. Hoover. 1993. Reproductive performance of territorial Ovenbirds occupying forest fragments and a contiguous forest in Pennsylvania. Conservation Biology 7:618–622.

Rappole, J. H., and A. R. Tipton. 1991. New harness design for attachment of radio transmitters to small passerines. Journal of Field Ornithology 62:335–337.

Refsnider, J. M., D. A. Warner, and F. J. Janzen. 2013. Does shade cover availability limit nest-site choice in two populations of a turtle with temperature-dependent sex determination? Journal of Thermal Biology 38:152–158.

Robinson, S. K. 1992. Population dynamics of breeding Neotropical migrants in a fragmented Illinois landscape. Pp. 408–418 in J. M. Hagan III and D. W. Johnston (editors), Ecology and conservation of Neotropical migrant landbirds. Smithsonian Institution Press, Washington, DC.

Robinson, S. K., and D. S. Wilcove. 1994. Forest fragmentation in the temperate zone and its effects on migratory songbirds. Bird Conservation International 4:223–249.

Rossell, C. R., S. C. Patch, and S. P. Wilds. 2003. Attributes of Golden-winged Warbler territories in a mountain wetland. Wildlife Society Bulletin 31:1099–1104.

Roth, A. M., and S. Lutz. 2004. Relationship between territorial male Golden-winged Warblers in managed aspen stands in northern Wisconsin, USA. Forest Science 50:153–161.

Rush, T., and T. Post. 2008. Golden-winged Warbler (Vermivora chrysoptera) and Blue-winged Warbler (Vermivora pinus) surveys and habitat analysis on Fort Drum Military Installation. New York State Department of Environmental Conservation, Division of Fish and Wildlife, Albany, NY.

Rush, S. A., and B. J. M. Stutchbury. 2008. Survival of fledgling Hooded Warblers (Wilsonia citrina) in small and large forest fragments. Auk 125:183–191.

Shaffer, T. L. 2004. A unified approach to analyzing nest success. Auk 121:526–540.

Shipley, A. A., M. T. Murphy, and A. H. Elzinga. 2013. Residential edges as ecological traps: postfledging survival of a ground-nesting passerine in a forested urban park. Auk 130:501–511.

Slaght, J. C., J. S. Horne, S. G. Surmach, and R. J. Gutiérrez. 2013. Home range and resource selection by animals constrained by linear habitat features: an example of Blakiston's Fish Owl. Journal of Applied Ecology 50:1350–1357.

Stephens, S. E., J. J. Rotell, M. S. Lindberg, M. L. Taper, and J. K. Ringelman. 2005. Duck nest survival in the Missouri Coteau of North Dakota: landscape effects at multiple spatial scales. Ecological Applications 15:2137–2149.

Streby, H. M., and D. E. Andersen. 2011. Seasonal productivity in a population of migratory songbirds: why nest data are not enough. Ecosphere 2:78.

Streby, H. M., and D. E. Andersen. 2013. Testing common assumptions in studies of songbird nest success. Ibis 155:327–337.

Streby, H. M., T. L. McAllister, S. M. Peterson, G. R. Kramer, J. A. Lehman, and D. E. Andersen. 2015. Minimizing marker mass and handling time when attaching radio-transmitters and geolocators to small songbirds. Condor 117:249–255.

Streby, H. M., S. M. Peterson, C. F. Gesmundo, M. K. Johnson, A. C. Fish, J. A. Lehman, and D. E. Andersen. 2013. Radio-transmitters do not affect seasonal productivity of female Golden-winged Warblers. Journal of Field Ornithology 84:316–321.

Streby, H. M., J. M. Refsnider, S. M. Peterson, and D. E. Andersen. 2014. Retirement investment theory explains patterns in songbird nest-site choice. Proceedings of the Royal Society of London B 281:20131834.

Vitz, A. C., and A. D. Rodewald. 2007. Vegetative and fruit resources as determinants of habitat use by mature-forest birds during the postbreeding period. Auk 124:494–507.

Weldon, A. J., and N. M. Haddad. 2005. The effects of patch shape on Indigo Buntings: evidence for an ecological trap. Ecology 86:1422–1431.

CHAPTER TEN

Management Implications of Brood Division in Golden-winged Warblers*

Sean M. Peterson, Henry M. Streby, and David E. Andersen

Abstract. Brood division in the postfledging period is a common avian behavior that is not well understood. Brood division has been reported in Golden-winged Warblers (*Vermivora chrysoptera*), but it is not known how common this behavior is, whether males and females exhibit different strategies related to parental care and habitat use, or how brood division might influence management strategies. We radiomarked fledglings and monitored divided broods of Golden-winged Warblers from fledging until independence from parental care at three sites in the western Great Lakes region from 2010 to 2012 to assess differences in strategies between male and female parents and to consider possible management implications. Male- and female-reared sub-broods exhibited different space use during the dependent postfledging period despite similar fledgling survival, cover-type use, and microhabitat use. By independence, female-reared sub-broods traveled over twice as far from the nest (mean = 461 ± 81 SE m) as male-reared sub-broods (164 ± 41 m).

Additionally, female-reared sub-broods traveled over three times as far from the natal patch edge (354 ± 72 m) as male-reared sub-broods (108 ± 36 m). Without accounting for differential space use by male- and female-reared sub-broods, we would have reported broods traveling 292 (± 46 m) from the nest and 214 (± 40 m) from the natal patch edge—distances that do not reflect how far females move sub-broods. Parental strategies differ between sexes with regard to movement patterns, and we recommend incorporating the differences in space use between sexes in future management plans for Golden-winged Warblers and other species that employ brood division. Specifically, management actions might be most effective when they are applied at spatial scales large enough to incorporate the habitat requirements of both sexes throughout the entire reproductive season.

Key Words: behavior, parental care, postfledging care, songbird, *Vermivora chrysoptera*.

Brood division is a widespread avian behavior that is characterized by adults provisioning and caring for a subset of their brood over all or a substantive portion of the postfledging period, after young leave the nest but before they reach independence from adult care, forming two stable "sub-broods" attended separately by male and female parents (Harper 1985, Leedman

* Peterson, S. M., H. M. Streby, and D. E. Andersen. 2016. Management implications of brood division in Golden-winged Warblers. Pp. 161–171 in H. M. Streby, D. E. Andersen, and D. A. Buehler (editors). Golden-winged Warbler ecology, conservation, and habitat management. Studies in Avian Biology (no. 49), CRC Press, Boca Raton, FL.

and Magrath 2003, Peterson 2014). Will (1986) reported that Golden-winged Warblers (*Vermivora chrysoptera*) exhibited brood division, but the extent to which this behavior occurs in Golden-winged Warblers and how it influences productivity and habitat use are unknown. Recently, Peterson (2014) observed division in 98.5% of broods in northern Minnesota and southern Manitoba, Canada, with equal parental effort where male and female parents cared for similar numbers of fledglings and provisioned fledglings at similar rates, suggesting that brood division is nearly obligate in Golden-winged Warblers.

Most studies of brood division in birds have focused on differences or similarities in adult care and the potential evolutionary benefits of employing this behavior (McLaughlin and Montgomerie 1985, Lessells 2002, Leedman and Magrath 2003, Draganoiu et al. 2005). In Golden-winged Warblers, parental care, such as provisioning of young and parental attendance, is similar between male and female parents (Peterson 2014), suggesting that both sexes are similarly capable of rearing young to independence. Furthermore, brood division in Golden-winged Warblers appears to most closely fit the evolutionary strategy proposed by Lessells (1998), in which brood division decreases the likelihood of survival for any individual parent when compared with brood abandonment, but increases survival of the other parent of a brood and for fledglings in both sub-broods (Peterson 2014).

An aspect of brood division that has potential conservation implications but has received little attention is how sub-broods partition space. Elsewhere, four studies of three species reported spatial patterns of division in broods (McLaughlin and Montgomerie 1985, Weatherhead and McRae 1990, Evans Ogden and Stutchbury 1997, Rush and Stutchbury 2008). Weatherhead and McRae (1990) observed that divided broods of American Robins (*Turdus migratorius*) traveled similar distances from the nest independent of parental sex. In contrast, female Hooded Warblers (*Setophaga citrina*; Evans Ogden and Stutchbury 1997, Rush and Stutchbury 2008) and female Lapland Longspurs (*Calcarius lapponicus*; McLaughlin and Montgomerie 1985) traveled farther from the nest with their sub-broods compared to males. How Golden-winged Warbler sub-broods partition space is unknown, but if female- and male-reared sub-broods exhibit different spatial use patterns, conservation strategies

may need to account for these differences to maximize their impact. In extreme cases, failure to recognize brood division may lead to flawed conservation plans for a species and failure to achieve the desired management effect.

We studied brood division in three populations of Golden-winged Warblers in the western Great Lakes region of central North America. Golden-winged Warblers are Neotropical migratory songbirds that breed in northeastern and north-central North America and spend the nonbreeding season in southern Central America and northern South America (Confer et al. 2011). Golden-winged Warbler populations outside of the core population in the western Great Lakes region are declining rapidly, resulting in Golden-winged Warblers being listed as Endangered, Threatened, or of high management concern in 10 states and Canada (Buehler et al. 2007). Golden-winged Warblers are single-brooded, fledging young from no more than one clutch per year, and each parent cares for both nestlings and fledglings (Confer et al. 2011).

Primary nesting habitat of the species consists of shrubby, early successional uplands, shrubby wetlands, and adjacent forests with dense undergrowth (Streby et al. 2014). Golden-winged Warbler postfledging habitat differs from that used for nesting, with birds selecting mature forest and mid-successional forest over other cover types, but also using early successional forests and wetlands (Chapter 8, this volume). Current Best Management Practices in the western Great Lakes region are based on a positive association of nesting Golden-winged Warblers with forest, shrubland, shrub-forest wetlands, and pasture-hay fields within 244 m of a managed upland patch and ≥50% forest cover within 2.4 km of a habitat patch (A. M. Roth et al., unpubl. plan). To provide available postfledging habitat, A. M. Roth et al. (unpubl. plan) suggested creating a dynamic forest landscape comprising different aged stands, maintaining 15%–20% of the total managed area in early seral stage, forested cover types.

Brood division during the 25-day postfledging period of parental care in Golden-winged Warblers appears to be nearly obligate, with rare observations of both parents provisioning the same fledglings (Will 1986, Peterson 2014). Sub-broods often form crèches with ≥1 conspecific sub-brood and fledglings of other species (Will 1986, H. M. Streby, unpubl. data). Similar to other songbirds (Ricklefs 1968, Anders et al. 1998, King et al. 2006, Streby and Andersen 2013a), Golden-winged Warbler fledgling survival is lowest in the first few days after fledging, with 75% of fledgling mortality occurring in the first three days after fledging from the nest (Chapter 8, this volume).

We monitored fledgling movements via radio-telemetry throughout the dependent postfledging period before independence to assess differences between male- and female-reared sub-broods of Golden-winged Warblers. We specifically compared patterns in movements, space use of locations on the landscape and within vegetation strata, structure and composition of vegetation used by fledglings, and fledgling survival. Based on the patterns we observed in habitat and space use by Golden-winged Warbler sub-broods, we suggest breeding habitat management implications for the western Great Lakes region.

METHODS

Study Sites

From 2010 to 2012, we studied Golden-winged Warbler breeding-season ecology at Tamarac National Wildlife Refuge (NWR) in Becker County, Minnesota (47.049°N, 95.583°W). In 2011 and 2012, we expanded our study to include sites at Rice Lake NWR in Aitkin County, MN (46.529°N, 93.338°W) and Sandilands Provincial Forest (PF) in southeastern Manitoba, Canada (49.637°N, 96.247°W). At each study site, we focused our efforts in 8–16 plots of 2.5–25 ha shrubby uplands and wetlands and the surrounding forest in a predominantly forested landscape. For a more detailed description of the landscapes present at our study sites, see Chapter 8 (this volume).

Field Methods

We located Golden-winged Warbler nests (n = 50) using nest-searching methods described by Martin and Geupel (1993). We found additional nests (n = 28) by passively mist-netting adult female Golden-winged Warblers after their arrival from spring migration but before most started nesting, and attaching a VHF radio transmitter that was ~4.1% of mean adult mass with a figure-eight harness (Rappole and Tipton 1991, Streby et al. 2015) and by subsequently monitoring the radiomarked females (Streby et al.

2013). Radio transmitters used in this study had no measureable impact on female productivity (Streby et al. 2013). We recorded nest locations using handheld Global Positioning System (GPS) units (GPSMAP 76 or eTrex Venture HC, Garmin Ltd., Schaffhausen, Switzerland), and we averaged locations using 100 points to achieve <5 m accuracy. We delineated cover types in ArcGIS 10.0 Geographic Information System (GIS) software (Environmental Systems Research Institute, Redlands, CA). For Tamarac NWR and Rice Lake NWR, we used 1-m resolution digital orthophoto quadrangles (2009; Minnesota Department of Natural Resources, St. Paul, MN). For Sandilands PF, we used 1-m resolution georeferenced satellite images obtained from Google Earth™ 6.2 (2010; Google Inc., Mountain View, CA). For each nest, we defined the natal patch as the contiguous patch of shrubland cover in which the nest was situated. For five nests located <25 m into mature forest (7%), we defined the natal patch as the nearest contiguous patch of shrubland cover. We excluded two nests that were >25 m into mature forest from this analysis. Natal patches ranged from 0.25 to 30 ha in area.

We monitored nests on 4-day intervals until near the estimated fledge date, when we visited nests daily. When nestlings were 6–9 days old (counting hatch day as day 1), we banded all nestlings in the nest with standard U.S. Geological Survey leg bands, measured their mass to the nearest 0.01 g with a digital scale (AWS-100, American Weigh Scales Inc., Norcross, GA), and attached a radio transmitter to 1–5 randomly selected nestlings using the figure-eight harness design (~4.6% of mean nestling mass; Rappole and Tipton 1991, Streby et al. 2015). In addition, we used mist nets to capture, band, and attach radio transmitters to fledglings from unknown nest locations (n = 14) detected by fledgling vocalization or adult behavior. We used these individuals only in analyses that did not require knowledge of nest location such as vegetation characteristics and minimum daily distance moved. We estimated the age of captured fledglings based on observed development of fledglings banded in the nest and subsequently monitored throughout the postfledging period.

We recorded daily fledgling locations using ground-based radiotelemetry, as described in Streby and Andersen (2013b). We avoided locating fledglings in inclement weather to minimize the chance that our activities would cause fledglings to move to locations where their survival might be compromised by exposure to excessive cold or moisture. For each fledgling, we recorded locations using handheld GPS units as described above for nest locations. For each encounter of a fledgling, we recorded vegetation strata as ground, understory, midstory, or canopy, and whether the location was in nonshrubland forest such as mature forest, forested wetland, and sapling-dominated clear-cuts. We defined the understory strata as the vegetation layer from ground to 2 m, midstory as the vegetation layer from >2–15 m, and the canopy as the vegetation layer >15 m. At each fledgling location, we estimated lateral vegetation density by recording the amount of vegetation obscuring a 2-m profile board (MacArthur and Macarthur 1961, Streby and Andersen 2013a, for more detailed methods see Chapter 8, this volume). We used ImageJ (National Institutes of Health, http://imagej.nih.gov/ij/) to derive percent vertical cover as the percent of sky covered by vegetation >2 m above the ground from digital photographs taken vertically from 2 m above the ground at fledgling locations. For each photograph, we divided color channels to differentiate between sky and vegetation, converted vegetation and sky to binary pixels, and measured the percentage of pixels occupied by vegetation. We used ArcGIS to calculate daily distance from nest for all fledglings for which we knew the location of the nest from which they fledged. We used ArcGIS to measure the daily distance from the edge of the natal patch to daily locations, with locations inside the natal patch considered to be negative distances from the edge of the patch.

We identified sex of the parent attending individual fledglings by plumage while observing adult and fledgling interactions. We excluded 50 broods from our analyses for which fledglings were depredated before we observed parental care, and two broods without brood division where fledglings were provisioned by both parents during multiple observations. We recorded observations daily over the 25-day period that Golden-winged Warbler fledglings are dependent upon adults (Will 1986).

Postfledging Periods

Our initial observations indicated that space use was similar between male- and female-reared sub-broods shortly after fledging from the nest, but space use appeared to differ between male- and female-reared sub-broods by the time fledglings became independent of adults (Peterson 2014). Changes in space use suggested a shift in parental movement strategies during the postfledging period. Peterson (2014) analyzed these data and identified a period of greater-than-expected directionality of movements on days 9 and 10 that indicated a change in female space use in relation to the nest site (Figure 10.1a). Therefore, to ensure that we compared differences in male and female parental behavior both before and after the apparent change in parental strategies occurred, we divided the postfledging period into two periods—an early postfledging period from days 1 to 8, and a late postfledging period from days 9 to 25.

Statistical Analyses

We tested for differences between male- and female-reared sub-broods in distance from nest and vegetation characteristics at daily fledgling locations including percent vertical cover, lateral vegetation density, nonshrubland forest use, and strata occupied by fledglings in broods for which we tracked both sub-broods via radiotelemetry (hereafter, "paired sub-broods"). We used data from this sample of broods to avoid potential bias from nonindependence of locations for sub-broods that we tracked without knowing the location of the sub-brood under the care of the other parent. After testing for differences between paired sub-broods, we used our entire sample of sub-broods to describe patterns over time for each variable as a function of parental sex. For sub-broods in which we monitored >1 fledgling, we used the mean value for all fledglings (usually two) in that sub-brood for each variable in analyses. All results are presented as mean ± SE.

Daily distance from a nest and natal patch edge are likely to be temporally autocorrelated because the distance from the nest one day is likely to be more similar to the distance from the nest the subsequent day than the distance from any randomly selected day. Thus, we used a sign test to assess differences between paired sub-broods in daily distance from nest and natal patch (Dixon and Mood 1946). We calculated mean distance from nest and natal patch edge near the end of the postfledging period using the mean distance

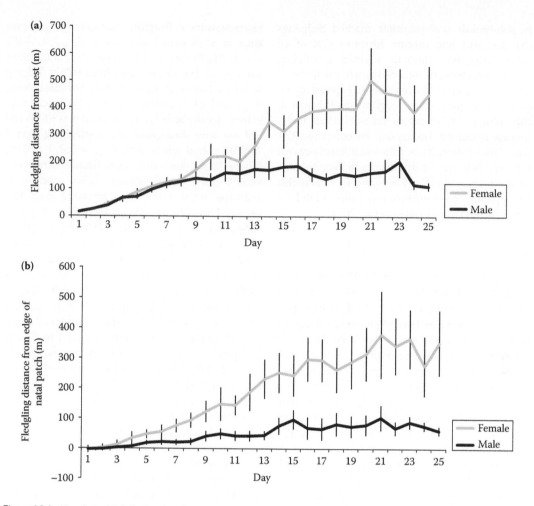

Figure 10.1. Trends in (a) daily distance from nest and (b) daily distance from the edge of the natal cover patch in male-reared sub-broods (Male) and female-reared sub-broods (Female) in three populations of Golden-winged Warblers in the western Great Lakes region 2010–2012 (data presented as mean ± SE).

of fledglings in each sub-brood over the last five days of the postfledging period prior to independence (i.e., days 21–25). We used a Pillai-M.S. Bartlett trace multivariate analysis of covariance (MANCOVA) to test for differences between male- and female-reared sub-broods in vegetation characteristics at fledgling locations using linear models in R (ver. 2.14.1, R Foundation for Statistical Computing, Vienna, Austria). Parental sex and fledgling age were explanatory variables and percent vertical cover, lateral vegetation density, nonshrubland forest use, and strata occupied by fledgling were response variables. For all MANCOVA tests that were statistically significant ($\alpha = 0.05$), we conducted an ANCOVA test on each dependent variable.

We could not compare survival between male- and female-reared sub-broods during the early postfledging period because in most cases, we were unable to identify the sex of the parental caregiver for fledglings that died during that period. In addition, our data indicated that space use was nearly identical for fledglings in male- and female-reared sub-broods in the early postfledging period, suggesting that predation pressure was likely similar between sub-broods. For the late postfledging period, we used the known fates module in program MARK (ver. 5.1, Colorado State University, Ft. Collins, CO) to estimate daily fledgling survival and 95% confidence intervals using the known fate module for a random fledgling per sub-brood (White and Burnham 1999).

In sub-broods with multiple marked fledglings that had different capture histories (2% of all sub-broods), we randomly selected a fledgling in that sub-brood to include in survival analyses. Fledgling survival during the late postfledging period was unrelated to fledgling age (Chapter 8, this volume); we therefore did not include fledgling age in our survival model. We compared 95% confidence intervals for male- and female-reared sub-broods to test for differences in fledgling survival. Previous models of fledgling survival in this study population determined that survival did not differ by site or year (Streby et al. 2014).

RESULTS

During the summers of 2010–2012, we monitored 66 Golden-winged Warbler fledglings from 60 sub-broods at Tamarac NWR, 30 fledglings from 28 sub-broods at Rice Lake NWR, and 27 fledglings from 24 sub-broods at Sandilands PF. In broods for which we monitored both sub-broods, both male- and female-reared sub-broods moved similar distances from the nest in the early postfledging period (sign test, n = 18, P = 0.82; Figure 10.1a). In contrast, female-reared sub-broods moved farther from the nest than male-reared sub-broods during the late postfledging period (sign test, n = 15, P = 0.04; Figure 10.1a). Female-reared sub-broods from broods in which we monitored both sub-broods were farther away from the natal patch edge than male-reared sub-broods throughout the entire postfledging period (sign test, n = 19, P = 0.02; Figure 10.1b). In the last five days of the postfledging period, female-reared sub-broods (n = 22) were 461 ± 81 m away from the nest and 354 ± 72 m from the natal patch edge, whereas male-reared sub-broods (n = 29) were 164 ± 41 m away from the nest and 108 ± 36 m away from the natal patch edge (Figure 10.1a,b). If we had not accounted for brood division and treated sub-broods as independent sample units (n = 51), we would have reported mean distance from the nest as 292 ± 46 m and mean distance from natal patch edge as 214 ± 40 m for the last five days of the postfledging period.

Male- and female-reared sub-broods used areas with similar vegetation characteristics (vertical cover, shrubland use, lateral vegetation density, and forest strata) in both the early postfledging period (n = 18 paired sub-broods, $F_{4, 221}$ = 0.95, P = 0.43) and the late postfledging period (n = 12 paired sub-broods, $F_{4, 262}$ = 1.92, P = 0.11). Vegetation

characteristics at fledgling locations changed over time (n = 19 paired sub-broods, $F_{4, 488}$ = 50.02, P < 0.001). Percentage of vertical cover at locations used by sub-broods increased from 40% at fledging to 60% at independence (n = 19 paired sub-broods, $F_{2, 499}$ = 12.23, P < 0.001; Figure 10.2a). Lateral vegetation density at fledgling locations was ~65% and did not vary throughout the postfledging period (n = 19 paired sub-broods, $F_{2, 537}$ = 2.8, P = 0.06; Figure 10.2b). Both male- and female-reared sub-broods used nonshrubland forest extensively, with use of this cover type increasing from 45% on day 1 to 85% by independence (n = 19 paired sub-broods, $F_{2, 541}$ = 25.47, P < 0.001; Figure 10.2c). We observed fledglings occupying vegetation strata farther from the ground with increasing frequency throughout the postfledging period. The frequency of fledglings perched on the ground declined from 35% on day 1 to 0% by day 13 (n = 19 paired sub-broods, $F_{2, 539}$ = 37.15, P < 0.001), frequency of use of midstory increased from 0% on day 1 to 15% by independence (n = 19 paired sub-broods, $F_{2, 539}$ = 18.25, P < 0.001), and frequency of use of canopy increased from 0% on day 1 to 25% by independence (n = 19 paired sub-broods, $F_{2, 539}$ = 30.2, P < 0.001) in the postfledging period (Figure 10.2d,e). Understory use did not change throughout the postfledging period, remaining at ~60% (n = 19 paired sub-broods, $F_{2, 539}$ = 1.305, P = 0.12). Fledglings in both male- and female-reared sub-broods had similarly high daily survival for the period from nine days after fledging until independence (male $\bar{x}$ = 0.9887, 95% CI = 0.9946–0.9766, n = 54; female $\bar{x}$ = 0.9873, 95% CI = 0.9939–0.9736, n = 46).

DISCUSSION

We observed a significant difference in space use between male- and female-reared sub-broods of Golden-winged Warblers during the postfledging period in the western Great Lakes region, and the difference has implications for management. Spatial patterns are present not only in relation to a single point at the nest, but also at a scale relevant to management such as a clearcut or shrubby wetland patch, as we observed that female-reared sub-broods traveled farther from the natal patch edge than male-reared sub-broods. Current Best Management Practices in the western Great Lakes region suggest active management at a radius of 244 m around a patch

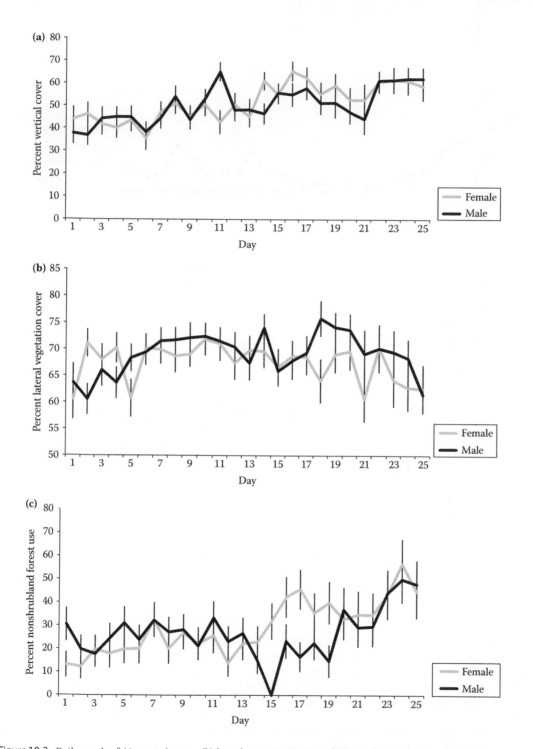

Figure 10.2. Daily trends of (a) vertical cover, (b) lateral vegetation cover, and (c) nonshrubland forest use in male-reared sub-broods and female-reared sub-broods.

(Continued)

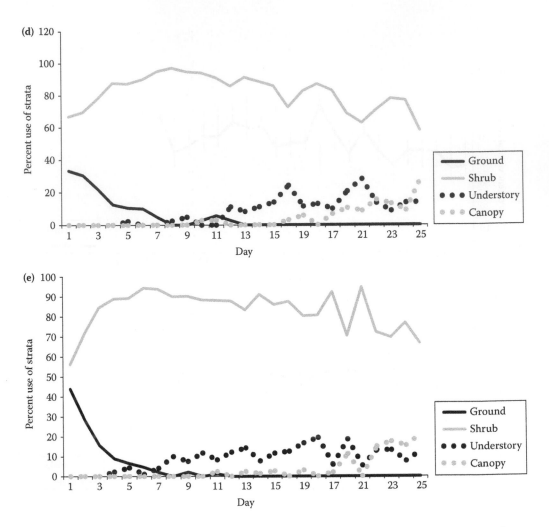

Figure 10.2. (*Continued*) Daily trends of strata use in (d) male-reared sub-broods and (e) female-reared sub-broods in three populations of Golden-winged Warblers in the western Great Lakes region 2010–2012 (data presented as mean ± SE; SE omitted from (d) and (e) for clarity).

(A. M. Roth et al., unpubl. plan). If we had not accounted for sex-related brood division in our study, we would have reported that broods move 201 m away from the natal patch edge and likely confirmed the current scale of management. However, both of these measurements significantly underestimate the extent of postfledging habitat used by female-reared sub-broods, which on average traveled 330 m from the natal patch edge before fledglings were independent from adult care.

To effectively manage breeding habitat of Golden-winged Warblers, it may be necessary to incorporate considerations for brood division and the different spatial scales used by male- and female-reared sub-broods during the postfledging period. Ignoring differential space use by male and female Golden-winged Warblers during the postfledging period may result in negative consequences for half the potential production of young to independence. We did not directly assess the consequences of failing to provide brood-rearing habitat at the spatial scaled used by female-reared sub-broods. However, landscapes that do not provide brood-rearing cover types at that spatial scale may provide lower quality habitat, and therefore productivity may be lower in such landscapes. Fledgling survival is a key component of full season productivity (Streby and Andersen 2011), and current recommendations underestimate the spatial scale of management needed to incorporate postfledging cover-type use (A. M. Roth et al., unpubl. plan). To provide adequate postfledging habitat for Golden-winged Warblers, mature forest or sapling-dominated clear-cuts should be maintained in close proximity to natal patches.

Habitat use of female-reared sub-broods is similar to habitat use of male-reared sub-broods throughout the postfledging period and managing for postfledging habitat from the edge of the natal patch to the scale that females use space will likely provide adequate habitat for both sub-broods, even as they move away from the natal patch. Current management recommendations need to be revised to consider cover types beyond what has been considered to be Golden-winged Warbler breeding habitat such as mature forest and sapling-dominated forest (Confer et al. 2011). Portions of the landscape not previously considered to be used by Golden-winged Warblers such as mature forest >244 m from shrublands may need to be reclassified as postfledging habitat and incorporated into management planning. Managing for postfledging cover types up to 700 m from a managed shrubland would provide postfledging habitat for 95% of the broods we monitored in our study.

Habitat characteristics at locations used by broods during the postfledging period likely influence survival (Streby and Andersen 2013a; Chapter 8, this volume). We observed no difference in survival between male- and female-reared sub-broods in the late postfledging period. Furthermore, although we did not directly test for differences in survival between male- and female-reared sub-broods in the early postfledging period, our observations that space use and vegetation characteristics at sub-brood locations were similar during this period suggest that predation pressures and, by extension, survival were likely also similar in male- and female-reared sub-broods. These observations suggest that both sexes are equally capable of rearing broods. Identification of parental care in our study populations often required multiple observations over several days. If the fledglings cared for by either parent were subjected to greater predation rates than fledglings cared for by the other parent during the early postfledging period, we would expect fewer fledglings observed under the care of that parental sex during the late postfledging period. Instead, we observed an equal proportion of parental care in males and females in this population (Peterson 2014), supporting the conclusions that predation pressures and survival rates were similar throughout the early postfledging period for both male- and female-reared sub-broods.

Last, our observations of differences in space use between male and female parents may influence mate choice and breeding habitat use. The mechanism of mate choice by female songbirds remains unclear, but mate choice in some species may be influenced more by territory quality than by male physical characteristics (Sirkiä and Laaksonen 2009, Temeles and Kress 2010, Hasegawa et al. 2012). We suggest that female Golden-winged Warblers likely use multiple criteria for choosing a breeding territory and mate, including male quality, potential nest success, and potential fledgling survival, all of which influence nest-site choice (Streby et al. 2014). Under such circumstances, many unoccupied areas that appear to humans structurally suitable as breeding territories may be unoccupied because they are not surrounded by adequate postfledging habitat at a scale used by females to rear their sub-broods. In addition, female-based postfledging habitat requirements may play a role in males remaining unpaired on seemingly suitable song territories. Managing at the territory scale, defined based on singing males, may be inadequate if females choose mates and associated breeding territories based on a much larger area or quality of adjacent habitats.

ACKNOWLEDGMENTS

We thank H. Saloka, P. Blackburn, W. Brininger, W. Faber, W. Ford, J. Loegering, M. McDowell, and R. Vallender for equipment and logistical support. We thank L. Arnold, S. Barlow, D. Bradshaw, J. Bruggman, R. Carr, M. Dawson, L. Deets, J. Feura, A. Fish, R. Franta, C. Gesmundo, A. Jensen, M. Johnson, G. Kramer, J. Lehman, T. McAllister, D. McNeil, E. Pokrivka, R. Poole, A. Rehmann, J. Refsnider, R. Refsnider, N. Seeger, E. Sinnot, and B. Vernasco for assistance with field data collection, D. Johnson for assistance with statistical analyses, L. Schofield, J. Loegering, and P. Wood for constructive comments on this manuscript, and two anonymous referees for comments improving the manuscript. Field data were collected during a project funded by the U.S. Fish and Wildlife Service, U.S. Geological Survey, and Minnesota Department of Natural Resources through Research Work Order no. 87 at the Minnesota Cooperative Fish and

Wildlife Research Unit. We collected data following Protocol #1004A80575, approved by the University of Minnesota Institutional Animal Care and Use Committee. Use of trade names does not imply endorsement by the U.S. Geological Survey, University of Minnesota, or any other organization supporting this research.

LITERATURE CITED

Anders, A. D., J. Faaborg, and F. R. Thompson III. 1998. Postfledging dispersal, habitat use, and home-range size of juvenile Wood Thrushes. Auk 115:349–358.

Buehler, D. A., A. M. Roth, R. Vallender, T. C. Will, J. L. Confer, R. A. Canterbury, S. B. Swarthout, K. V. Rosenberg, and L. P. Bulluck. 2007. Status and conservation priorities of Golden-winged Warbler (*Vermivora chrysoptera*) in North America. Auk 124:1439–1445.

Confer, J. L., P. Hartman, and A. Roth. 2011. Golden-winged Warbler (*Vermivora chrysoptera*). In A. Poole (editor), The birds of North America online. Cornell Lab of Ornithology, Ithaca, NY.

Dixon, W. J., and A. M. Mood. 1946. The statistical sign test. Journal of the American Statistical Association 41:557–566.

Draganoiu, T. I., L. Nagle, R. Musseau, and M. Kreutzer. 2005. Parental care and brood division in a songbird, the Black Redstart. Behaviour 142:1495–1514.

Evans Ogden, L. J., and B. J. M. Stutchbury. 1997. Fledgling care and male parental effort in the Hooded Warbler (*Wilsonia citrina*). Canadian Journal of Zoology 75:576–581.

Harper, D. G. C. 1985. Brood division in robins. Animal Behaviour 33:466–480.

Hasegawa, M., E. Arai, M. Watanabe, and M. Nakamura. 2012. Female mate choice based on territory quality in Barn Swallows. Journal of Ethology 30:143–150.

King, D. I., R. M. Degraaf, M. L. Smith, and J. P. Buonaccorsi. 2006. Habitat selection and habitat-specific survival of fledgling Ovenbirds (*Seiurus aurocapilla*). Journal of Zoology 269:414–421.

Leedman, A. W., and R. D. Magrath. 2003. Long-term brood division and exclusive parental care in a cooperatively breeding passerine. Animal Behaviour 65:1093–1108.

Lessells, C. M. 1998. A theoretical framework for sex-biased parental care. Animal Behaviour 56:395–407.

Lessells, C. M. 2002. Parentally biased favouritism: why should parents specialize in caring for different offspring? Philosophical Transactions of the Royal Society of London B 357:381–403.

MacArthur, R. H., and J. W. MacArthur. 1961. On bird species diversity. Ecology 42:594–598.

Martin, T. E., and G. R. Geupel. 1993. Nest-monitoring plots: methods for locating nests and monitoring success. Journal of Field Ornithology 64:507–519.

McLaughlin, R. L., and R. D. Montgomerie. 1985. Brood division by Lapland Longspurs. Auk 102:687–695.

Peterson, S. M. 2014. Landscape productivity and the ecology of brood division in Golden-winged Warblers in the western Great Lakes region. M.S. thesis, University of Minnesota, Minneapolis, MN.

Rappole, J. H., and A. R. Tipton. 1991. New harness design for attachment of radio transmitters to small passerines. Journal of Field Ornithology 62:335–337.

Ricklefs, R. E. 1968. The survival rate of juvenile Cactus Wrens. Condor 70:388–389.

Rush, S. A., and B. J. M. Stutchbury. 2008. Survival of fledgling Hooded Warblers (*Wilsonia citrina*) in small and large forest fragments. Auk 125:183–191.

Sirkiä, P. M., and T. Laaksonen. 2009. Distinguishing between male and territory quality: females choose multiple traits in the Pied Flycatcher. Animal Behaviour 78:1051–1060.

Streby, H. M., and D. E. Andersen. 2011. Seasonal productivity in a population of migratory songbirds: why nest data are not enough. Ecosphere 2:78.

Streby, H. M., and D. E. Andersen. 2013a. Survival of fledgling Ovenbirds: influences of habitat characteristics at multiple spatial scales. Condor 115:403–410.

Streby, H. M., and D. E. Andersen. 2013b. Movements, cover-type selection, and survival of fledgling Ovenbirds in managed deciduous and mixed-coniferous forests. Forest Ecology and Management 287:9–16.

Streby, H. M., T. L. McAllister, S. M. Peterson, G. R. Kramer, J. A. Lehman, and D. E. Andersen. 2015. Minimizing marker mass and handling time when attaching radio-transmitters and geolocators to small songbirds. Condor 117:249–255.

Streby, H. M., S. M. Peterson, C. F. Gesmundo, M. K. Johnson, A. C. Fish, J. A. Lehman, and D. E. Andersen. 2013. Radio-transmitters do not affect seasonal productivity of female Golden-winged Warblers. Journal of Field Ornithology 84:316–321.

Streby, H. M., J. M. Refsnider, S. M. Peterson, and D. E. Andersen. 2014. Retirement investment theory explains patterns in songbird nest-site choice. Proceedings of the Royal Society of London B 281:20131834.

Temeles, E. J., and W. J. Kress. 2010. Mate choice and mate competition by a tropical hummingbird at a floral resource. Proceedings of the Royal Society of London B 277:1607–1613.

Weatherhead, P. J., and S. B. McRae. 1990. Brood care in American Robins: implications for mixed reproductive strategies by females. Animal Behaviour 39:1179–1188.

White, G. C., and K. P. Burnham. 1999. Program MARK: survival estimation from populations of marked animals. Bird Study 46 (Suppl.):120–138.

Will, T. C. 1986. The behavioral ecology of species replacement: Blue-winged and Golden-winged Warblers in Michigan. Ph.D. dissertation, University of Michigan, Ann Arbor, MI.

Nonbreeding Season and Migration

CHAPTER ELEVEN

Conservation Implications of Golden-winged Warbler Social and Foraging Behaviors during the Nonbreeding Season*

Richard B. Chandler, Sharna Tolfree, John Gerwin, Curtis Smalling,
Liliana Chavarría-Duriaux, Georges Duriaux, and David I. King

Abstract. We used radiotelemetry and observations of color-banded birds in Costa Rica and Nicaragua to characterize the social system and foraging behavior of Golden-winged Warblers (*Vermivora chrysoptera*) at the nonbreeding grounds, and we assessed how these behaviors affected intraspecific spacing and home-range size. Golden-winged Warblers spent the majority of their time associating with mixed-species flocks composed of migrant and resident species. Males were territorial, responding aggressively to broadcast vocalizations and exhibiting a high degree of within- and among-season site fidelity. We rarely observed males flocking with other male Golden-winged Warblers, and there was little overlap of neighboring male home ranges. In contrast, female home ranges overlapped extensively with neighboring male home ranges. Home-range sizes did not differ between sexes but were larger in Costa Rica (8.77 ± 0.92 ha) than in Nicaragua (4.09 ± 1.30 ha). Home ranges were larger than reports of most other migratory parulids, and we hypothesize that large home-range size and high propensity to join mixed-species flocks result from the species' specialized foraging behaviors. The predominant foraging behavior involved probing hanging dead leaves and epiphytes for arthropods. Although this foraging strategy can be highly effective, it is noisy and reduces vigilance, which may explain the propensity for joining mixed-species flocks because group living can reduce predation risk. Our results indicate that the nonbreeding season behaviors of Golden-winged Warbler have important conservation implications because mixed-species flocks can be disrupted by habitat loss and fragmentation, and because specialized foraging requirements, large home ranges, and territorial behavior reduce the potential density at which the species can occur.

Key Words: behavioral ecology, mixed-species flocks, site fidelity, social system.

Little is known about the social and foraging behaviors of long-distance migratory birds during the nonbreeding season, yet these behaviors have important conservation implications because they affect space use, energy expenditure, susceptibility to predation, habitat

* Chandler, R. B., S. Tolfree, J. Gerwin, C. Smalling, L. Chavarría-Duriaux, G. Duriaux, and D. I. King. 2016. Conservation implications of Golden-winged Warbler social and foraging behaviors during the nonbreeding season. Pp. 175–192 in H. M. Streby, D. E. Andersen, and D. A. Buehler (editors). Golden-winged Warbler ecology, conservation, and habitat management. Studies in Avian Biology (no. 49), CRC Press, Boca Raton, FL.

requirements, and hence population density and survival (Crook 1970, Rappole and Morton 1985, Sutherland 1998, Rappole et al. 2003, Morton and Stutchbury 2005). A lack of information on basic behavioral ecology limits efforts to conserve these species, many of which are declining, possibly due to events occurring on the nonbreeding grounds (Rappole et al. 2003, Sherry et al. 2005, King et al. 2006, Sauer et al. 2008, Calvert et al. 2009).

Greenberg and Salewski (2005) summarized the literature on Neotropical–Nearctic migrant social systems during the nonbreeding season and classified species according to the following aspects of sociality: regional movements, local tenacity, territoriality, group size, and tendency to occur in mixed-species flocks. High variability exists among species with respect to each of these components, and within species, variation exists among populations and between sexes. Important factors thought to explain variation in social systems include foraging behavior and predator avoidance behavior (Buskirk 1976, Pulliam and Millikan 1982).

Research on the influences of foraging behavior on sociality has revealed several common patterns in migratory birds during the nonbreeding season. In general, most studies support hypotheses predicting that species exploiting rapidly renewing, defensible resources should have smaller home ranges and be more territorial than species exploiting ephemeral resources (Brown 1969, Pulliam and Millikan 1982). For example, many frugivorous migrants such as Swainson's Thrushes (Catharus ustulatus) and Eastern Kingbirds (Tyrannus tyrannus) move over large regions in search of preferred fruit (Rappole and Morton 1985). In contrast, many insectivorous migrants such as Black-throated Blue Warblers (Setophaga caerulescens) and Ovenbirds (Seiurus aurocapilla) maintain small territories within seasons and exhibit high site fidelity among seasons (Rappole and Warner 1980, Wunderle 1992, Sherry and Holmes 1996). Several frugivores also maintain territories (Latta and Faaborg 2002; Brown and Sherry 2008; Townsend et al. 2010, 2012), which suggests that social behavior is influenced less by resource type than it is by the abundance and temporal variability of resources. In addition, most published evidence supports theory predicting that species exploiting pulsed resources should occur in larger groups than species that

rely on stable resources (Zahavi 1971, Greenberg and Salewski 2005).

Predation risk may be at least as important as resource exploitation strategies in influencing group size and propensity to occur in mixed-species flocks (Morse 1977, King and Rappole 2000). Survival probability can be lower for solitary individuals than for individuals in flocks (Page and Whitacre 1975), and it has been noted that flock formation is rare on islands lacking predators (Willis 1972). However, the importance of predation pressure does not diminish other benefits of flocking such as increased food intake via social enhancement (Krebs et al. 1972, Sridhar et al. 2012), and it seems likely that many migrant birds that participate in mixed-species flocks do so for multiple reasons.

Golden-winged Warblers (Vermivora chrysoptera) are one of the most rapidly declining Neotropical–Nearctic migrants (Chapter 1, this volume), yet little is known about its nonbreeding season ecology. Recent studies suggest that Golden-winged Warblers are patchily distributed and occur at low densities throughout their nonbreeding range (Chapter 1, this volume). Golden-winged Warblers can occur in multiple forest types, but they are strongly associated with specific microhabitat characteristics such as hanging dead leaves and epiphytes (Chandler and King 2011). Prior anecdotal observations suggested that Golden-winged Warblers forage by probing these substrates while in mixed-species flocks (Buskirk et al. 1972, Morton 1980, Tramer and Kemp 1980); however, until now no quantitative studies of Golden-winged Warbler nonbreeding season foraging and social behaviors have been conducted. Without this information, it is impossible to understand the mechanisms governing spatial variation in density during the nonbreeding season, which is necessary for effective conservation planning (Sutherland 1998). The objectives of this study were to (1) quantify Golden-winged Warbler social system and foraging behavior in terms of within- and among-season site fidelity, territoriality, and mixed-species flock participation; and (2) assess how these behaviors are related to home-range size and overlap, which are important determinants of population density. When possible, we assessed whether males and females differed with respect to each of these behaviors because such differences could influence sexual segregation among habitats (Marra 2000).

METHODS

Field Methods

We studied social and foraging behaviors of Golden-winged Warblers in Costa Rica during three nonbreeding seasons: December 2006–March 2007, October 2007–March 2008, and January–March 2009, and during two nonbreeding seasons in Nicaragua: January–March 2012 and February–March 2013. We used radiotelemetry to collect data on site fidelity, home-range size, and foraging behavior. Telemetry was necessary because Golden-winged Warblers are generally too silent and move too rapidly to track without telemetry in the structurally complex habitats in which they occur, as can also be true on the breeding grounds (Streby et al. 2012; Chapter 5, this volume). We captured individuals using mist nets, broadcast vocalizations, and decoys. We banded each individual with a unique U.S. Geological Survey metal band and two or three unique combinations of color bands. We determined sex and age using plumage characteristics, but we could not reliably age several individuals, and we therefore excluded age from our analyses.

We began tracking each bird one day after attaching a VHF radio transmitter using the backpack harness design of Rappole and Tipton (1991). Most transmitters weighed 0.43 g (Holohil Systems Ltd., Ontario, Canada) with an expected battery life of 21 days, although we used some 0.27-g Holohil units and 0.35-g Blackburn units (Blackburn Transmitters, Nacogdoches, TX) in Nicaragua. All transmitters weighed <5% of body mass, which averaged 8.7 g in both countries. We relocated birds every 1–2 days, and we tracked individuals for 2–4 hr per day. During our first season in Costa Rica, we recorded bird locations with a GPS unit (Garmin Ltd., Olathe, KS) every 30 min, but only when we saw the bird. However, the resulting data did not adequately reflect space use because some birds were difficult to see while in dense vegetation or high in the canopy. Therefore, in the latter two seasons, we recorded locations every 30 min regardless of whether we saw the bird. When we did not see birds, we determined approximate locations by triangulation or signal strength, which we calibrated from visual observations. We stayed >5 m from birds in an effort to avoid influencing their behavior. In Nicaragua, we initially recorded locations every hour, but after the first four birds removed their transmitters after four days, we recorded locations ≥20 min apart. At both study sites, we continued tracking each individual until battery failure or bird mortality. The only mortality event we observed in Costa Rica involved depredation by a striped palm pit viper (*Bothriechis lateralis*) that consumed a bird with an active transmitter. In Nicaragua, we recovered one dead individual, but the cause of mortality was unknown.

Home-Range Size and Overlap

We estimated home-range size using bivariate normal kernel density estimators (Worton 1989). Kernel density estimators yield utilization distributions, which are the relative probability of an individual occurring at each location in its home range. We characterized home-range size using 50% and 95% intensity levels. We regarded 50% kernels as core areas (Townsend et al. 2010). We excluded individuals for which we recorded <15 locations because it was not possible to estimate the kernel bandwidth for those individuals.

As a measure of territoriality and social tolerance, we computed the overlap of 50% kernel home ranges using the volume intersection index described in Fieberg and Kochanny (2005) and implemented in the R package adehabitat (Calenge 2006). For 50% kernels, the index ranges from 0.0 (no overlap) to 0.5 (complete overlap). We used 2 × 2 factorial ANOVAs to test if home-range size differed between sexes and between Costa Rica and Nicaragua. We used a similar approach to test for differences in home-range overlap between male–male and male–female neighbors.

Social System

We assessed site fidelity using our telemetry data and by monitoring color-banded individuals over multiple nonbreeding seasons. We considered an individual to exhibit high within-season site fidelity if it maintained a home range during the course of tracking and resighting. We considered individuals located during multiple seasons to exhibit high among-season site fidelity if they were located ≤100 m from their previous home range. In Costa Rica, we searched for birds that had been color banded in previous years by making monthly visits to three locations within each home range, and we broadcast Golden-winged Warbler songs and chip notes for 30 min or until we encountered the marked bird. We did not conduct resighting efforts in Nicaragua.

We assessed territoriality by observing the response of birds to broadcast vocalizations and decoys (clay mounts) of conspecifics used to capture birds. Birds that attacked the mount were considered to be territorial. Other behaviors in response to call broadcast and decoys that we considered to reflect territoriality, as opposed to mere curiosity, included chipping and singing, rapid position switching, and feather raising (Rappole and Warner 1980). Chipping and singing were never heard without the call broadcast stimuli.

In Costa Rica, we recorded data on mixed-species flock participation and composition at 30-min intervals while radio-tracking. We classified Golden-winged Warblers as flocking, not flocking, or associating with mixed-species flocks. Following Hutto (1987), we classified individuals as flocking if they were ≤25 m from groups of other species and moving in the same direction. We classified Golden-winged Warblers as "associating" with mixed-species flocks if they occurred ≤25 m from groups of other species but were not moving with the flock. For instance, we occasionally observed Golden-winged Warblers associating with large flocks of frugivores that spent >20 min in a single tree. We collected mixed-species flock composition data continuously during each tracking period, and we compiled a list of all species that were observed flocking with Golden-winged Warblers.

Foraging Behavior

We recorded foraging observations exclusively on radiomarked individuals in Costa Rica and Nicaragua. In Costa Rica, we recorded data on the first foraging maneuver we observed during each 30-min interval. Foraging data included the height that the bird was above the ground; the height of the tree; the foraging maneuver, including glean, probe, sally, hawk, or flush; and the substrate on which the maneuver occurred: open live leaf, rolled live leaf, dead leaf, flower, bark, moss or lichen, bromeliad, or miscellaneous epiphyte. We were unable to adequately measure foraging rate because it was difficult to watch radiomarked birds for prolonged periods due to their rapid movements within dense vegetation. In Nicaragua, we recorded the height that the bird was above the ground but not the other foraging variables.

To determine if foraging behavior differed between sexes or among the three flocking states, we used mixed-effects models that allowed for inference about variation within and among individuals. We modeled height at which birds foraged and tree height as normally distributed response variables and treated variation among individuals as a random effect. Models were equivalent to two-way ANOVA models with an additional error term for random variation among individuals. Foraging maneuver is a categorical variable, but because 99% of observations were probes or gleans, we modeled maneuver as a binomial response variable. Models were fit in Program R (ver. 3.0.1; R Development Core Team 2013) using the lme4 package (Bates et al. 2013). We tested the effects of sex and flocking state using likelihood ratio tests applied to models with and without each effect. We used a two-sample t-test to test for differences between sexes.

RESULTS

Home-Range Size and Overlap

We radio-tracked a total of 39 Golden-winged Warblers in Costa Rica (n = 26) and Nicaragua (n = 13, Figure 11.1). In Costa Rica, the battery life of transmitters ranged from two to 26 days (median = 12 days). Premature battery failure prevented us from acquiring >15 location points to calculate kernel-based utilization distributions for six individuals. Of the 20 individuals with a sufficient number of locations for home-range analysis, 17 were males and three were females, and we recorded an average of 31.6 locations for each of these individuals. In Nicaragua, 10 of the 13 radiomarked individuals had sufficient data for analysis, and we recorded an average of 30.1 locations for each of eight males and two females.

Home-range size was similar between sexes in both countries (Figure 11.2); however, our sample size was small, with only five females included in the analysis. Estimates of 50% home-range size were 0.44–2.52 ha, and we found no effect of sex ($F_{1,26}$ = 0.49, P = 0.49) or of the interaction of sex and country ($F_{1,26}$ = 0.041, P = 0.84). However, there was an effect of country alone ($F_{1,26}$ = 8.78, P < 0.01), with home-range sizes in Nicaragua averaging less than half of those in Costa Rica (Table 11.1).

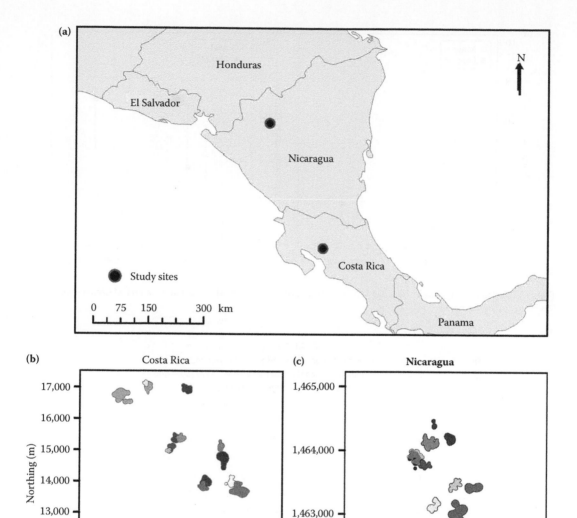

Figure 11.1. The two study sites (a), and the distribution of Golden-winged Warbler home ranges during the nonbreeding season in Costa Rica (b) and Nicaragua (c).

We obtained similar results for 95% home-range sizes with an area of 2.3–19.5 ha in Costa Rica and 2.48–10.82 ha in Nicaragua (Table 11.1).

Home ranges of neighboring males did not overlap extensively (Figure 11.3). The overlap of 50% core areas was higher for male–female neighbors than for male–male neighbors. Even with a sample of only three male–female neighbors and two male–male neighbors in Costa Rica, home-range overlap was different between the two groups ($t_2 = 3.78$,

$P = 0.031$). For the male–male neighbors, overlap occurred only in the outer extremes of the home range, and not in the core areas. We found similar results in Nicaragua. Core areas did not overlap in one set of neighboring males, but as in Costa Rica, male–female home-range overlap was extensive, ranging from 21.7% to 51.3% (Figure 11.3). In Nicaragua, we observed one instance of female–female home-range overlap, but one of the individuals was not radiomarked. Relying on resighting data

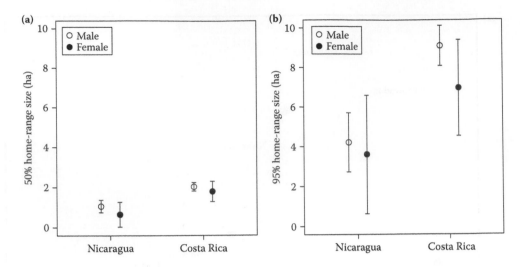

Figure 11.2. Home-range size (±1 SE) for 50% contour (a) and 95% contour (b) by sex and country of Golden-winged Warblers during the nonbreeding season.

TABLE 11.1

Summary statistics (mean, standard deviation, minimum, and maximum) of Golden-winged Warbler home-range sizes (50% and 95% kernel home-range percentiles).

Country	Percentile (%)	Mean	SD	Min	Max
Costa Rica	50	1.99	0.95	0.44	4.00
	95	8.77	4.69	2.31	19.50
Nicaragua	50	0.98	0.60	0.49	2.52
	95	4.09	2.48	2.22	10.82

Twenty individuals were tracked during three nonbreeding seasons 2006–2009 in Costa Rica, and 10 individuals were tracked during two seasons in Nicaragua. Home-range size units are hectares.

from a nonradiomarked individual for home-range delineation is problematic due to potentially low detection probability, but we include the overlap information in Figure 11.3 as anecdotal information.

Site Fidelity and Territoriality

In both Costa Rica and Nicaragua, male and female Golden-winged Warblers maintained home ranges characterized by one or two core areas where most activity was concentrated (Figure 11.4). In some instances, activity centers shifted slightly among days, with extensive inter-day overlap (Figure 11.5). Patterns were consistent for birds tracked both early and late in the season, suggesting that Golden-winged Warblers exhibit high site fidelity throughout the entire season.

In Costa Rica, three Golden-winged Warblers exhibited movement patterns inconsistent with the general patterns described earlier. We relocated one individual, a male, the day following capture and then never saw it again despite three days of searching the surrounding area using radiotelemetry and broadcast vocalizations. Probability of detection with playback is extremely high (0.97; Chandler and King 2011), and it is unlikely that this individual remained within the study area. If the bird moved to another location within the study area, the radio must have been defective because we were able to detect transmitter signals at distances >1 km and the entire study area was included within that range. Alternatively, the individual may have been depredated and the transmitter destroyed.

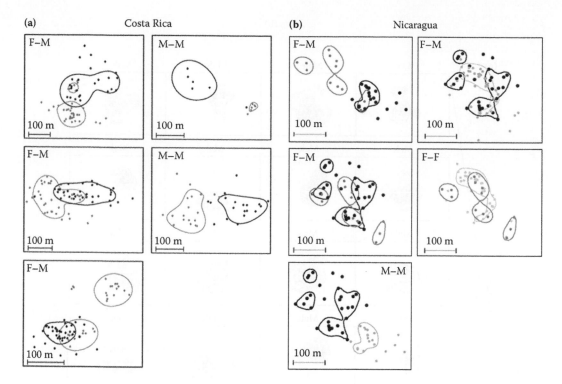

Figure 11.3. Overlap of 50% home ranges for neighboring Golden-winged Warblers during the nonbreeding season in Costa Rica and Nicaragua. The sex of members of neighboring pairs is indicated by F–F, M–F, or M–M. Only neighbors that were radio-tracked simultaneously and had ≥5 recorded locations are shown.

Two other males made brief long-distance forays. In each case, the bird moved from patches of secondary forest to contiguous forest ~2 km away and then returned to its home range within 24 hr. Foray locations were not included in home-range size calculations. In Nicaragua, three Golden-winged Warblers exhibited similarly uncharacteristic movement patterns. One male traveled 369 m and one female 478 m but returned to their respective home ranges within 24 hr. The other male may have been an early passage migrant or a floater that covered an area much larger than the size of an average home range (Brown and Long 2007). The male was captured at the end of March and frequently moved >150 m between sequential locations and was often difficult to locate. After two days of tracking, we were no longer able to detect the signal from that individual's transmitter in the study area.

Systematic visits to home ranges of color-banded birds throughout all three field seasons in Costa Rica indicated that all relocated individuals remained on their home ranges until the onset of migration. Furthermore, we found all five individuals that we were able to relocate in subsequent years ≤200 m from their original capture location, including three individuals that we observed during three consecutive nonbreeding seasons. During our two field seasons in Nicaragua, we encountered four of 28 individuals that had been banded prior to our study. Two of the radiomarked males in Nicaragua were originally banded two and four years before, and we recaptured them ≤100 m from their initial capture locations.

Both male and female Golden-winged Warblers showed aggressive responses to broadcast vocalizations and decoys. In Costa Rica, we captured 23 of 26 birds as part of our radiotelemetry study using these stimuli. Each of these individuals flew at the decoy, occasionally making direct contact. We caught the other three individuals (two males and one female) during constant-effort mist netting as part of a separate study (Chandler et al. 2013). Warblers captured without stimuli in systematic mist netting also maintained home ranges suggesting that our sample was not biased toward territorial individuals. In Nicaragua, we captured all birds using broadcast vocalizations, often without decoys.

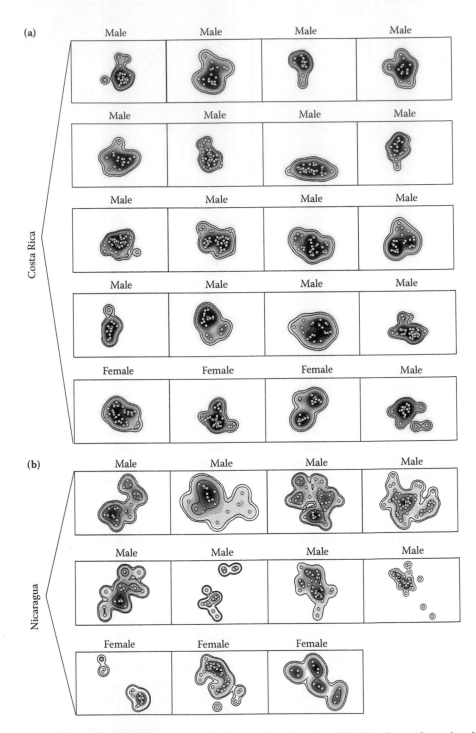

Figure 11.4. Kernel utilization distributions and location points for Golden-winged Warblers during the nonbreeding season in (a) Costa Rica and (b) Nicaragua. Contour lines represent kernel home-range percentiles.

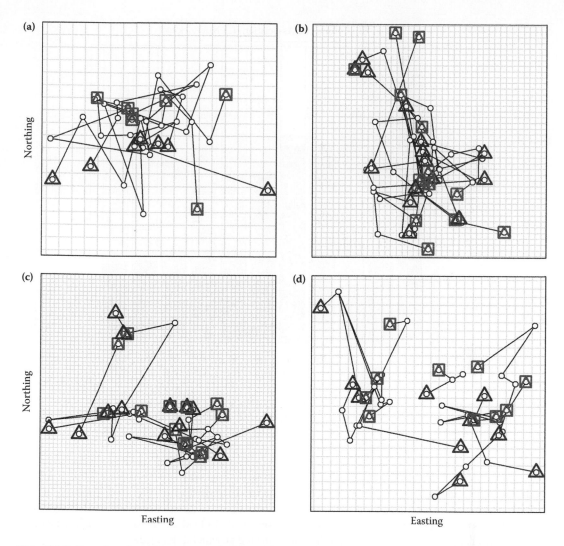

Figure 11.5. Examples of movement patterns for four Golden-winged Warblers (a–d) during the nonbreeding season in Costa Rica. Each tracking episode is represented by segments starting from a triangle and ending at a triangle within a square. Segments represent 30-min intervals. A 10-m grid is provided for scale.

Mixed-Species Flock Participation

We collected mixed-species flock data for 26 Golden-winged Warblers observed on 214 occasions totaling 562 hr. These individuals spent an average of 59% of their time with mixed-species flocks (Figure 11.6), which typically included Common Bush-Tanagers (*Chlorospingus ophthalmicus*) as the nuclear species (Table 11.2). Marked Golden-winged Warblers spent an additional 26% of their time associated with other species in loose flocks without obvious movement cohesion or nuclear species. Thus, we only observed marked Golden-winged Warblers away from flocks 15% of the time. We observed 88 species flocking with Golden-winged Warblers in cohesive flocks. No species was ubiquitously present in these flocks, and both resident and migratory species were common participants (Table 11.2). Flock participation was not related to sex of warblers ($t_9 = 0.148$, P = 0.89), although only three females were included in our sample.

Foraging Behavior

Of 293 foraging maneuvers we recorded for 24 color-banded Golden-winged Warblers, 72% were

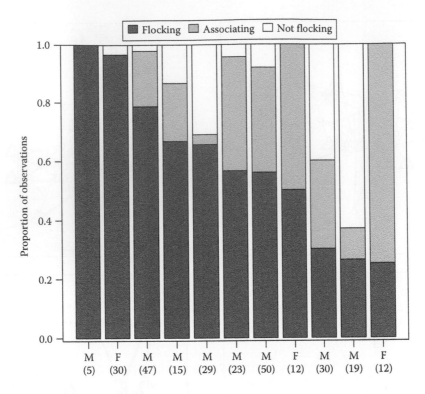

Figure 11.6. Proportion of observations in which 11 individual Golden-winged Warblers were flocking in Costa Rica during the nonbreeding season. Associating refers to cases where an individual occurred with other species but was not moving together in concert. Individuals are ranked by propensity for flocking and referenced by sex (M, male; F, female) with sample size of observations in parentheses.

probes and 27% were gleans (Figure 11.7). Sallies, hawks, hover-gleans, and flush-dives constituted <2% of observations. The most commonly probed substrate was hanging dead leaves (40%), although moss, bark, rolled leaves, bromeliads, and flowers were used to a lesser extent (Figure 11.7). Golden-winged Warblers probe in a specialized fashion that involves inserting the beak and opening it to pry open the leaf or flake off bark. Most of the bark foraging we observed occurred in *Psidium guajava*, which like many other members of Myrtaceae has thin flaking bark. Golden-winged Warblers were the only bird species observed using this resource. The longest foraging maneuvers (>1 min) occurred on individual *Cecropia* sp. leaves. Although *Cecropia* sp. was never a dominant plant species in home ranges of marked Golden-winged Warblers, the large leaves with hooked petioles are easily caught in the canopy. The leaves form tight curls upon desiccation and often host diverse arthropod assemblages (Rosenberg 1997). The only marked warbler that regularly foraged ≥20 m above the

ground almost exclusively used *Ocotea* sp. It was not possible to closely observe the foraging behavior of this individual due to the height above the ground at which it foraged.

We found no evidence that any of the foraging variables, including foraging height, tree height, and maneuver differed between sexes or among the three flocking states of flocking, associating, or solitary in Costa Rica (all P-values from mixed effects models >0.05). However, in Nicaragua male foraging height (13.1 ± 1.5 m) was higher than female foraging height (4.2 ± 2.4 m; $t_9 = 3.19$, P = 0.01). In addition to differences between sexes, we found substantial variation in foraging heights within and among individuals in Costa Rica and Nicaragua (Figure 11.8).

DISCUSSION

We documented the aspects of behavioral ecology during the nonbreeding season, which have important conservation implications for Golden-winged

TABLE 11.2

Co-occurrence probabilities for bird species observed flocking with 26 Golden-winged Warblers on >5% of observation days in Costa Rica.

Species	Resident/ migrant	Co-occurrence probability
Myioborus miniatus	R	0.37
Chlorospingus ophthalmicus	R	0.36
Cardellina pusilla	M	0.36
Setophaga virens	M	0.36
Setophaga pensylvanica	M	0.35
Mniotilta varia	M	0.34
Oreothlypis peregrina	M	0.29
Tangara icterocephala	R	0.18
Vireo philadelphicus	M	0.16
Myiarchus tuberculifer	R	0.16
Mionectes olivacea	R	0.14
Vireo flavifrons	M	0.12
Basileuterus culcivorous	R	0.10
Saltator maximus	R	0.09
Hylophilus decurtatus	R	0.08
Turdus grayi	R	0.08
Basileuterus tristriatus	R	0.08
Elaenia frantzii	R	0.07
Basileuterus rufifrons	R	0.07
Ramphocelus passerinii	R	0.07
Thraupis episcopus	R	0.06
Xiphorhynchus erythropygius	R	0.06
Euphonia hirundinacea	R	0.06
Phlogothraupis sanguinolenta	R	0.05
Piranga rubra	M	0.05
Premnoplex brunnescens	R	0.05

Probabilities represent the mean proportion of telemetry occasions during which a species was observed in flocks with radiomarked Golden-winged Warblers. Anecdotal observations from Nicaragua suggest that flocks there contained similar species.

Warblers. Golden-winged Warblers appear to have larger nonbreeding season area requirements than other long-distance migratory passerines, which may limit their ability to persist in small forest fragments. In Costa Rica, home-range size averaged 8.8 ha, almost 10 times larger than the average of 0.78 ha for Ovenbirds studied by Brown and Sherry (2008). Similarly, Rappole and Warner (1980) reported home-range sizes <1 ha

for 10 species of wintering long-distance migrants. Wood Thrushes (*Hylocichla mustelina*) and Bicknell's Thrushes (*Catharus bicknelli*), both larger-bodied migrant species, maintained smaller nonbreeding season territories (0.44 and 1.41 ha, respectively) than those of Golden-winged Warblers in either Costa Rica or Nicaragua (Winker et al. 1990, Townsend et al. 2010). Golden-winged Warbler home ranges were smaller in Nicaragua than in Costa Rica, and additional research is needed to determine the causes of geographic variation in home-range size.

In addition to large area requirements, male Golden-winged Warblers appeared to be highly territorial during the nonbreeding season. We rarely observed more than one male in mixed-species flocks, and neighboring male home ranges had little overlap. Males also responded aggressively to playback and decoys. Large home-range size and territorial behavior may explain why Golden-winged Warblers are not reported to be abundant anywhere throughout their known nonbreeding range (Johnson 1980, Morton 1980, Orejuela et al. 1980, Powell et al. 1992, Komar 1998, Blake and Loiselle 2000; Chapter 1, this volume).

Territoriality may also affect how Golden-winged Warblers respond to habitat loss because limited habitat can lead to competitive interactions resulting in losers that cannot defend territories. For Ovenbirds in Jamaica, where predation pressure is low, costs and benefits appear to be associated with territorial and nonterritorial social systems (Brown and Sherry 2008). Territorial individuals can access stable resources and minimize space use and energy expenditure, whereas nonterritorial individuals can exploit temporary resources more effectively. Wood Thrushes, in contrast, conform to an ideal despotic population model in which territory owners occur in primary forest whereas floaters occur in lower quality areas (Fretwell and Lucas 1969, Winker et al. 1990). Floaters wander over large areas and incur higher mortality costs (Rappole et al. 1989). We encountered only one Golden-winged Warbler that could have been described as a floater, suggesting that few individuals adopt this strategy at our study sites.

We found some evidence that tolerance of sharing space was higher between sexes than within sexes. Home-range overlap was higher for male–female neighbors than for male–male neighbors. Generally, within zones of

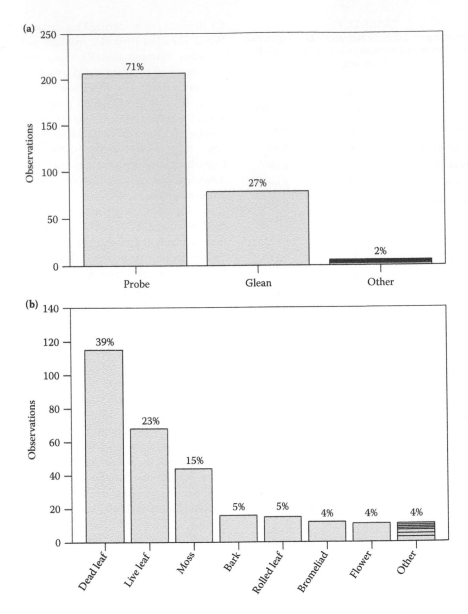

Figure 11.7. Bar plots of foraging maneuvers (a) and foraging substrates (b) of Golden-winged Warblers during the non-breeding season. Data are from Costa Rica, 2006–2009.

home-range overlap, male–female neighbors did not occur near each other, although one male in Costa Rica consistently foraged ≤5m from an unbanded female without displaying aggressive behaviors. In one area in Nicaragua, we identified four Golden-winged Warblers (two males and two females) with varying levels of home-range overlap. We resighted one female, which was not radiomarked, multiple times, typically ≤10 m from one of the other radiomarked individuals.

We found no evidence of sexual differences in the foraging behavior of Golden-winged Warblers in Costa Rica. However, males tended to forage higher above the ground than females in Nicaragua. With data on only five females in Costa Rica and two females in Nicaragua, a larger sample is necessary before conclusive statements can be made regarding sex-specific foraging behavior. Furthermore, even with the use of transmitters, it was difficult to record behavioral observations when Golden-winged Warblers were

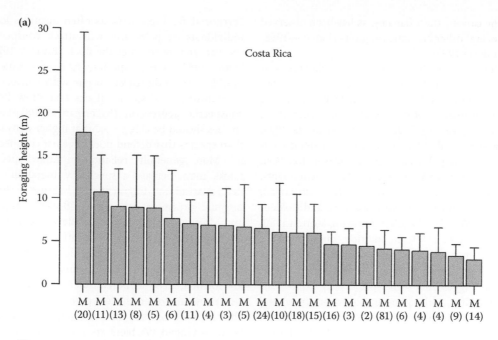

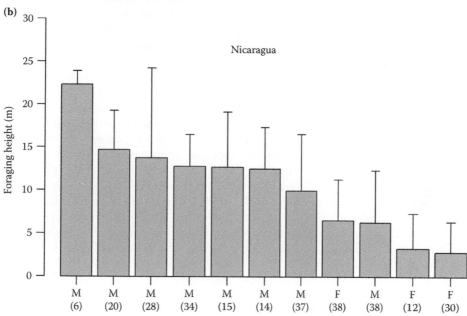

Figure 11.8. Mean foraging heights (+1 SE) for (a) 20 male and three female Golden-winged Warblers in Costa Rica and (b) eight males and three females in Nicaragua monitored using radiotelemetry during the nonbreeding season. Sample size and sex are indicated below each individual's bar.

high in the canopy or in low thickets. Low detectability may have introduced bias because two of the three females we studied in Costa Rica and all three of the females we studied in Nicaragua used dense understory vegetation, and we were only able to record foraging behavior when birds emerged from thickets. Future work could overcome this problem by estimating the probability of detecting individuals in different types of cover as a function of the height of foraging locations in the canopy. Corrections for detectability would shed light on whether males tend to forage higher in the canopy than females, as has been observed in several other Nearctic migrants (Rappole 1988, Wunderle 1992).

Our finding that male–female tolerance was higher than male–male tolerance differs from studies demonstrating sexual habitat segregation and dominance (Morton et al. 1987, Marra 2000, Latta and Faaborg 2002, Studds and Marra 2005, Townsend et al. 2012). However, male–female tolerance during the nonbreeding season has been observed for many Neotropical–Nearctic migrants, including Prothonotary Warblers (*Protonotaria citrea*), Canada Warblers (*Cardellina canadensis*), and Golden-cheeked Warblers (*Setophaga chrysoparia*; Morton 1980, Rappole et al. 1999). In addition, observations of pairs of Philadelphia Vireos (*Vireo philadelphicus*), Blue-headed Vireos (*Vireo solitarius*), and Gray Vireos (*Vireo vicinior*) suggest that some species may exhibit pair territoriality, as do Stonechats (*Saxicola torquata*) and White Wagtails (*Motacilla alba*; Zahavi 1971, Tramer and Kemp 1982, Gwinner et al. 1994, E. Morton, unpubl. data). In other species, males and females may be randomly distributed within a forest type (Brown and Sherry 2008), or may have horizontally overlapping home ranges, but stratify vertically (Rappole 1988, Wunderle 1992). The reason for higher intersexual tolerance in Golden-winged Warblers is unclear, but we found some evidence that males foraged higher above the ground than females in Nicaragua. However, given our small sample sizes, more research is needed to assess the possibility of vertical resource partitioning. The primary conservation implication of the absence of sexual habitat segregation is that differential rates of habitat loss would not lead to biased sex ratios, an important concern with species such as American Redstarts (*Setophaga ruticilla*) and Bicknell's Thrushes (Marra 2000, Townsend et al. 2012). In spite of high tolerance and overlapping home ranges, an apparent sex ratio bias was observed within our Costa Rica study area (Chandler and King 2011), indicating that males and females may segregate geographically.

Golden-winged Warblers occurred in mixed-species flocks 85% the time and were highly territorial. Many Neotropical resident species that regularly join mixed-species flocks are also territorial and may either defend the flock against conspecifics (Munn and Terborgh 1979) or have distinct territory boundaries and drop out of the flock when these boundaries are crossed (Powell 1979, Munn 1985, King and Rappole 2001).

Territorial flock participants often occur as lone individuals or pairs and will attack conspecifics that attempt to enter the flock (Buskirk 1976, Hutto 1987). Other species, such as Cerulean Warblers (*Setophaga cerulea*), occur with conspecifics within mixed-species flocks and show little conspecific aggression (Bakermans 2008). Such species should be able to occur at higher densities than species that defend flocks against conspecifics. More generally, reliance on mixed-species flocks may increase a species' vulnerability to deforestation and fragmentation because anthropogenic processes can disrupt flocks (Rappole and Morton 1985, Stouffer and Bierregaard 1995, Stratford and Stouffer 1999, Stouffer et al. 2006). Dependence on mixed-species flocks might also suggest that nuclear species, around which flocks are formed, need to be considered in conservation plans; however, the low co-occurrence probabilities we observed in our study suggest that Golden-winged Warblers are not reliant on particular nuclear species.

Our results support the hypothesis that social systems develop as an outcome of resource availability and foraging behavior. Golden-winged Warblers exhibit a specialized foraging strategy during the nonbreeding season in which they primarily probe hanging dead leaves and epiphytes. A gleaning or probing foraging strategy is shared by many resident Neotropical species from several families (Capitonidae, Formicaridae, Furnariidae, Troglodytidae) and some Neotropical–Nearctic migrants (Morton 1980, Remsen and Parker 1984, Greenberg 1987, Rosenberg 1993). Dead leaves provide habitat for numerous large-bodied arthropods, especially roaches (Blattaria), spiders (Araneae), and Orthopterans (Gradwohl and Greenberg 1982, Rosenberg 1993, R.B. Chandler, unpubl. data). Prey density and biomass can be much higher in dead leaves than in live leaves, due to the larger body size and different taxonomic compositions among arthropods (fewer Hymenopterans and Dipterans, Rosenberg 1997). Accessing these food resources, however, requires behaviors and morphological traits that nonspecialized species do not possess (Rosenberg 1993). For example, many dead-leaf foragers, including Golden-winged Warblers, lack rictal bristles that could interfere with probing.

The benefits of accessing abundant food resources in dead leaves are associated with two important costs. First, dead leaves are much less abundant than live leaves and are patchily distributed (Remsen and Parker 1984). Dead-leaf foragers might therefore be required to travel farther than live-leaf foragers, which may partially explain the large home ranges of Golden-winged Warblers during the nonbreeding season. Second, dead-leaf foraging may increase predation risk because it is a noisy process and precludes vigilance because the entire head of a foraging bird is often inside a curled leaf (Morton 1980). The predator avoidance benefits of mixed-species flocks might therefore explain why virtually all regular dead-leaf foragers participate in mixed-species flocks (Remsen and Parker 1984, Rosenberg 1997).

The reliance on a high-quality, patchily distributed food resource may also explain Golden-winged Warbler territoriality during the nonbreeding season. Arthropod populations in dead leaves can be quickly diminished by avian insectivores, but colonization rate is also high (Gradwohl and Greenberg 1982, Rosenberg 1993). Therefore, successfully defending an area with many dead-leaf clusters could ensure an adequate food supply throughout the nonbreeding season. In accordance with this hypothesis, most dead-leaf-foraging resident species occur as single individuals or pairs in mixed-species foraging flocks and actively defend territories against conspecifics during the nonbreeding season (Munn and Terborgh 1979, Powell 1979).

Our results suggest that the distinctive social and foraging behaviors of Golden-winged Warblers explain why the species exhibits a patchy distribution and low density during the nonbreeding season. These behaviors may also make Golden-winged Warblers vulnerable to habitat loss and fragmentation because they have relatively large area requirements and depend on flocks, which are more common in contiguous forest. However, Golden-winged Warblers use forest fragments and advanced secondary forests that contain vine tangles and hanging dead leaves (Chandler and King 2011). Future research is needed to compare nonbreeding season behavior and survival between fragmented and contiguous forests to assess the quality of these landscapes. Direct energetic measurements and their influences on body condition would also be helpful in identifying suitable areas for targeted conservation and management.

ACKNOWLEDGMENTS

We thank the many landowners in Costa Rica who allowed us to work on their properties. C. Chandler helped design and implement the study in Costa Rica. J. Wolfe, S. Beaudreault, N. Hazlet, A. Anderson, J. Ritterson, J. Wells, M. Gonzales, C. Delgado, O. Rodriguez, M. Siles, and M. Mietzelfeld provided excellent assistance in the field. Comments by two anonymous reviewers and the volume editors led to substantial improvements in our manuscript.

LITERATURE CITED

Bakermans, M. H. 2008. Demography and habitat use of Cerulean Warblers on breeding and wintering grounds. Ph.D. dissertation, Ohio State University, Columbus, OH.

Bates, D., M. Maechler, B. Bolker, and S. Walker. 2013. Lme4: linear mixed-effects models using Eigen and S4, R package version 1.0-5. <http://CRAN.R-project.org/package=lme4> (1 February 2013).

Blake, J. G., and B. A. Loiselle. 2000. Diversity of birds along an elevational gradient in the Cordillera Central, Costa Rica. Auk 117:663–686.

Brown, D. R., and J. A. Long. 2007. What is a winter floater? Causes, consequences, and implications for habitat selection. Condor 109:548–565.

Brown, D. R., and T. W. Sherry. 2008. Alternative strategies of space use and response to resource change in a wintering migrant songbird. Behavioral Ecology 19:1314–1325.

Brown, J. L. 1969. Territorial behavior and population regulation in birds: a review and re-evaluation. Wilson Bulletin 81:293–329.

Buskirk, W. H. 1976. Social systems in a tropical forest avifauna. American Naturalist 110:293–310.

Buskirk, W. H., G. V. N. Powell, J. F. Wittenberger, R. E. Buskirk, and T. U. Powell. 1972. Interspecific bird flocks in tropical highland Panama. Auk 89:612–624.

Calenge, C. 2006. The package adehabitat for the R software: a tool for the analysis of space and habitat use by animals. Ecological Modelling 197:516–519.

Calvert, A. M., S. J. Walde, and P. D. Taylor. 2009. Nonbreeding-season drivers of population dynamics in seasonal migrants: conservation parallels across taxa. Avian Conservation and Ecology 4:5.

Chandler, R. B., and D. I. King. 2011. Habitat quality and habitat selection of Golden-winged Warblers in Costa Rica: an application of hierarchical models for open populations. Journal of Applied Ecology 48:1038–1047.

Chandler, R. B., D. I. King, C. C. Chandler, R. Raudales, R. Trubey, and V. J. Arce Chavez. 2013. A small-scale land-sparing approach to conserving biodiversity in tropical agricultural landscapes. Conservation Biology 27:785–795.

Crook, J. H. 1970. Social organization and the environment: aspects of contemporary social ethology. Animal Behavior 18:197–209.

Fieberg, J., and C. O. Kochanny. 2005. Quantifying home-range overlap: the importance of the utilization distribution. Journal of Wildlife Management 69:1346–1359.

Fretwell, S. D., and H. L. Lucas. 1969. On territorial behavior and other factors influencing habitat distribution in birds. Acta Biotheoretica 19:16–36.

Gradwohl, J., and R. Greenberg. 1982. The effect of a single species of avian predator on the arthropods of aerial leaf litter. Ecology 63:581–583.

Greenberg, R. 1987. Seasonal foraging specialization in the Worm-eating Warbler. Condor 89:158–168.

Greenberg, R., and V. Salewski. 2005. Ecological correlates of wintering social systems in New World and Old World migratory passerines. Pp 336–358 in R. Greenberg and P. P. Marra (editors), Birds of two worlds: the ecology and evolution of migration. John Hopkins University Press, Baltimore, MD.

Gwinner, E., T. Rödl, and H. Schwabl. 1994. Pair territoriality of wintering Stonechats: behaviour, function and hormones. Behavioral Ecology and Sociobiology 34:321–327.

Hutto, R. L. 1987. A description of mixed-species insectivorous bird flocks in western Mexico. Condor 89:282–292.

Johnson, T. B. 1980. Resident and North American migrant bird interactions in the Santa Marta highlands, northern Colombia. Pp. 239–247 in A. Keast and E. S. Morton (editors), Migrant birds in the Neotropics: ecology, behavior, distribution, and conservation. Smithsonian Institution Press, Washington, DC.

King D. I., and J. H. Rappole. 2000. Winter flocking of insectivorous birds in montane pine-oak forests in Middle America. Condor 102:664–672.

King, D. I., and J. H. Rappole. 2001. Mixed-species bird flocks in dipterocarp forests of north-central Burma (Myanmar). Ibis 143:380–390.

King, D. I., J. H. Rappole, and J. P. Buonaccorsi. 2006. Long-term population trends of forest-dwelling Nearctic-Neotropical migrant birds: a question of temporal scale. Bird Populations 7:1–9.

Komar, O. 1998. Avian diversity in El Salvador. Wilson Bulletin 110:511–533.

Krebs, J. R., M. H. MacRoberts, and J. M. Cullen. 1972. Flocking and feeding in the Great Tit Parus major L.—an experimental study. Ibis 114:507–530.

Latta, S. C., and J. Faaborg. 2002. Demographic and population responses of Cape May Warblers wintering in multiple habitats. Ecology 83:2502–2515.

Marra, P. P. 2000. The role of behavioral dominance in structuring patterns of habitat occupancy in a migrant bird during the nonbreeding season. Behavioral Ecology 11:299–308.

Morse, D. H. 1977. Feeding behavior and predator avoidance in heterospecific groups. Bioscience 27:332–339.

Morton, E. S. 1980. Adaptations to seasonal change by migrant land birds in the Panama Canal zone. Pp. 437–453 in A. Keast and E. S. Morton (editors), Migrant birds in the Neotropics: ecology, behavior, distribution, and conservation. Smithsonian Institution Press, Washington, DC.

Morton, E. S., J. F. Lynch, K. Young, and P. Mehlhop. 1987. Do male Hooded Warblers exclude females from nonbreeding territories in tropical forest? Auk 104:133–135.

Morton E. S., and B. J. Stutchbury. 2005. The significance of mating system and nonbreeding behavior to population and forest patch use by migrant birds. Pp. 285–289 in C. J. Ralph and T. D. Rich (editors), Bird conservation implementation and integration in the Americas. U.S. Department of Agriculture, Pacific Southwest Research Station, Albany CA.

Munn, C. A. 1985. Permanent canopy and understory flocks in Amazonia: species composition and population density. Ornithological Monographs 36:683–712.

Munn, C. A., and J. W. Terborgh. 1979. Multi-species territoriality in Neotropical foraging flocks. Condor 81:338–347.

Orejuela, J. E., R. J. Raitt, and H. Alvarez. 1980. Differential use by North American migrants of three types of Colombian forests. Pp. 253–264 in A. Keast and E. S. Morton (editors), Migrant birds in the Neotropics: ecology, behavior, distribution, and conservation. Smithsonian Institution Press, Washington, DC.

Page, G., and D. F. Whitacre. 1975. Raptor predation on wintering shorebirds. Condor 77:73–83.

Powell, G. V. N. 1979. Structure and dynamics of interspecific flocks in a Neotropical mid-elevation forest. Auk 96:375–390.

Powell, G. V. N., J. H. Rappole, and S. A. Sader. 1992. Neotropical migrant landbird use of lowland Atlantic habitats in Costa Rica: a test of remote sensing for identification of habitat. Pp. 287–298 in J. M. Hagan and D. W. Johnson (editors), Ecology and conservation of Neotropical migrant landbirds. Smithsonian Institution Press, Washington, DC.

Pulliam, H. R., and G. C. Millikan. 1982. Social organization in the nonreproductive season. Avian Biology 6:169–197.

R Development Core Team. 2013. R: a language and environment for statistical computing. R Foundation for Statistical Computing, Vienna, Austria. <http://www.R-project.org/>.

Rappole, J. H. 1988. Intra- and intersexual competition in migratory passerine birds during the non-breeding season. Proceedings of the International Ornithological Congress 19: 2308–2317.

Rappole, J. H., D. I. King, and W. C. Barrow Jr. 1999. Winter ecology of the endangered Golden-cheeked Warbler. Condor 101:762–770.

Rappole, J. H., D. I. King, and J. Diez. 2003. Winter- vs. breeding-habitat limitation for an endangered avian migrant. Ecological Applications 13:735–742.

Rappole, J. H., and E. S. Morton. 1985. Effects of habitat alteration on a tropical avian forest community. Ornithological Monographs 36:1013–1021.

Rappole, J. H., M. A. Ramos, and K. Winker. 1989. Wintering Wood Thrush movements and mortality in southern Veracruz. Auk 106:402–410.

Rappole, J. H., and A. R. Tipton. 1991. New harness design for attachment of radio transmitters to small passerines. Journal of Field Ornithology 62:335–337.

Rappole, J. H., and D. W. Warner. 1980. Ecological aspects of migrant bird behavior in Veracruz, Mexico. Pp. 353–393 in A. K. Keast and E. S. Morton (editors), Migrant birds in the Neotropics: ecology, behavior, distribution, and conservation. Smithsonian Institution Press, Washington, DC.

Remsen, J. V., and T. A. Parker. 1984. Arboreal dead-leaf-searching birds of the Neotropics. Condor 86:36–41.

Rosenberg, K. V. 1993. Diet selection in Amazonian antwrens: consequences of substrate specialization. Auk 110:361–375.

Rosenberg, K. V. 1997. Ecology of dead-leaf foraging specialists and their contribution to Amazonian bird diversity. Ornithological Monographs 48:673–700.

Sauer, J. R., J. E. Hines, and J. Fallon. [online]. 2008. The North American Breeding Bird Survey, results and analysis 1966–2007, version 5.15.2008. USGS Patuxent Wildlife Research Center, Laurel, MD. <http://www.mbr-pwrc.usgs.gov/bbs/bbs.html> (1 April 2014).

Sherry, T. W., and R. T. Holmes. 1996. Winter habitat quality, population limitation, and conservation of Neotropical-Nearctic migrant birds. Ecology 77:36–48.

Sherry, T. W., M. D. Johnson, and A. M. Strong. 2005. Does winter food limit populations of migratory birds? Pp. 414–425 in R. Greenberg and P. P. Marra (editors), Birds of two worlds. Smithsonian Institution Press, Washington, DC.

Sridhar, H., U. Srinivasan, R. A. Askins, J. C. Canales-Delgadillo, C. Chen, D. N. Ewert, G. A. Gale, E. Goodale, W. K. Gram, P. J. Hart, K. A. Hobson, R. L. Hutto, S. W. Kotagama, J. L. Knowlton, T. M Lee, C. A. Munn, S. Nimnuan, B. Z. Nizam, G. Péron, V. V. Robin, A. D. Rodewald, P. G. Rodewald, R. L. Thomson, P. Trivedi, S. L. Van Wilgenburg, and K. Shanker. 2012. Positive relationships between association strength and phenotypic similarity characterize the assembly of mixed-species bird flocks worldwide. American Naturalist 180:777–790.

Stouffer, P. C., and R. O. Bierregaard. 1995. Use of Amazonian forest fragments by understory insectivorous birds. Ecology 76:2429–2445.

Stouffer, P. C., R. O. Bierregaard, C. Strong, and T. E. Lovejoy. 2006. Long-term landscape change and bird abundance in Amazonian rainforest fragments. Conservation Biology 20:1212–1223.

Stratford, J. A., and P. C. Stouffer. 1999. Local extinctions of terrestrial insectivorous birds in a fragmented landscape near Manaus, Brazil. Conservation Biology 13:1416–1423.

Streby, H. M., J. P. Loegering, and D. E. Andersen. 2012. Spot-mapping underestimates song-territory size and use of mature forest by breeding Golden-winged Warblers in Minnesota, USA. Wildlife Society Bulletin 36:40–46.

Studds, C. E., and P. P. Marra. 2005. Nonbreeding habitat occupancy and population processes: an upgrade experiment with a migratory bird. Ecology 86:2380–2385.

Sutherland, W. 1998. The importance of behavioural studies in conservation biology. Animal Behaviour 56:801–809.

Townsend, J. M., C. C. Rimmer, and K. P. McFarland. 2010. Winter territoriality and spatial behavior of Bicknell's Thrush (Catharus bicknelli) at two ecologically distinct sites in the Dominican Republic. Auk 127:514–522.

Townsend, J. M., C. C. Rimmer, K. P. McFarland, and J. E. Goetz. 2012. Site-specific variation in food resources, sex ratios, and body condition of an overwintering migrant songbird. Auk 129:683–690.

Tramer, E. J., and T. R. Kemp. 1980. Foraging ecology of migrant and resident warblers and vireos in the highlands of Costa Rica. Pp. 285–296 in A. Keast and E. S. Morton (editors), Migrant birds in the Neotropics: ecology, behavior, distribution, and conservation. Smithsonian Institution Press, Washington, DC.

Tramer, E. J., and T. R. Kemp. 1982. Notes on migrants wintering at Monteverde, Costa Rica. Wilson Bulletin 94:350–354.

Willis, E. O. 1972. Do birds flock in Hawaii, a land without predators? California Birds 3:1–8.

Winker, K., J. H. Rappole, and M. A. Ramos. 1990. Population dynamics of the Wood Thrush in southern Veracruz, Mexico. Condor 92:444–460.

Worton, B. J. 1989. Kernel methods for estimating the utilization distribution in home-range studies. Ecology 70:164–168.

Wunderle, J. M. 1992. Sexual habitat selection in wintering Black-throated Blue Warblers in Puerto Rico. Pp. 299–307 in J. M. Hagan and D. W. Johnston (editors), Ecology and conservation of Neotropical migrant landbirds. Smithsonian Institution Press, Washington, DC.

Zahavi, A. 1971. The social behaviour of the White Wagtail *Motacilla alba alba* wintering in Israel. Ibis 113:203–211.

CHAPTER TWELVE

Golden-winged Warbler Migratory Connectivity Derived from Stable Isotopes*

Keith A. Hobson, Steven L. Van Wilgenburg, Amber M. Roth, Ruth E. Bennett, Nicholas J. Bayly, Liliana Chavarría-Duriaux, Gabriel J. Colorado, Pablo Elizondo, Carlos G. Rengifo, and Jeffrey D. Ritterson

Abstract. Establishing migratory connectivity between breeding and nonbreeding sites of populations of migratory birds is crucial to their effective management, but the use of conventional tracking tags is not an option for most small passerines. For declining species like Golden-winged Warblers (*Vermivora chrysoptera*), it is especially important to determine the relative impact of factors on breeding and nonbreeding areas and those experienced *en route* to address means of mitigating or reversing such declines. We used measurements of naturally occurring hydrogen isotopes (δ^2H) in feathers of Golden-winged Warblers sampled in Honduras (n = 68), Nicaragua (n = 19), Costa Rica (n = 65), Colombia (n = 16), and Venezuela (n = 3) and used the isotopic data to assign probable breeding or natal origins of individuals to a North American feather δ^2H isoscape. We found considerable structure in migratory connectivity, with Honduran birds generally deriving from more southern portions of the breeding distribution and those in Nicaragua and Costa Rica from more northern regions. Birds in Venezuela and Colombia were generally from the Appalachians and more southern portions of their breeding distribution, suggesting the existence of a migratory divide between Appalachian and Great Lakes breeding-distribution segments in North America. Our results have important implications for the design of ongoing sampling of this species on the nonbreeding areas and the interpretation of population trajectories of birds from various regions of their breeding distribution.

Key Words: breeding ground, deuterium, migration, nonbreeding ground.

Effective conservation of migratory birds requires an understanding of geographical connections between breeding and nonbreeding populations, and migratory stopover sites used *en route* (Webster and Marra 2005). Information on migratory connectivity serves to highlight critical habitats used by a species during most of the annual cycle and to integrate

* Hobson, K. A., S. L. Van Wilgenburg, A. M. Roth, R. E. Bennett, N. J. Bayly, L. Chavarría-Duriaux, G. J. Colorado, P. Elizondo, C. G. Rengifo, and J. D. Ritterson. 2016. Golden-winged Warbler migratory connectivity derived from stable isotopes. Pp. 193–203 in H. M. Streby, D. E. Andersen, and D. A. Buehler (editors). Golden-winged Warbler ecology, conservation, and habitat management. Studies in Avian Biology (no. 49), CRC Press, Boca Raton, FL.

conservation efforts across hemispheres and political boundaries (Faaborg et al. 2010). For Neotropical–Nearctic migrants, habitat requirements and biogeography of populations are better understood at the breeding grounds, although it remains generally unclear which breeding areas produce the most recruits into the adult breeding population (Hobson et al. 2006, 2009). In contrast, the distribution and habitat needs of migratory birds at nonbreeding sites in Mexico, the Caribbean, and Central and South America remain poorly known, despite considerable loss of forest and other vital cover types in those areas (FAO and JRC 2012). Establishing key migratory connections between breeding and nonbreeding populations and identifying which areas make the greatest contributions to continental breeding populations are two vital goals in migratory bird conservation. However, these goals have remained largely elusive due to the practical difficulties in tracking movements of small songbirds (Hobson and Norris 2008).

The development of the use of intrinsic markers, such as ratios of naturally occurring stable isotopes in tissues, has greatly advanced understanding of animal migration (Hobson and Wassenaar 2008). In particular, the use of stable-hydrogen isotope measurements (δ^2H) of tissues such as feathers or claws that are metabolically inert following formation allows a probabilistic assignment of individuals and populations to origins due to the occurrence of strong isotopic gradients of hydrogen isotope ratios in food webs at a continental scale (Hobson and Wassenaar 2008). The establishment of calibration algorithms linking δ^2H values in feathers (δ^2H_f) with amount-weighted δ^2H in growing-season precipitation (δ^2H_p) allows for the creation of δ^2H_f isoscapes that can be used to infer potential origins where feathers were grown. Most, but not all Neotropical migrant songbirds breeding in North America grow their feathers on or close to their breeding grounds before migration (Rohwer et al. 2005). That, together with the fact that the δ^2H_f isoscape for North America is well established compared to those of nonbreeding areas results in the isotope approach being particularly well suited to inferring breeding or natal origins of Neotropical migrants sampled at their nonbreeding locations (Hobson and Wassenaar 1997, Hobson et al. 2004, Rubenstein and Hobson 2004, Boulet et al. 2006, Norris et al. 2006).

Golden-winged Warblers (*Vermivora chrysoptera*) breed primarily in the northcentral region of the United Sates (Figure 12.1) with populations centered in the Appalachian Mountains from southern New York to northwestern Georgia and in the Great Lakes from Quebec to Minnesota extending northwest into Manitoba (Chapter 1, this volume). Golden-winged Warblers generally molt at their breeding grounds prior to migration (Pyle et al. 1997, Confer et al. 2011), and spend the nonbreeding season primarily in Central America and northern South America, from central Guatemala and northern Honduras south to northwestern Venezuela and western Colombia (Confer et al. 2011; Chapter 1, this volume). Most conservation efforts and concern for Golden-winged Warblers have focused on breeding ecology (Buehler et al. 2007, Thogmartin 2010), where population size has declined at an annual rate of about 2.5%–3% (Sauer et al. 2008) and where the species hybridizes with congeneric Blue-winged Warblers (*Vermivora cyanoptera*; Vallender et al. 2007, 2009). No bands have been recovered from the nonbreeding grounds and virtually nothing is known about migratory connectivity in Golden-winged Warblers. The small body mass of ~9 g has resulted in geolocators only recently becoming small enough to be tested on Golden-winged Warblers (H. M. Streby, pers. comm.) making the species a candidate for stable-isotope investigation. The stable-isotope approach is not biased with respect to origin because only one capture is required. Information on origin can be derived for all birds sampled and the approach does not influence the prior migratory behavior of the bird up to the point of sampling (Hobson 2005, Hobson and Wassenaar 2008). Last, with extensive geographic sampling, estimates of the most productive breeding regions can be delineated by examining estimated origins of hatch-year or second-year (SY) birds. Here, we present results of our investigation into migratory connectivity in Golden-winged Warblers by analyzing δ^2H in feathers (δ^2H_f) from individuals captured at sites used during the nonbreeding period. We augmented our study with a species-specific feather isoscape based on a calibration between feathers and precipitation derived from after-second year (ASY) birds that have the highest probability of being associated with a set of known breeding locations in North America. Our objectives were to examine the evidence for spatial structure in

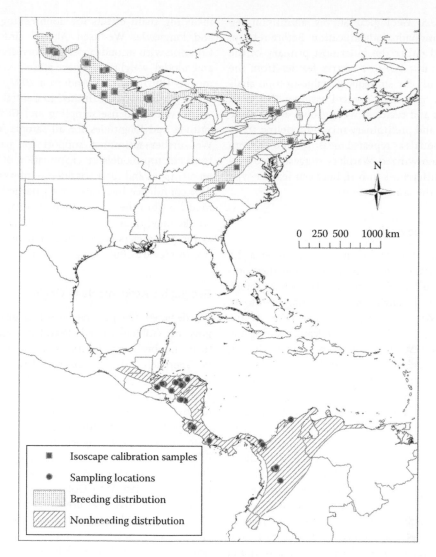

Figure 12.1. Distribution of feather collection sites (circles) within the Golden-winged Warbler nonbreeding distribution (hatched shading) and locations used to calibrate a model of expected δ^2H in Golden-winged Warbler feathers grown on the breeding distribution (stippling).

migratory connectivity versus panmixis, and establish a protocol for Golden-winged Warblers that could be used for future sampling at nonbreeding sites. Scant information is currently available and we had no *a priori* predictions concerning patterns of migratory connectivity.

METHODS

Field Methods

We obtained feather samples from Golden-winged Warblers captured at 74 locations across much of their nonbreeding distribution, grouped

at 22 distinct localities (Figure 12.1). We captured one to eight individuals per site in Honduras ($n = 68$), Nicaragua ($n = 19$), Costa Rica ($n = 65$), Colombia ($n = 16$), and Venezuela ($n = 3$) between 2003 and 2012 (Figure 12.1). We collected 171 samples at nonbreeding sites: 139 from males, 16 from females, and 16 from individuals of unknown sex. We captured birds using mist nets and call playback of male type I and type II songs during the period when most individuals are territorial, with the exception of Colombia where birds were likely captured during migration (Chandler 2011). At capture, we banded birds with a U.S. Geological Survey aluminum leg band for individual identification. Before release, we pulled either the innermost primary of the wing (P1; n = 92) or a rectrix feather from the tail (n = 61) for subsequent stable-isotope analyses. Our sampling protocol assumes that both primaries and rectrices were grown at the same location, and preliminary mixed modeling treating individuals as repeated measures of δ^2H_f from 147 Golden-winged Warblers suggested no systematic differences in δ^2H_f between feather type grown at the same location.

Stable-Isotope Analysis

We cleaned feathers of surface oils using a 2:1 chloroform:methanol solvent rinse and then prepared them for δ^2H analysis at the Stable Isotope Laboratory of Environment Canada, Saskatoon, Canada. We determined the δ^2H value of the nonexchangeable hydrogen of feathers using the method described by Wassenaar and Hobson (2003) using two calibrated keratin hydrogen-isotope reference materials. We performed hydrogen isotopic measurements on H_2 gas derived from high-temperature (1350°C) flash pyrolysis of 350 ± 10 µg feather (distal vane) subsamples in a Eurovector elemental analyzer (Milan, Italy) coupled with an VG Isoprime mass spectrometer (Manchester, UK) using continuous-flow isotope-ratio mass spectrometry. Measurements of the two keratin laboratory reference materials (CBS, KHS) corrected for linear instrumental drift were both accurate and precise with typical within-run mean $\delta^2H \pm SD$ values of −197‰ ± 0.79‰ (n = 5) and −54.1‰ ± 0.33‰ (n = 5), respectively. We report all results for nonexchangeable H expressed in the typical delta notation, in units per mil (‰), and normalized on the Vienna Standard Mean Ocean Water-Standard Light Antarctic Precipitation standard scale.

Data Analysis

We tested for spatial segregation of breeding populations on their nonbreeding grounds using general linear models in which we considered four a priori candidate models. Specifically, we used linear models to examine if birds from different breeding locations (as inferred from δ^2H_f) segregated on the nonbreeding range. Our candidate models included an intercept-only model and models including main effects for nonbreeding latitude and longitude. We used Akaike's Information Criterion with second-order bias correction (AICc) and model weights to select among competing models (Burnham and Anderson 2002). Data for linear modeling were limited to Central America where we had dense sampling and GPS-derived latitude and longitudes for all sample locations. We further examined support for parameters from our top models by examination of parameter estimates and their confidence intervals. If the 85% confidence intervals for a parameter from a model within the "top models" overlapped zero, we excluded these models from further inference as the parameters were considered noninformative (Anderson 2008, Arnold 2010).

Geographic Assignments to Origin

We depicted the putative breeding-ground origins of Golden-winged Warblers sampled at their nonbreeding locations in a probabilistic approach using a likelihood-based assignment technique (Hobson et al. 2009, Wunder 2010, Van Wilgenburg and Hobson 2011). The technique entailed the creation of a feather δ^2H isoscape using an amount-weighted precipitation-to-feather calibration algorithm ($\delta^2H_f = -45.7 + 0.74 * \delta^2H_p$) derived from work using 93 known (between-year recaptures) or presumed origin (from unbanded ASY) Golden-winged Warblers distributed across the breeding range (Figure 12.1). We used the regression parameters of our algorithm to convert δ^2H_p estimates provided by Bowen et al. (2005) into a δ^2H_f isoscape.

We estimated the likelihood that a cell (pixel) within the δ^2H_f isoscape (resolution of 0.33°) represented a potential origin for a sample by using a normal probability density function to estimate the likelihood function based on the observed δ^2H_f (Hobson et al. 2009, Wunder 2010, Van Wilgenburg and Hobson 2011). We depicted the likely origins of sampled Golden-winged Warblers by assigning individuals to the δ^2H_f isoscape one at a time. Following Hobson et al. (2009), we accomplished assignment by determining the odds that an assigned origin was correct relative to the odds that it was incorrect. We recoded the set of raster cells that defined the upper 67% of estimated "probabilities of origin" for each individual and coded those as 1, and all others as 0, resulting in one binary map per assigned individual, which was consistent with 2:1 odds of being correct versus incorrect (Hobson et al. 2009, Van Wilgenburg et al. 2012). We then summed the results of individual assignments over all individuals by addition of the predicted surfaces (Hobson et al. 2009, Van Wilgenburg and Hobson 2011). We made geographic assignments to origin using functions within the R statistical computing environment (R Core Team 2012) using scripts employing the raster package (Hijmans and Van Etten 2012).

RESULTS

Golden-winged Warblers differed in their inferred breeding-ground origins depending on their capture location (Figure 12.2). Birds tended to be more enriched in ^{2}H the farther south they were captured on the nonbreeding grounds (Figure 12.2). However, for birds captured in Central America, our linear models suggested a more complex pattern of connectivity with the

top two models of variation in δ^2H_f receiving ~98% of the summed model weights (Table 12.1). The latitude-only model received <2% of model weight of the models in our candidate set (see w_i; Table 12.1), and so we did not consider this model to be competitive. The second-most parsimonious model (ΔAICc = 1.94) included parameters for both latitude and longitude (Table 12.1); however, the 85% confidence interval for the latitude parameter estimate overlapped zero (β = −0.4, 85% CI = −1.7 to 0.9‰), and thus was a noninformative parameter and we did not consider it further. As a result, we limited all inference to the top model, which only included a longitude covariate (Table 12.1). The parameter estimate for nonbreeding longitude from our top model (β = −4.3, 85% CI = −5.8 to −2.9‰) suggested that birds captured at our westernmost site within Central America (88.633°W) were ~24.5‰ more enriched in ^{2}H on average than birds from the easternmost site (85.333°W; Figure 12.3).

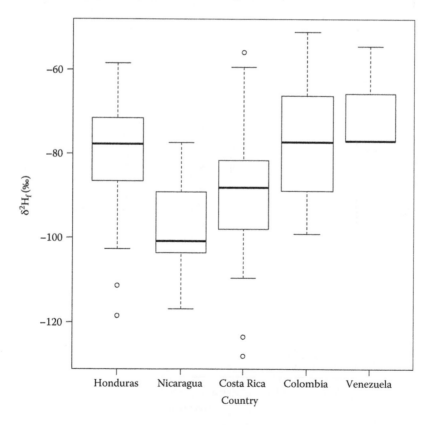

Figure 12.2. Variation in δ^2H_f within Golden-winged Warblers sampled in Honduras (n = 68), Nicaragua (n = 19), Costa Rica (n = 65), Colombia (n = 16), and Venezuela (n = 3) between 2003 and 2012. Dark solid line represents the median, the box indicates the interquartile range, whiskers are 1.5 times the interquartile range, and dots indicate extreme values.

TABLE 12.1

Selection among linear models exploring spatial (latitude and longitude) variation on the nonbreeding grounds in δ^2H_f from breeding-ground-grown feathers of Golden-winged Warblers captured in Costa Rica (n = 65), Honduras (n = 68), and Nicaragua (n = 19).

Model	K	−2 ln L	ΔAICc	w_i
Longitude	3	1214.02	0.00	0.71
Longitude + latitude	4	1213.84	1.94	0.27
Latitude	3	1221.04	7.03	0.02
Null	2	1231.72	15.62	0.00

K denotes the number of estimated parameters, −2 ln L denotes deviance (i.e., −2 times log-likelihood), ΔAICc denotes difference in Akaike's Information Criterion adjusted for sample size relative to the most parsimonious model, and w_i is the Akaike weight.

Golden-winged Warblers sampled in Honduras included individuals assigned locations from across the species' breeding distribution, but the majority of birds was consistent with the more southerly portions of their breeding distribution (Figures 12.3 and 12.4). We assigned most Golden-winged Warblers sampled in Honduras to breeding-ground origins consistent with a region including Wisconsin through Michigan in the west and south along the Appalachian Mountains to the extreme southern edge of their breeding distribution. Consistent with the results of our linear modeling (Figure 12.3), the proportion of Golden-winged Warblers consistent with origins in the northern portion of the breeding distribution was greater in eastern Honduras than western Honduras (Figure 12.4a,b). Up to 70% of the birds captured in eastern Honduras (east of 87°W) were consistent with breeding origins south of 45°N (Figure 12.4a). Between 3% and 40% of the birds sampled in eastern Honduras were consistent with regions north of 45°N (Figure 12.4a). In contrast, up to 93% of the birds sampled in Honduras west of 87°W were assigned to breeding origins south of 45°N. Only ~6%–26% of the sample was isotopically consistent with areas north of 45°N, including the southern Manitoba

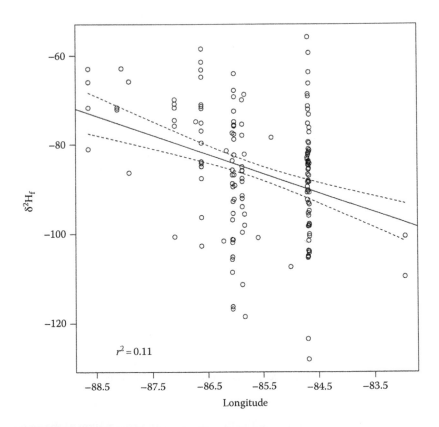

Figure 12.3. Relationship between δ^2H_f and longitude for Golden-winged Warblers sampled in Costa Rica (n = 65), Honduras (n = 68), and Nicaragua (n = 19). Solid line is least squares fit, with dashed lines representing 85% confidence intervals.

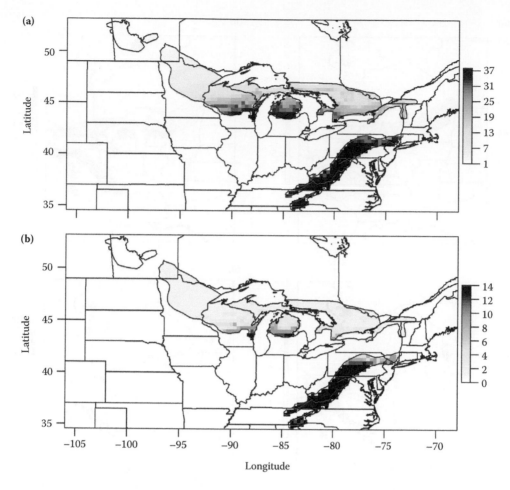

Figure 12.4. Geographic distribution of assigned origins for 68 Golden-winged Warblers sampled in (a) eastern Honduras (east of 87°W, $n = 53$) and (b) western Honduras (west of 87°W, $n = 15$) inferred from δ^2H_f. Assignments were based on likelihood-based comparison of observed δ^2H_f against predicted δ^2H_f from a species-specific calibration of the isoscape of Bowen et al. (2005); see "Methods" section for details. Numbers on legend indicate the number of individuals in the sample that were isotopically consistent with similarly colored portions of the map at our selected (2:1) odds ratio.

breeding population, whereas none of the sample was consistent with the north-westernmost breeding population (Figure 12.4b).

In contrast to birds captured in Honduras (Figure 12.4), birds captured in Nicaragua and Costa Rica were assigned origins that were generally farther north (Figure 12.5a,b). Most of the Golden-winged Warblers sampled in Nicaragua were consistent with origins in a region from New York state through Ontario and the Great Lakes region and into Minnesota and southern Manitoba in the northwest (Figure 12.5a). High variation in feather isotope ratios (δ^2H_f) among individuals captured in Costa Rica (Figure 12.2) implied origins from throughout the breeding distribution (Figure 12.5b); however, the majority of Costa

Rican samples were assigned to midlatitudes of the species breeding distribution, from Minnesota in the west to Cape Cod and Pennsylvania in the east (Figure 12.5b). Several individuals captured in Costa Rica were also consistent with origins in the Appalachian Mountains and the northwestern extreme of the species breeding distribution (Figure 12.5b).

Golden-winged Warblers captured in South America were largely consistent with origins in the southern extremes of the breeding distribution (Figure 12.5c,d). A majority of samples from Colombia derived from origins in the Appalachian Mountains or portions of the breeding distribution abutting the southern end of Lake Michigan, whereas ≤5 birds (≤31%) were assigned to upper

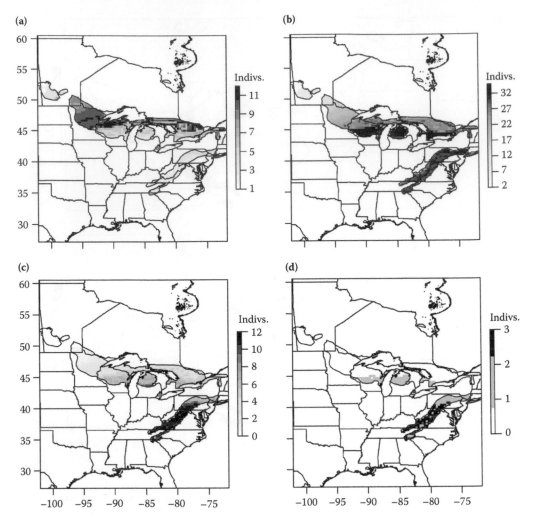

Figure 12.5. Geographic distribution of assigned origins for Golden-winged Warblers sampled in (a) Nicaragua (n = 19), (b) Costa Rica (n = 65), (c) Colombia (n = 16), and (d) Venezuela (n = 3) inferred from $\delta^2 H_f$. Assignments were based on likelihood-based comparison of observed $\delta^2 H_f$ against predicted $\delta^2 H_f$ from a species-specific calibration of the isoscape of Bowen et al. (2005); see "Methods" section for details. Numbers on legend indicate the number of individuals in the sample that were isotopically consistent with similarly colored portions of the map at our selected (2:1) odds ratio.

New York State and regions north of the Great Lakes (Figure 12.5c). All three birds captured in Venezuela were consistent with the southernmost portion of the breeding distribution along the Appalachian Mountains; however, southern portions of the disjunct northern distribution could also represent potential origins for these individual birds (Figure 12.5d).

DISCUSSION

Despite modest sample sizes from some locations in Central America, our study provides

evidence for structure in the migratory connectivity between breeding sites in North America and nonbreeding sites in Central and South America for Golden-winged Warblers. Samples of birds from the southernmost portion of the nonbreeding distribution in Colombia and Venezuela were among the highest $\delta^2 H_f$ values representing origins in the southernmost regions of the breeding distribution, whereas those from Central America were generally lower, suggesting more northerly origins. Within Central America, we further discovered that a greater proportion of birds in more easterly locations originated from more northerly breeding locations compared to those in western Central America, which derived from more southern origins.

Although our results are not definitive, we suggest a potential migratory connectivity scenario operating in Golden-winged Warblers involving three migratory pathways. Some birds from southern breeding populations may cross the Caribbean Sea, make landfall in northern Central America, and fly overland to Honduras. Samples collected in Venezuela and Colombia were associated with the most southern breeding origins consistent with the southern Appalachian Mountains. The most parsimonious migratory routes for birds migrating to South America would appear to be via trans-Gulf of Mexico migration or following a trans-Caribbean Sea path through the eastern Caribbean Islands to arrive on nonbreeding areas via eastern Venezuela. Fall migration records from eBird (2012) include sparse observations along the Caribbean Island chain, which provide partial support for the eastern Caribbean as a potential migratory route to South America. In contrast, birds from more northern and northwestern breeding areas may migrate across the Gulf of Mexico and settle directly in eastern Honduras, Nicaragua, and Costa Rica. As eBird migration records for Golden-winged Warbler are sparse in Mexico north of Veracruz, we predict that circum-Gulf of Mexico migration is relatively rare in Golden-winged Warblers.

The putative migratory structure we have described largely corresponds to breeding latitude, which also corresponds to populations experiencing differential population trends because southern breeding populations are known to be experiencing the steepest rates of decline (Buehler et al. 2007). In addition to the loss of breeding habitat in the Appalachian Mountains, the compounding stresses of long migration and scarcity of fueling sites could negatively affect populations of birds from more southerly breeding origins that migrate to Colombia and Venezuela. Bayly et al. (2012, 2013) have convincingly shown the importance of migratory refueling sites in forested remnants of coastal Colombia for trans-Caribbean Sea migrants, and these habitats are declining in extent due to ongoing deforestation. Possibly, southern-breeding Golden-winged Warblers have declined most precipitously due to factors operating at both ends of the migratory route.

Our study has demonstrated the strength of using δ^2H_f analyses coupled with a detailed species-specific δ^2H_f breeding-ground isoscape for describing the origins of Neotropical migrants captured at sites used during the boreal winter. Stable isotopes have clear advantages over conventional mark–recapture techniques that have generally failed to describe migratory connectivity for migratory songbirds. Our results should also guide future research on Golden-winged Warblers at sites used during both the breeding and nonbreeding seasons. For example, renewed sampling efforts in Venezuela and Colombia would improve estimates of the proportion of those populations that migrate there from the southeastern portion of their breeding distribution. Further isotopic sampling in Central America may improve understanding of how different populations settle along a latitudinal gradient. For example, the northern edges of the nonbreeding distribution in Guatemala and Belize have yet to be sampled.

Linking regional population trends on the breeding grounds with differential migratory routes or ultimate nonbreeding destinations is an important goal for the conservation of Golden-winged Warblers. Rubenstein et al. (2002) used stable-isotope methods to show that declining, southern breeding populations of Black-throated Blue Warblers (Setophaga caerulescens) spent the nonbreeding season in the eastern Antilles, unlike more stable northern breeding populations. That result provided evidence that habitat loss on the nonbreeding grounds may be an important driver of regional differences in breeding population trends. Similarly, and again using stable-isotope approaches, strong migratory connectivity has been demonstrated for Wilson's Warblers (Cardellina pusilla; Kelly et al. 2002) and American Redstarts (Setophaga ruticilla; Norris et al. 2006), which may allow population-specific conservation. As geolocators become small enough to be used on Golden-winged Warblers without adversely affecting their migratory behavior, we suggest that these tags be deployed to test the hypotheses we have generated here based on stable-isotope data. Last, with more sampling across sites used during the nonbreeding period, it may be possible to identify those regions of the breeding distribution contributing most to the recruitment of young birds (Hobson et al. 2009).

ACKNOWLEDGMENTS

We are extremely grateful to M. I. Moreno, A. Paez, G. Suarez, and J. Sanabria for collecting samples on the wintering grounds. We thank U.S. Forest Service International Programs and The Nature Conservancy for financial support and Universidad Nacional de Colombia for collaboration. We are also grateful to the following individuals who contributed breeding ground samples for isoscape calibration: K. Aldinger, J. Chernek, J. Confer, D. Flaspohler, M. Fowlds, M.-F. Julien, J. Larkin, K. Percy, C. Smalling, H. Streby, and R. Vallender. We are thankful for assistance supplied by the Honduran Ornithological Association, Mancomunidad MAPANCE, J. Wilson, T. Daulton, D. King, J. Guevara, F. Elizondo Camacho, A. el Socorro, The Nature Conservancy, Universidad Nacional de Colombia, CORANTIOQUIA, and the San Vito Bird Club for their assistance. B. M. Alvarez, M. Allen, A. Crosby, and C. Gryba prepared feathers for stable-isotope analysis. Funding was provided by grants from the U.S. Fish and Wildlife Service Region 3, the Inter-Departmental Recovery Fund, the MTU Ecosystem Science Center, U.S. Forest Service International Programs, Neotropical Migratory Bird Conservation Act, the National Science Foundation, and an operating grant to KAH from Environment Canada.

LITERATURE CITED

Anderson, D. R. 2008. Model based inference in the life sciences: a primer on evidence. Springer, New York, NY.

Arnold, T. W. 2010. Uninformative parameters and model selection using Akaike's Information Criterion. Journal of Wildlife Management 74:1175–1178.

Bayly, N., C. Gomez, K. Hobson, A. Gonzalez, and K. Rosenberg. 2012. Fall migration of the Veery (*Catharus fuscescens*) in northern Colombia: determining the energetic importance of a stopover site. Auk 129:449–459.

Bayly, N., C. Gomez, and K. A. Hobson. 2013. Energy reserves stored by migrating Gray-cheeked Thrushes *Catharus minimus* at a spring stopover site in northern Colombia are sufficient for a long-distance flight to North America. Ibis 155:271–283.

Boulet, M., H. Gibbs, and K. A. Hobson. 2006. Integrated analysis of genetic, stable isotope, and banding data reveal migratory connectivity and flyways in the northern Yellow Warbler (*Dendroica petechia*; aestiva group). Ornithological Monographs 61:29–78.

Bowen, G. J., L. I. Wassenaar, and K. A. Hobson. 2005. Global application of stable hydrogen and oxygen isotopes to wildlife forensics. Oecologia 143:337–348.

Buehler, D., A. Roth, R. Vallender, T. Will, J. Confer, R. Canterbury, S. B. Swarthout, K. Rosenberg, and L. Bulluck. 2007. Status and conservation priorities of Golden-winged Warbler (*Vermivora chrysoptera*) in North America. Auk 124:1439–1445.

Burnham, K. P., and D. R. Anderson. 2002. Model selection and multimodel inference: a practical information-theoretic approach. Springer-Verlag, New York, NY.

Chandler, R. B. 2011. Avian ecology and conservation in tropical agricultural landscapes with emphasis on *Vermivora chrysoptera*. Ph.D. dissertation, University of Massachusetts Amherst, Amherst, MA.

Confer, J., P. Hartman, and A. Roth. 2011. Golden-winged Warbler (*Vermivora chrysoptera*). In A. Poole (editor), The birds of North America online. Cornell Lab of Ornithology, Ithaca, NY.

eBird. [web application]. 2012. eBird: an online database of bird distribution and abundance. eBird, Cornell Lab of Ornithology, Ithaca, NY. <http://www.ebird.org> (30 May 2016).

Faaborg, J., R. T. Holmes, A. D. Anders, K. L. Bildstein, K. M. Dugger, S. A. Gauthreaux, P. Heglund, K. A. Hobson, A. E. Jahn, D. H. Johnson, S. C. Latta, D. J. Levey, P. P. Marra, C. L. Mekord, E. Nol, S. I. Rothstein, T. W. Sherry, T. S. Sillett, F. R. Thompson III, and N. Warnock. 2010. Conserving migratory land birds in the New World: do we know enough? Ecological Applications 20:398–418.

FAO and JRC. 2012. Global forest land-use change 1990–2005, by E. J. Lindquist, R. D'Annunzio, A. Gerrand, K. MacDicken, F. Achard, R. Beuchle, A. Brink, H. D. Eva, P. Mayaux, J. San-Miguel-Ayanz, and H.-J. Stibig. FAO Forestry Paper No. 169. Food and Agriculture Organization of the United Nations and European Commission Joint Research Centre, Rome, Italy.

Hijmans, R. J., and J. Van Etten. [online]. 2012. Raster: geographic analysis and modeling with raster data, R package version 2.0-12. <http://CRAN.R-project.org/package=raster> (1 April 2014).

Hobson, K. A. 2005. Stable isotopes and the determination of avian migratory connectivity and seasonal interactions. Auk 122:1037–1048.

Hobson, K. A., Y. Aubry, and L. I. Wassenaar. 2004. Migratory connectivity in Bicknell's Thrush: locating missing populations with hydrogen isotopes. Condor 106:905–909.

Hobson, K. A., and D. R. Norris. 2008. Animal migration: a context for using new techniques and approaches. Pp. 1–19 in K. A. Hobson and L. I. Wassenaar (editors), Tracking animal migration using stable isotopes. Academic Press, London, UK.

Hobson, K. A., S. Van Wilgenburg, L. I. Wassenaar, H. Hands, W. P. Johnson, M. O'Meilia, and P. Taylor. 2006. Using stable hydrogen isotope analysis of feathers to delineate origins of harvested Sandhill Cranes in the Central Flyway of North America. Waterbirds 29:137–147.

Hobson, K. A., and L. I. Wassenaar. 1997. Linking breeding and wintering grounds of Neotropical migrant songbirds using stable hydrogen isotopic analysis of feathers. Oecologia 109:142–148.

Hobson, K. A., and L. I. Wassenaar. 2008. Tracking animal migration using stable isotopes. Academic Press-Elsevier, Amsterdam, Netherlands.

Hobson, K. A., M. B. Wunder, S. L. Van Wilgenburg, R. G. Clark, and L. I. Wassenaar. 2009. A method for investigating population declines of migratory birds using stable isotopes: origins of harvested Lesser Scaup in North America. PLoS One 4:e7915.

Kelly, J., V. Atudorei, Z. Sharp, and D. Finch. 2002. Insights into Wilson's Warbler migration from analyses of hydrogen stable-isotope ratios. Oecologia 130:216–221.

Norris, D., P. Marra, G. Bowen, L. Ratcliff, J. Royle, and T. Kyser. 2006. Migratory connectivity of a widely distributed songbird, the American Redstart (Setophaga ruticilla). Auk 61:14–28.

Pyle, P., S. Howell, D. DeSante, R. Yunick, and M. Gustafson. 1997. Identification guide to North American birds. Slate Creek Press, Ann Arbor, MI.

R Core Team. 2012. R: a language and environment for statistical computing. The R Foundation for Statistical Computing, Vienna, Austria.

Rohwer, S., L. K. Butler, and D. R. Froehlich. 2005. Ecology and demography of east-west differences in molt scheduling of Neotropical migrant passerines. Pp. 87–105 in R. Greenberg and P. P. Marra (editors), Birds of two worlds: the ecology and evolution of migration. Johns Hopkins University Press, Baltimore, MD.

Rubenstein, D. R., C. P. Chamberlain, R. T. Holmes, M. P. Ayres, J. R. Waldbauer, G. R. Graves, and N. C. Tuross. 2002. Linking breeding and wintering ranges of a migratory songbird using stable isotopes. Science 295:1062–1065.

Rubenstein, D. R., and K. A. Hobson. 2004. From birds to butterflies: animal movement patterns and stable isotopes. Trends in Ecology and Evolution 19:256–263.

Sauer, J., J. Hines, and J. Fallon. 2008. The North American Breeding Bird Survey; results and analysis 1966–2007, version 5.15.2008. Patuxent Wildlife Research Center, Laurel, MD.

Thogmartin, W. 2010. Modeling and mapping Golden-winged Warbler abundance to improve regional conservation strategies. Avian Conservation and Ecology 5:12.

Vallender, R., R. Robertson, V. Friesen, and I. Lovette. 2007. Complex hybridization dynamics between Golden-winged and Blue-winged Warblers (Vermivora chrysoptera and Vermivora pinus) revealed by AFLP, microsatellite, intron and mtDNA markers. Molecular Ecology 16:2017–2029.

Vallender, R., S. L. Van Wilgenburg, L. Bulluck, R. Roth, R. Canterbury, J. Larkin, R. Fowlds, and I. Lovette. 2009. Extensive rangewide mitochondrial introgression indicates substantial cryptic hybridization in the Golden-winged Warbler (Vermivora chrysopetra). Avian Conservation and Ecology 4:4.

Van Wilgenburg, S., K. Hobson, K. Brewster, and J. Welker. 2012. Assessing dispersal in threatened migratory birds using stable hydrogen isotope (δD) analysis of feathers. Endangered Species Research 16:17–29.

Van Wilgenburg, S. L., and K. A. Hobson. 2011. Combining stable-isotope (δD) and band recovery data to improve probabilistic assignment of migratory birds to origin. Ecological Applications 21:1340–1351.

Wassenaar, L. I., and K. A. Hobson. 2003. Comparative equilibrium and online technique for determination of non-exchangeable hydrogen of keratins for use in animal migration studies. Isotopes in Environmental and Health Studies 39:211–217.

Webster, M. S., and P. P. Marra. 2005. The importance of understanding migratory connectivity and cross-seasonal interactions. Pp. 199–209 in R. Greenberg and P. P. Marra (editors), Birds of two worlds: the ecology and evolution of migration. Johns Hopkins University Press, Baltimore, MD.

Wunder, M. B. 2010. Using isoscapes to model probability surfaces for determining geographic origins. Pp. 251–270 in G. J. Bowen, J. B. West, K. P. Tu, and T. E. Dawson (editors), Isoscapes: understanding movement, pattern, and process on earth through isotope mapping. Springer, Dordrecht, Netherlands.

Synthesis and Future Directions

CHAPTER THIRTEEN

Conservation Perspectives*

REVIEW OF NEW SCIENCE AND PRIMARY THREATS TO GOLDEN-WINGED WARBLERS

Ronald W. Rohrbaugh, David A. Buehler, Sara Barker Swarthout, David I. King, Jeffery L. Larkin, Kenneth V. Rosenberg, Amber M. Roth, Rachel Vallender, and Tom Will

Abstract. In this penultimate chapter, we examine new perspectives on ecology of Golden-winged Warblers (*Vermivora chrysoptera*), review primary population-level threats, and offer conservation recommendations. Adequate forest cover and patch-level habitat configuration are important for successful reproduction and to buffer against negative interactions with Blue-winged Warblers (*V. cyanoptera*). We recommend landscape-scale forest cover of 50%–100% and meso-scale (500-m radius) habitat designs that provide nesting habitat bounded by a mosaic of structurally diverse, multiple age-class forest or connected to such forest by dispersal corridors <200 m in length. The primary threat to breeding and nonbreeding Golden-winged Warbler populations is land-use change, resulting in forest conversion to human development and agriculture. In the Great Lakes breeding-distribution segment, which holds 95% of the global breeding population, we recommend protection and improvement of existing habitat, whereas we recommend critically needed habitat creation in the Appalachian Mountains breeding-population segment. At the nonbreeding grounds, we recommend protection of humid forest at 700–1,400 m elevation, establishment of a system of national forest reserves, and promotion of agroforestry, such as Integrated Open Canopy Coffee. Given that Golden-winged Warblers likely use a migration pathway across the Gulf of Mexico, which is similar to many other Neotropical migrants, we recommend a general strategy of protecting coastal Gulf of Mexico stopover locations. Last, protection of inland migration pathways such as ridge tops and riparian forests along major river systems could also confer benefits to Golden-winged Warblers.

INTRODUCTION

Devising conservation strategies for Neotropical migrant landbirds, such as Golden-winged Warblers (*Vermivora chrysoptera*), is complicated by a remarkable life cycle that includes a latitudinal migration covering thousands of kilometers and separate breeding and nonbreeding distributions with different ecological conditions and threats. The diverse science presented in this Studies in

* Rohrbaugh, R. W., D. A. Buehler, S. B. Swarthout, D. I. King, J. L. Larkin, K. V. Rosenberg, A. M. Roth, R. Vallender, and T. Will. 2016. Conservation perspectives: Review of new science and primary threats to Golden-winged Warblers. Pp. 207–215 in H. M. Streby, D. E. Andersen, and D. A. Buehler (editors). Golden-winged Warbler ecology, conservation, and habitat management. Studies in Avian Biology (no. 49), CRC Press, Boca Raton, FL.

Avian Biology volume, ranging from regional studies of postfledgling survival to broad inquiries about migratory connections that potentially stretch from Manitoba to Colombia, underscores the complexities of understanding Golden-winged Warbler ecology and conservation, not only at one point in space or time, but across all stages of the annual life cycle—breeding, migrating, and nonbreeding.

During the 1990s and early 2000s, Golden-winged Warbler conservation efforts were opportunistic, site-specific, and largely focused on the Appalachian Mountains. Researchers and managers, often working at local scales, protected and maintained human-modified (e.g., utility rights-of-way and surface mines) and natural (e.g., forested wetlands) habitats occupied by breeding Golden-winged Warblers (Canterbury and Stover 1999, Confer and Pascoe 2003, Kubel and Yahner 2008). Recently, the first guide to Best Management Practices (Bakermans et al. 2011) and comprehensive conservation plan (A. M. Roth et al., unpubl. plan) have taken a more systematic, distribution-wide approach to Golden-winged Warbler conservation. A. M. Roth et al. (unpubl. plan) provided a strategic, spatially explicit approach to setting population and habitat goals, identifying threats, delineating high-priority focal areas, and making habitat management recommendations. Guidance provided by these publications has been used across the breeding distribution to improve and create habitat. For example, the Golden-winged Warbler is one of seven species targeted for conservation via the U.S. Department of Agriculture, Natural Resource Conservation Service's Working Lands for Wildlife program. In the Appalachian Mountains during 2012–2015, the working lands program created 3,700 ha of Golden-winged Warbler habitat on private lands by using science-based habitat management prescriptions based on Bakermans et al. (2011, 2015) and A. M. Roth et al. (unpubl. plan).

The new science in this volume provides fresh viewpoints on the challenges facing Golden-winged Warbler populations and the strategies required to implement effective conservation practices. Continued steep population declines in the Appalachian Mountains breeding-distribution segment and broadening declines in the Great Lakes breeding-distribution segment emphasize the urgency in addressing known threats. In this chapter, we (1) examine results from this volume that yield new perspectives on Golden-winged Warbler conservation needs, including suggestions for updating current management prescriptions, and (2) review primary population-level threats and offer strategies for ameliorating these in the context of full life-cycle conservation planning. There is growing need to better understand the connectivity among different life-cycle phases. We have organized this chapter around the breeding, nonbreeding, and migrating stages, and identify potential linkages where possible. We conclude with a discussion of the need for quantitative, full life-cycle models to identify spatiotemporal population constraints and more efficiently direct conservation resources for the greatest positive impact on the global Golden-winged Warbler population.

BREEDING

New Conservation Perspectives

Status and Distribution

Effective conservation planning requires detailed knowledge about the distribution and abundance of target organisms on the landscape, especially as populations rapidly decline or undergo geographic shifts in response to environmental changes (Faaborg et al. 2010). Rosenberg et al. (Chapter 1, this volume) point out the inadequacy of available monitoring programs to track current population trends and distribution shifts in patchily distributed Golden-winged Warbler populations. The North American Breeding Bird Survey is now ineffective at monitoring Golden-winged Warbler population trends and distribution in the Appalachian Mountains breeding-distribution segment because birds are detected on too few survey routes. The same issue is a concern in portions of the Great Lakes breeding-distribution segment, and the problem will likely become more systemic in coming years as fewer Golden-winged Warblers are available to be detected on each route. A related issue is the lack of knowledge about what is driving population dynamics and shifts in distribution of Blue-winged Warblers (*V. cyanoptera*). Hybridization and competitive exclusion by Blue-winged Warblers may be drivers of Golden-winged Warbler population declines (Buehler et al. 2007). However, few research projects and no targeted monitoring projects for Blue-winged Warblers have been undertaken in the past decade.

This basic lack of data is an obstacle to understanding Golden-winged Warbler ecology and population dynamics at a distribution-wide scale where it is difficult to compare the relative influence of habitat loss, nonbreeding-season survival, and climate change against the influence of interactions with Blue-winged Warblers.

Rohrbaugh et al. (2011) developed a spatially balanced, occupancy-based monitoring program for Golden-winged Warblers in the Appalachian Mountains breeding-distribution segment. We recommend establishment of a similar standardized, distribution-wide program capable of tracking Golden-winged and Blue-winged Warbler breeding populations at multiple spatial scales. A standardized protocol will allow biologists, managers, and ultimately policy makers to measure population response to management actions and refine conservation strategies.

Habitat Requirements and Reproduction

Emerging science documents the importance of adequate forest cover at all spatial scales for successful Golden-winged Warbler reproduction and possibly as a buffer against negative interactions with Blue-winged Warblers (Streby et al. 2014; Chapters 3, 5, 8, and 9, this volume). Crawford et al. (Chapter 3, this volume) reported that landscape-scale settlement patterns of Golden-winged Warblers were positively associated with increasing forest cover and higher elevations but negatively associated with increasing agriculture and human development. In contrast, Blue-winged Warblers showed the opposite relationship with the same covariates and were more commonly associated with agriculture. These new findings are important in understanding and mitigating for land-use patterns at landscape scales, which are driving interspecific spatial interactions and may facilitate competitive exclusion or hybridization. For example, in the absence of a strong elevational gradient, does an increase in agricultural cover in the landscape facilitate co-occurrence or increased frequency of hybridization between Golden-winged and Blue-winged Warblers?

A. M. Roth et al. (unpubl. plan) recommended focusing Golden-winged Warbler breeding habitat management on sites with 50%–75% forest cover within 2.5 km. Although this recommendation is consistent with results in Crawford et al. (Chapter 3, this volume), we see no reason to cap the range at 75%, as occupancy has not been shown to decline above this percentage (Thogmartin 2010; Chapter 3, this volume). Moreover, a recent study that examined Golden-winged Warbler habitat use in New York and Pennsylvania suggests working in landscapes with >70% forest cover (Wood et al. 2016). Also, in light of a positive correlation between Blue-winged Warblers and agricultural and human-developed landscapes, conservation projects in landscapes (2.5-km scale) with >25% combined agricultural and urban cover should be considered lower priority relative to those in landscapes with <25% of these cover types. One exception to this recommendation would be projects focused on reforesting large portions of open land-use types, such as reclaimed surface mines that are widespread in some portions of the Appalachian Mountains breeding-distribution segment.

Forest that is more mature than that used for primary nesting sites is essential, not only in the larger landscape, but also adjacent to Golden-winged Warbler breeding territories, where it has been shown to be used by adults and is important for fledgling survival (Streby et al. 2014; Chapters 5, 8, and 9, this volume). The results underscore the importance of considering not only cover-type composition, but also configuration within and among Golden-winged Warbler management sites. To better incorporate adult and postfledging habitat requirements, mesoscale habitat designs (e.g., within 500-m radius) should provide potential nesting habitat in shrubland or young forest that is bounded by structurally diverse, multiple age-class forest (at least one home-range size or ≥6 ha, see Chapter 5, this volume) or connected to such forests by dispersal corridors no longer than 200 m (Chapters 5 and 10, this volume). Functionally, conservation plans should maintain dynamic forested landscapes with a shifting mosaic of forest-patch sizes and ages, where the interior of clustered young forest patches is <200 m from the edge of surrounding older age-class forest.

Primary Threats and Conservation Actions

The primary threat to Golden-winged Warbler breeding populations is land-use change through conversion of forest to agriculture or other human development, resulting in habitat loss and reduction in habitat quality by creating inappropriate landscape and patch-level configurations. These landscape changes facilitate negative interactions with Blue-winged Warblers (Chapters 3 and 4, this volume) and impair reproduction and fledgling survival (Chapters 8 and 9, this volume). Compounding the problem of land-use change, an overall maturing of forest within significant parts of the Golden-winged Warbler's breeding distribution, combined with lack of natural disturbances and forest management, has reduced regeneration of early successional habitats used by Golden-winged Warblers for nesting.

A. M. Roth et al. (unpubl. plan) set a goal of restoring the global Golden-winged Warbler population from the current estimate of 414,000 to 621,000 individuals by 2050. Currently, the Great Lakes breeding-distribution segment is estimated to hold 95% of the global breeding population, with only 5% in the Appalachian Mountains breeding-distribution segment (Chapter 1, this volume). The imbalance is increasing as populations are declining more rapidly in the Appalachian Mountains breeding-distribution segment than in the Great Lakes segment. Data from the U.S. Forest Service's Forest Inventory Analysis indicate that opportunities to create new habitat are greater in the Appalachian Mountains segment, where presently only about 7% of forest is suitable for Golden-winged Warblers compared with 19% suitable forest in the Great Lakes segment (A. M. Roth et al., unpubl. data). A. M. Roth et al. (unpubl. plan) suggested that landscape-scale forest management should strive to perpetually keep 15%–20% of forest in an early successional stage for Golden-winged Warblers and associated bird species. Following this guidance, the proportion of early successional forest in the Appalachian Mountains breeding-distribution segment could be more than doubled before the 15%–20% threshold is met, whereas the Great Lakes segment is already near the recommended maximum. Given the imbalance in initial population size and opportunities for habitat and population growth, different conservation strategies are required to address the threat of land-use change in each breeding-distribution segment.

The most important conservation actions for Golden-winged Warblers in the Great Lakes breeding-distribution segment are to:

1. Improve the quality of existing habitat by altering forest configuration and structure, following recommendations in Roth et al. (2014, unpubl. plan) and the new science presented in this volume.

2. Prevent a net loss of habitat, especially on private lands, which support about 70% of the current breeding distribution (North American Bird Conservation Initiative 2013).

3. Protect significant populations on public land, particularly in places where surrounding unprotected land is likely to undergo extensive land-use change.

4. Increase cooperation among conservation groups to ensure management for associated species, such as American Woodcock (*Scolopax minor*) also benefits Golden-winged Warblers (Bakermans et al. 2015).

5. Work collaboratively with Canadian officials to facilitate population expansion in northern latitudes where genetic purity of subpopulations is still high (Vallender et al. 2009; Chapter 4, this volume).

Conservation actions in the Appalachian Mountains breeding-distribution segment are necessary to maintain the species within portions of its historical breeding distribution (Chapter 1, this volume). Nonbreeding season loss of individuals from the vanishingly small Appalachian Mountains breeding-distribution segment to rapid habitat loss or stochastic events could hasten local extirpations and hinder population recovery. The most important immediate actions in the Appalachian Mountains breeding-distribution segment are to:

1. Protect key, high-elevation populations that have historically withstood co-occurrence with Blue-winged Warblers and that remain largely unaffected by hybridization (Dabrowski et al. 2005, Vallender et al. 2009).

2. Increase subpopulations by creating and maintaining biologically meaningful amounts of habitat within Golden-winged Warbler focal areas (A. M. Roth et al., unpubl. plan) where Blue-winged Warbler co-occurrence is unlikely.

3. Act on new knowledge about linkages between breeding and nonbreeding populations to enhance nonbreeding survival and condition of the Appalachian Mountains breeding-distribution segment where it occurs during the nonbreeding season (Chapters 12 and 14, this volume).

NONBREEDING

New Conservation Perspectives

Findings from this volume inform conservation of Golden-winged Warblers at nonbreeding areas in a number of important ways. Rosenberg et al. (Chapter 1, this volume) provided the first detailed description of the nonbreeding distribution, including areas where the species is most concentrated during the nonbreeding season; these results are being used to identify focal areas as part of the ongoing development of a nonbreeding grounds conservation plan. King et al. (Chapter 2, this volume) reported that Golden-winged Warblers in Costa Rica, Honduras, and Nicaragua occupied both regenerating and primary forest within a suitable range of elevation and moisture conditions, bounded by dry forests at lower elevations and cloud forest at upper elevations. Edges and canopy gaps with abundant dead leaves and vine tangles for foraging were important determinants of Golden-winged Warbler presence within their nonbreeding distribution (Chapter 2, this volume).

Observations concerning other aspects of Golden-winged Warbler nonbreeding ecology have conservation implications. Chandler et al. (Chapter 11, this volume) reported that Golden-winged Warblers spent most of their time within mixed-species flocks, within which they defended territories from conspecifics. Because individual flocks use large areas (~9 ha), and each flock may only support a single Golden-winged Warbler, nonbreeding densities are necessarily low, and thus a large area is required to support their nonbreeding populations.

Development of a comprehensive conservation strategy for the nonbreeding grounds is a daunting task because of the number of countries involved (Chapter 2, this volume). Landscape-scale conservation on the nonbreeding grounds will require partnering with governments, industries, and nongovernmental organizations throughout the Golden-winged Warbler's nonbreeding distribution. There is urgent need to involve the "Protected Areas Management" departments of Latin American governments. With partnerships, government officials can incorporate conservation recommendations into their protected areas management plans and train their field biologists to recognize and manage for Golden-winged Warbler habitat.

In addition to working within protected areas, it is important to provide Golden-winged Warbler habitat in working lands. Shade coffee represents a potential strategy for accomplishing conservation on working lands because it involves the retention of trees within areas of coffee production, which support more biodiversity than other forms of agriculture. Golden-winged Warblers are reported from shade-coffee farms, but these farms may provide suboptimal habitat because microhabitat features required by Golden-winged Warblers are seldom present in coffee farms. Moreover, the cohesion of the mixed species flocks on which Golden-winged Warblers depend is not maintained in these farms (Pomara et al. 2007; Chapter 11, this volume). Perhaps a more effective option is Integrated Open Canopy (IOC) Coffee, where coffee is grown with sparse or no shade adjacent to forest patches of equivalent or greater size that provide habitat for Golden-winged Warblers (Chapter 2, this volume). In addition to promoting the conservation of forest habitat required by Golden-winged Warblers and other species, IOC coffee increases income to farmers by increasing yields and providing a market-based incentive for forest conservation.

Primary Threats and Conservation Actions

Quantitative data on Golden-winged Warbler population limiting factors during the nonbreeding season are lacking, but we presume that the primary threat is loss of forested habitat from conversion to agriculture and other land uses, similar to known limiting factors for better studied Neotropical migrant species (Johnson et al. 2006). In the middle-elevation humid forest zones occupied throughout the nonbreeding distribution of Golden-winged Warblers (Chapter 1, this volume), forest conversion for commodity production is particularly acute, with rapid recent loss of forest occurring in Guatemala, Honduras, and Nicaragua (Cherrington et al. 2011). In Colombia and Venezuela, most forest conversion occurred in the 1980s and 1990s; thus, reduction in nonbreeding Golden-winged Warblers populations in South America may reflect past loss of forest and may be linked to population declines in the Appalachian Mountains breeding-distribution segment (Chapter 1, this volume). In Costa Rica, forest loss has slowed in the past decade and forested corridors are being regenerated in an attempt to reconnect a system of isolated national parks and preserves (Cherrington et al. 2011). Clearly, any assessment of threats to nonbreeding populations must account for regional variation in land-use change. As part of developing a nonbreeding grounds conservation strategy for Golden-winged Warblers, a country-by-country threats analysis was recently completed within the focal areas identified in Rosenberg et al. (Chapter 1, this volume).

Even with limited knowledge, strategies for mitigating threats and increasing survival of nonbreeding Golden-winged Warblers can be derived from recent studies of nonbreeding distribution, ecology, and habitat use (Chapters 1, 2, and 11, this volume). These actions fall into four broad areas:

1. Protect remaining primary and secondary humid forests between 700 and 1,400 m in the Central American highlands and northern Andes by establishing national reserves, municipal watershed protection, and conservation easements. Protected areas should be large enough to support cohesive mixed-species flocks within which Golden-winged Warblers can maintain nonbreeding territories (Chapter 11, this volume).

2. Restore and regenerate forest patches and corridors within focal areas of the nonbreeding grounds, especially near occupied Golden-winged Warbler nonbreeding sites.

3. Promote agroforestry practices and other land uses that are compatible with nonbreeding Golden-winged Warblers, especially IOC coffee, which encourages retention of intact forest patches in the landscape (Chapter 2, this volume).

MIGRATING

New Conservation Perspectives

Of the three life-cycle stages for Golden-winged Warblers, migration is the least studied and consequently has the largest number of knowledge gaps (Chapter 14, this volume). Given that the migration period may be when mortality for migratory birds is greatest (Sillett and Holmes 2002), comprehensive full life-cycle conservation

planning for Golden-winged Warblers will be hampered until knowledge gaps during this stage are addressed.

Stable isotope analysis has provided the first look at migratory connectivity for Golden-winged Warblers (Chapter 12, this volume). The geographic resolution of the connections is broad, owing to small sample sizes and the uncertainty inherent in interpreting isotope data. Nevertheless, these results provide the first chance to link regional Golden-winged Warbler population declines with not only breeding ground attributes, but also specific locations and corresponding habitat conditions at nonbreeding areas. Understanding connectivity is a critical step in making linkages among life cycle stages and discerning the interdependence of each stage.

New technologies, such as light-level and GPS-enabled geolocators, provide great promise for developing fine-scale maps of migratory connectivity and stopover sites. These devices have been used to successfully map migratory connectivity of Neotropical migrants such as Wood Thrushes (*Hylocichla mustelina*) and Purple Martins (*Progne subis*) (Fraser et al. 2012, McKinnon et al. 2013). Current research with geolocators provides promising results that may soon generate meaningful insights into connectivity patterns in Golden-winged Warbler populations (Chapter 14, this volume).

Primary Threats and Conservation Actions

Two potential threats to migrating Golden-winged Warblers are loss and degradation of migratory stopover habitats and fatal collisions with anthropogenic structures (Arnold and Zink 2011, Loss et al. 2014). Rosenberg et al. (Chapter 1, this volume) and Hobson et al. (Chapter 12, this volume) speculated that Golden-winged Warblers mainly use a migration pathway across the Gulf of Mexico during spring and fall. A large number of Neotropical migrant species use a similar pathway (La Sorte et al. 2014). Thus, a general strategy of protecting coastal stopover sites that have been identified as being critical for other Neotropical migrants along the Gulf Coast of the U.S., Mexico, and Central America, and known inland migration pathways such as ridge tops and riparian forests along major river systems, could enable Golden-winged Warbler migration. For Golden-winged Warblers and other migrants that spend the nonbreeding season in South America, stopover areas and corridors on the north coast of Colombia also might be critical for successful long-distance migration (Bayly et al. 2012).

Collision mortality might pose a significant threat during migration for Golden-winged Warblers. Arnold and Zink (2011) identified Golden-winged Warblers as having a greater than expected incidence of collision at communications towers, and Loss et al. (2014) estimated that Golden-winged Warblers had a collision risk with buildings (windows) that was 35.3 times greater than a migrant species with average risk. Moreover, Loss et al. (2014) suggested that building collisions may contribute to or exacerbate overall Golden-winged Warbler population declines.

Lacking specific information for Golden-winged Warblers, the most straightforward strategy to help ensure safe migration passage is to support existing efforts that are already protecting critically important stopover habitats and influencing policies focused on reducing collision risk with buildings and towers. The most relevant initiatives include the following:

1. The Joint Venture network operated by U.S. Fish and Wildlife Service and American Bird Conservancy, especially the Gulf Coast, East Gulf Coastal Plain, and Appalachian Mountains Joint Ventures

2. The Fatal Light Awareness Program and other similar programs, which focus on reducing collisions with buildings and collecting data on fatal bird strikes

3. The American Bird Conservancy's "Bird Smart" programs aimed at policy reforms to reduce collision risk at wind turbines, buildings, and communications towers

4. The Midwest Landbird Migration Monitoring Network and other regional initiatives focused on improved understanding of migratory landbird ecology and conservation

CONCLUSIONS

Call for Full Life-Cycle Conservation

As evidenced from the extensive field projects conducted by a diversity of research partners in this volume, the Golden-winged Warbler is becoming one of the most studied Neotropical migrant warbler species. Yet, we still lack necessary information about the extent to which the Golden-winged Warbler's population is being limited in each stage of its life-cycle. Conservation biologists have recognized the value in considering all segments of a migratory species' annual cycle to develop full life-cycle population models and conservation strategies (Sherry and Holmes 1995, Faaborg et al. 2010). This full life-cycle approach has illuminated biological and geographical relationships between nonbreeding grounds and breeding grounds. For example, Norris et al. (2004) found that variation in site quality of American Redstart (*Setophaga ruticilla*) nonbreeding locations carried over into the breeding season, causing impacts to reproductive fitness. New understanding about population limitations at various stages throughout the full life cycle opens novel opportunities to create dynamic conservation plans that identify and mitigate for population constraints in both space and time. Traditional conservation strategies for migratory birds have mostly relied on the broad scheme of protecting and creating habitat within the breeding distribution. The traditional approach can increase breeding population density and perhaps fecundity, but source–sink population models (Donovan and Thompson 2001) have shown that passerine populations are most sensitive to adult and juvenile mortality, which is often highest during migration (Sillett and Holmes 2002). Full life-cycle strategies could help resolve this potential conservation disconnect by pinpointing the locations and periods when management action would be most effective, thereby hastening population recovery and maximizing the impact of limited conservation funds (Berlanga et al. 2010).

Recent, large-scale collaborative research on Golden-winged Warblers has generated new demographic and movement data required for developing quantitative, full life-cycle population models. The Golden-winged Warbler research community has collected useful data for such models in both the breeding and nonbreeding periods, and ongoing research (Chapters 12 and 14, this volume) will soon provide data to parameterize the demographics of migration. Given new information, and the urgent need to reverse Golden-winged Warbler population declines, we suggest a collaborative research effort to begin developing quantitative full life-cycle models that can adaptively inform existing conservation plans.

In the meantime, the lack of such models should not preclude immediate conservation action to address known threats in each life-cycle stage. Conservation actions can protect existing populations, reduce mortality, and increase reproduction, while building capacity, infrastructure, and partnerships to more quickly and effectively take additional action when spatiotemporal population limits have been identified through full life-cycle population modeling.

LITERATURE CITED

Arnold, T. W., and R. M. Zink. 2011. Collision mortality has no discernable effect on population trends of North American birds. PLoS One 6:e24708.

Bakermans, M. H., J. L. Larkin, B. W. Smith, T. M. Fearer, and B. C. Jones. 2011. Golden-winged Warbler habitat best management practices for forestlands in Maryland and Pennsylvania. American Bird Conservancy, The Plains, VA.

Bakermans, M. H., B. W. Smith, B. C. Jones, and J. L. Larkin. 2015. Stand and within-stand factors influencing Golden-winged Warbler use of regenerating stands in the central Appalachian Mountains. Avian Conservation and Ecology 10:10.

Bakermans, M., C. Zeigler, and J. L. Larkin. 2015. American Woodcock and Golden-winged Warbler abundance and associated vegetation in managed habitats. Northeastern Naturalist 22:690–703.

Bayly, N. J., Gómez, C., Hobson, K., and K. V. Rosenberg. 2012. Fall migration of the Veery (Catharus fuscescens) in northern Colombia: determining the importance of a stopover site. Auk 129:449–459.

Berlanga, H., J. A. Kennedy, T. D. Rich, M. C. Arizmendi, C. J. Beardmore, P. J. Blancher, G. S. Butcher, A. R. Couturier, A. A. Dayer, D. W. Demarest, W. E. Easton, M. Gustafson, E. Inigo-Elias, E. A. Krebs, A. O. Panjabi, V. Rodriguez Contreas, K. V. Rosenberg, J. M. Ruth, E. Santana Castellon, R. Ma. Vidal, and T. Will. 2010. Saving our shared birds: Partners in Flight tri-national vision for landbird conservation. Cornell Lab of Ornithology, Ithaca, NY.

Buehler, D. A., A. M. Roth, R. Vallender, T. C. Will, J. L. Confer, R. A. Canterbury, S. B. Swarthout, K. V. Rosenberg, and L. P. Bulluck. 2007. Status and conservation priorities of Golden-winged Warbler (Vermivora chrysoptera) in North America. Auk 124:1439–1445.

Canterbury, R. A., and D. M. Stover. 1999. The Golden-winged Warbler: an imperiled migrant songbird of the southern West Virginia coalfields. Green Lands 29:44–51.

Cherrington, E. A., B. E. Hernandez, B. C. Garcia, M. O. Oyuela, and A. H. Clemente. 2011. Land cover change and deforestation in Central America. CATHALAC, Panama City, Panama. <http://issuu.com/cathalac/docs/servir_land_cover_eng> (1 January 2013).

Confer, J. L., and S. M. Pascoe. 2003. Avian communities on utility rights-of-way and other managed shrublands in the northeastern United States. Forest Ecology and Management 185:193–205.

Dabrowski, A., R. Fraser, J. L. Confer, and I. J. Lovette. 2005. Geographic variability in mitochondrial introgression among hybridizing populations of Golden-winged (Vermivora chrysoptera) and Blue-winged (V. pinus) Warblers. Conservation Genetics 6:843–853.

Donovan, T., and F. R. Thompson. 2001. Modeling the ecological trap hypothesis: a habitat and demographic analysis for migrant songbirds. Ecological Applications 11:871–882.

Faaborg, J., R. T. Holmes, A. D. Anders, K. L. Bildstein, K. M. Dugger, S. A. Gauthreaux, P. Heglund, K. A. Hobson, A. E. Jahn, D. H. Johnson, S. C. Latta, D. J. Levey, P. P. Marra, C. L. Merkord, E. Nol, S. I. Rothstein, T. W. Sherry, T. S. Sillett, F. R. Thompson, and N. Warnock. 2010. Conserving migratory land birds in the New World: do we know enough? Ecological Applications 20:398–418.

Fraser, K. C., B. J. M. Stutchbury, C. Silverio, P. Kramer, J. Barrow, D. Newstead, N. Mickle, B. F. Cousens, J. C. Lee, D. M. Morrison, T. Shaheen, P. Mammenga, K. Applegate, and J. Tautin. 2012. Continent-wide tracking to determine migratory connectivity and tropical habitat associations of a declining aerial insectivore. Proceedings of the Royal Society of London B 279:4901–4906.

Johnson, M. D., T. W. Sherry, R. T. Holmes, and P. P. Marra. 2006. Assessing habitat quality for a migratory songbird wintering in natural and agricultural habitats. Conservation Biology 20:1433–1444.

Kubel, J. E., and R. H. Yahner. 2008. Quality of anthropogenic habitats for Golden-winged Warblers in central Pennsylvania. Wilson Journal of Ornithology 120:801–812.

La Sorte, F. A., D. Fink, W. M. Hochachka, A. Farnsworth, A. D. Rodewald, K. V. Rosenberg, B. L. Sullivan, D. W. Winkler, C. Wood, and S. Kelling. 2014. The role of atmospheric conditions in the seasonal dynamics of North American migration flyways. Journal of Biogeography 41:1685–1696.

Loss, S. R., T. Will, S. S. Loss, and P. Marra. 2014. Bird-building collisions in the United States: estimates of annual mortality and species vulnerability. Condor 116:8–23.

McKinnon, E. A., K. C. Fraser, and B. J. M. Stutchbury. 2013. New discoveries in landbird migration using geolocators, and a flight plan for the future. Auk 130:211–222.

Norris, R. D., P. P. Marra, T. K. Kyser, T. W. Sherry, and L. M. Ratcliffe. 2004. Tropical winter habitat limits reproductive success on the temperate breeding grounds in a migratory bird. Proceedings of the Royal Society of London B 271:59–64.

North American Bird Conservation Initiative, U.S. Committee. 2013. The state of the birds 2013 report on private lands. U.S. Department of Interior, Washington, DC.

Pomara, L.Y., R. J. Cooper, and L. J. Petit. 2007. Modeling the flocking propensity of passerine birds in two Neotropical habitats. Oecologia 153:121–133.

Rohrbaugh, R.W., S. B. Swarthout, D. L. Crawford, M. D. Piorkowski, and J. D. Lowe. 2011. Golden-winged Warbler conservation initiative: year 3 breeding grounds monitoring throughout the Appalachian Region. Final Report to U.S. Fish and Wildlife Service (Cooperative Agreement No. 501819M859). Cornell Lab of Ornithology, Ithaca, NY.

Roth, A. M., D. J. Flaspohler, and C. R. Webster. 2014. Legacy tree retention in young aspen forest improves nesting habitat quality for Golden-winged Warbler (*Vermivora chrysoptera*). Forest Ecology and Management 321:61–70.

Sherry, T. W., and R. T. Holmes. 1995. Summer versus winter limitation of populations: what are the issues and what is the evidence? Pp. 85–120 in T. E. Martin and D. M. Finch (editors), Ecology and management of Neotropical migratory birds. Oxford University Press, New York, NY.

Sillett, T. S., and R. T. Holmes. 2002. Variation in survivorship of a migratory songbird throughout its annual cycle. Journal of Animal Ecology 71:296–308.

Streby, H. M., J. M. Refsnider, S. M. Peterson, and D. E. Andersen. 2014. Retirement investment theory explains patterns in songbird nest-site choice. Proceedings of the Royal Society of London B 281:20131834.

Thogmartin, W. E. 2010. Modeling and mapping Golden-winged Warbler abundance to improve regional conservation strategies. Avian Conservation and Ecology 5:12.

Vallender, R., S. L. Van Wilgenburg, L. P. Bulluck, A. Roth, R. Canterbury, J. Larkin, R. M. Fowlds, and I. J. Lovette. 2009. Extensive rangewide mitochondrial introgression indicates substantial cryptic hybridization in the Golden-winged Warbler. Avian Conservation and Ecology 4:4.

Wood, E. M., S. E. Barker Swarthout, W. M. Hochachka, J. L. Larkin, R. W. Rohrbaugh, K. V. Rosenberg, and A. D. Rodewald. 2016. Intermediate habitat associations by hybrids may exacerbate genetic introgression in a songbird. Journal of Avian Biology 47:508–520.

CHAPTER FOURTEEN

Research on Golden-winged Warblers*

RECENT PROGRESS AND CURRENT NEEDS

*Henry M. Streby, Ronald W. Rohrbaugh, David A. Buehler,
David E. Andersen, Rachel Vallender, David I. King, and Tom Will*

Abstract. Considerable advances have been made in knowledge about Golden-winged Warblers (*Vermivora chrysoptera*) in the past decade. Recent employment of molecular analysis, stable-isotope analysis, telemetry-based monitoring of survival and behavior, and spatially explicit modeling techniques have added to, and revised, an already broad base of published knowledge. Here, we synthesize findings primarily from recent peer-reviewed literature on Golden-winged Warblers, from this volume and elsewhere, and we identify some of the substantial remaining research needs. We have organized this synthesis by stages of the Golden-winged Warbler annual cycle. First, we discuss the relatively well-studied breeding-grounds ecology including nesting and post-fledging ecology and hybridization with closely related Blue-winged Warblers (*Vermivora cyanoptera*). Second, we discuss the much-less-studied, non-breeding-grounds ecology, including the first empirical studies of non-breeding-grounds cover-type associations and spatial and social behavioral ecology. Third, we address migratory connectivity and migration ecology, for which little is known and research has only just begun. Last, we close with cautious optimism that current knowledge is adequate to inform initial conservation and management plans for Golden-winged Warblers, and with a sobering acknowledgement of the quantity of research still needed.

Key Words: annual cycle, breeding ecology, hybridization, migration, nonbreeding ecology, *Vermivora cyanoptera*.

BREEDING-GROUNDS ECOLOGY

Nesting-Habitat Associations and Nesting Ecology

The nesting ecology of Golden-winged Warblers has been the focus of most research on the species prior to this volume of Studies in Avian Biology (Ficken and Ficken 1968, Klaus and Buehler 2001, Martin et al. 2007, Vallender et al. 2007a, Confer et al. 2010) and within this volume (Chapters 7 and 9, this volume). Across their breeding distribution, Golden-winged Warblers are typically associated

* Streby, H. M., R. W. Rohrbaugh, D. A. Buehler, D. E. Andersen, R. Vallender, D. I. King, and T. Will. 2016. Research on Golden-winged Warblers: Recent progress and current needs. Pp. 217–227 in H. M. Streby, D. E. Andersen, and D. A. Buehler (editors). Golden-winged Warbler ecology, conservation, and habitat management. Studies in Avian Biology (no. 49), CRC Press, Boca Raton, FL.

with landscapes dominated by forest in later seral stages (hereafter, later successional forest) with openings of shrub–sapling and herbaceous vegetation that can be shrub-dominated uplands or wetlands or very young forest (hereafter, shrublands or early successional areas). However, beyond this broad-stroke description of the breeding landscape structure, results reported here and elsewhere suggest there is no one-size-fits-all breeding habitat description at the stand-level scale (Chapter 9, this volume) or the territory or nest-site scale (Confer et al. 2003, Bulluck and Buehler 2008, Aldinger and Wood 2014; Chapter 7, this volume) that is consistently associated with nesting habitat selection and nest success of Golden-winged Warblers.

Perhaps not surprising, considering the tremendous variation in plant communities across the breeding distribution of Golden-winged Warblers, the relative importance of small-scale vegetation characteristics such as percent ground cover by certain species or vegetation types to nesting habitat selection and nest success is not generalizable among regions or even among sites within a region (Chapter 7, this volume). Studies incorporating data from many sites primarily in the Appalachian Mountains region support the results of previous single-site studies (Confer et al. 2003, 2010; Bulluck and Buehler 2008; Kubel and Yahner 2008; Roth et al. 2014), indicating fine-scale vegetation associations of nesting Golden-winged Warblers are mostly site-specific. Furthermore, spatially explicit models of full-season productivity, or young raised to independence from adult care, indicate that a single management action can affect productivity differently depending on the cover-type composition of the surrounding landscape (Chapter 10, this volume). Therefore, it is unlikely there is any single stand-level management action, such as increasing area of shrubland, that generally increases Golden-winged Warbler productivity, at least in the western Great Lakes region. In fact, Chapter 10 (this volume) demonstrated that in some areas that already host highly productive populations, reducing the area of later successional forest can have a net negative impact on population productivity, even if breeding density increases. As frustrating as it might be that there is no evidence for a single management prescription that can be applied broadly to benefit breeding Golden-winged Warblers, the knowledge that management and conservation plans must be flexible and locally informed is an equally important research outcome.

Postfledging Survival and Habitat Associations

For two decades, a growing body of literature has demonstrated that postfledging survival and habitat associations are critical components of breeding-grounds ecology in migratory songbirds (Anders et al. 1997, Streby and Andersen 2011). Several species that nest in later successional forest use early successional areas during the postfledging period (Anders et al. 1998, Pagen et al. 2000, Marshall et al. 2003, Vitz and Rodewald 2007, Streby et al. 2011) and in some species, a mid-season switch in cover-type associations can benefit body condition (Stoleson 2013) and fledgling survival (Streby and Andersen 2013a, Vitz and Rodewald 2013). To our knowledge, recent postfledging research on Golden-winged Warblers represents the first report of the opposite pattern, in which a species most commonly associated with nesting in early successional areas selects later successional forest for raising young (Streby et al. 2014a, 2015a; Chapters 8 through 10, this volume).

In the western Great Lakes region, Golden-winged Warblers choose later successional forest over shrublands for raising recently fledged young (Streby et al. 2014a; Chapter 8, this volume). Parents of both sexes raise fledglings in later successional forest, but adult females lead fledglings hundreds of meters away from natal stands before young are independent from adult care (Chapter 10, this volume). In addition, fledglings choose later successional forest over all other cover types after independence from adult care (Streby et al. 2015a). Postfledging research is currently underway in populations within the Appalachian Mountains region (J. A. Lehman, unpubl. data), and those studies could provide at least a partial explanation for the substantial differences in population growth between regions with similar fledgling production from nests and apparently similar adult annual survival. Indeed, pilot research in Tennessee suggests that fledgling survival there might be low (J. A. Lehman, unpubl. data).

Limitations of Current Breeding-Grounds Knowledge

A tremendous amount of research has focused on breeding Golden-winged Warblers in the past few decades and that research has provided considerable new insights. However, many important gaps remain in our understanding of the breeding-grounds ecology of Golden-winged Warblers. For example, attempts to relate food availability to foraging habitat selection (Chapter 6, this volume) are complicated by limited available data on the diet of Golden-winged Warblers (Streby et al. 2014b). The published and anecdotal descriptions of diet in Golden-winged Warblers (summarized in Confer et al. 2011, Streby et al. 2014b) suggest their diet may be too specialized for food availability to be sampled with standard methods. Leaf-roller caterpillars (*Archips* spp.) constitute 89% of the diet of nestling and fledgling Golden-winged Warblers in Minnesota and Manitoba (Streby et al. 2014b), but initial attempts to develop efficient prey sampling methods indicated the required sampling effort might be a prohibitive challenge (B. Vernasco, unpubl. data). Regardless, recent evidence suggests patterns of differential arthropod abundance and distribution among shrub and tree species influence territory placement by Golden-winged Warblers in central Pennsylvania (Chapter 6, this volume). Quantifying diet across the entire distribution and throughout the breeding season, and developing methods for sampling prey abundance and phenology, is necessary precursors for future work regarding potential range shifts, predator–prey mismatches, and inter- and intraspecific resource competition.

Many remaining gaps in knowledge of the breeding ecology of Golden-winged Warblers are related to the question of transferability or the geographic range of inference from observations made at only one or a few locations. For example, are the radiotelemetry-based observations that song territories are considerably larger than those based on spot mapping (Streby et al. 2012), that breeding adults forage in later successional forest (Streby et al. 2012; Chapter 5, this volume), and that a considerable proportion of females nest in later successional forest adjacent to early successional stands (Streby et al. 2014a; Chapter 10, this volume) consistent in other portions of the breeding distribution? How widespread is the problem of misidentifying nest fates based on traditional monitoring methods (Streby and Andersen 2013b), and what impact does that bias have on local and regional productivity estimates? What are the factors underlying female mate choice, as song variation (Harper et al. 2010) and plumage phenotype of social mates (Vallender et al. 2007a) do not appear to be important? Are the positive

relationships between residual tree retention after forest harvest and breeding-male density and pairing success observed in Wisconsin (Roth et al. 2014) also present in the Appalachian region where breeding densities are relatively low, or in areas of Minnesota where pairing success is near 100%? Do patterns of postfledging cover-type selection observed in the western Great Lakes region (Chapter 8, this volume) hold true in the rest of the Golden-winged Warbler breeding distribution? If not, what are the cover types and habitat characteristics associated with high fledgling survival in populations outside the western Great Lakes region?

The vast majority of research on breeding Golden-winged Warblers has focused on the nesting season in the Appalachian population segment, which is declining and represents ~5% of the global breeding population (Chapter 1, this volume). Current breeding-distribution-wide conservation plans are based on published and unpublished data of which nearly all represent the Appalachian Mountains region (A. M. Roth et al., unpubl. plan), and little data represent the other 95% of the global breeding population of Golden-winged Warblers. The precipitous decline of the Appalachian population segment necessitates intensive monitoring and research designed to identify the causes of the decline and develop strategies and plans to reverse it. However, there is an inherent risk of bias when deriving most of the knowledge about a species from small and declining populations, especially if that knowledge is intended to inform management and conservation for the species as a whole. Until recently, relatively little research had focused on Golden-winged Warblers in the western Great Lakes region where a majority of the species breeds (Streby et al. 2012, Roth et al. 2014; Chapters 9 and 10, this volume). More information is needed on breeding-grounds ecology in this core, with respect to population density, of the Golden-winged Warbler breeding distribution, where population numbers are relatively stable (Chapter 1, this volume), or apparently recently populated and increasing, such as parts of Manitoba (Sauer et al. 2012). Currently, the only information available about postfledging ecology of Golden-winged Warblers is from the western Great Lakes region. Considering the importance of fledgling survival to full-season productivity and population growth of songbirds (Anders et al. 1997, Faaborg et al. 2010a, Streby and

Andersen 2011), and of Golden-winged Warblers in particular (Streby et al. 2014a; Chapter 8, this volume), research on postfledging ecology is a pressing need in the Appalachian Mountains breeding-distribution segment.

Rapidly Changing Knowledge of Breeding-Grounds Cover-Type Associations

In a special section of The Wildlife Society Bulletin summarizing research on Golden-cheeked Warblers (*Setophaga chrysoparia*), Morrison et al. (2012) made a unique and valuable contribution that also rings true with Golden-winged Warblers. Morrison et al. (2012) described how a prevailing paradigm can hinder progress in current research and conservation. In Golden-cheeked Warblers, a limited and biased understanding of the distribution, habitat associations, and status of the species led to research unconsciously designed to perpetuate the prevailing paradigm, which reduced the efficacy of conservation efforts (Morrison et al. 2012). Similar biased paradigm perpetuation is evident in Golden-winged Warbler breeding-grounds research. Golden-winged Warblers are commonly described as an early successional shrubland specialist throughout the published literature (Confer and Knapp 1981, Confer et al. 2011). The current paradigm acknowledges that early successional stands or patches must be within a later successional forested matrix to host breeding Golden-winged Warblers, and a positive association between Golden-winged Warblers and primarily forested landscapes has been demonstrated (Thogmartin 2010). However, the degree to which later successional forest is used by Golden-winged Warblers throughout the breeding season is only starting to be appreciated.

The early successional specialist paradigm is so entrenched in Golden-winged Warbler research that early successional or open shrubby areas are commonly described as "study sites" bounded by the edges of later successional forest (Ficken and Ficken 1968; Confer et al. 2011; Roth et al. 2014; Chapter 7, this volume). Early successional areas are certainly a critical component of breeding habitat, and Golden-winged Warblers are usually not observed breeding in landscapes that lack at least some forest openings. However, recent research published in this volume and elsewhere demonstrates that later successional forest is used throughout the breeding season for nesting,

foraging, and raising young. Adult Golden-winged Warblers usually include later successional forest in territories and home ranges (Streby et al. 2012; Chapter 5, this volume), especially later in the day when territorial defense subsides (Streby et al. 2012). The use of later successional forest by territorial males goes undetected when forest edges are assumed to be study-site boundaries (Streby et al. 2012). In addition, tracking radio-marked females to nest sites reveals that some females nest in later successional forest ≤100 m away from early successional areas (Streby et al. 2014a; Chapter 9, this volume). It is not possible to locate nests in areas not searched in traditional nest-finding protocols, and the prevailing paradigm results in study designs that only consider early successional shrublands, which can result in a biased distribution of nest sites and a biased estimate of nest success (S. M. Peterson, unpubl. data). Golden-winged Warbler nests in later successional forest can be found during standard nest-finding methods if search areas are expanded to include later successional forest (H. M. Streby, unpubl. data).

It remains to be seen if the nesting and postfledging use of later successional forest by Golden-winged Warblers in the western Great Lakes region is also common in the Appalachian breeding-distribution segment. However, the recent observations that Golden-winged Warblers use at least some later successional forest during the nesting season in two areas in the central Appalachians (Chapter 5, this volume), and a radiomarked adult male that traveled into forest with fledglings in West Virginia (Chapter 5, this volume), suggest the early successional specialist description is likely inappropriate across the breeding distribution. A more appropriate categorization for Golden-winged Warblers might be diverse-forest obligate or dynamic-forest specialist. Incorporating full-season habitat associations in breeding-grounds management and conservation planning will be crucial to the success of conservation plans for Golden-winged Warblers and likely other forest-associated songbirds.

Hybridization of Golden-winged Warbler × Blue-winged Warbler

Hybridization between Golden-winged and Blue-winged Warblers (*Vermivora cyanoptera*) is thought to be one of the principal drivers behind widespread population declines in Golden-winged

Warblers. Indeed, a typical pattern of replacement of the Golden-winged Warbler phenotype by the Blue-winged Warbler phenotype has been documented in several regions throughout the breeding distribution, and replacement tends to occur within 50 years of contact between the two species. The hybridization system has interested researchers for many years (Gill and Murray 1972; Gill 1980, 1987) but has only been quantitatively examined since the application of genetic markers in several studies since the late 1990s (Gill 1997; Shapiro et al. 2004; Dabrowski et al. 2005; Vallender et al. 2007a,b, 2009). There is variation across the breeding distribution—perhaps dependent on the length of time that the two species have been in contact with one another in a given region—with a well-established pattern of bidirectional gene flow between Blue-winged Warblers and Golden-winged Warblers. To date, patterns in gene flow have been established using genetic markers derived from the mitochondrial genome, as attempts to quantify hybridization using markers from the nuclear genome have been largely uninformative (Vallender et al. 2007b). The lack of data from the nuclear genome is the most significant challenge to elucidating the true impact of hybridization on Golden-winged Warblers and to enabling researchers to make predictions about the future of this species pair.

The most immediate short-term research need with respect to hybridization between Golden-winged and Blue-winged Warblers is the development of informative genetic markers from the nuclear genome for each species. Once available, genetic markers can be applied to the large number of samples already collected from across the breeding distributions of both species. As detailed in the chapter on hybridization of these species (Chapter 4, this volume), next-generation sequencing likely provides the most promise in this regard. A second short-term need is a better understanding of the ecology of Blue-winged Warblers and the impact of hybridization on allopatric populations of this poorly studied species, and a better understanding of the mate-choice dynamics within this hybrid complex (Vallender et al. 2007a, Hartman et al. 2012). Over the long term, the continued application of genetic techniques using markers derived from both the mitochondrial and nuclear genomes to better reveal patterns and extent of hybridization between Golden-winged Warblers and Blue-winged

Warblers will inform conservation plans for both species. Some have suggested that Golden-winged Warblers may become very rare or even extinct by 2080 (Gill 1980). Continuing to document the progression and impact of hybridization is paramount for determining where genetically pure populations of Golden-winged Warblers remain and for applying appropriate on-the-ground conservation measures.

Chapter 3 (this volume) reported that landscape-scale settlement patterns of Golden-winged Warblers in the Appalachian Mountains region are positively associated with increasing percent forest and high elevations, and negatively correlated with increasing agriculture and human development, whereas Blue-winged Warblers showed the opposite relationship with these covariates. In fact, Blue-winged Warblers were positively associated with agriculture. The findings may be important to better understanding the land-use patterns that are driving spatial interactions between Golden-winged and Blue-winged Warblers and in describing potential distribution shifts for both species. More work, however, is needed to understand how landscape-scale changes in distribution and abundance of cover types used by Golden-winged Warblers are related to population trends in both species and if these changes have a role in mediating hybridization.

Breeding-Grounds Research Needs

In addition to the needs discussed above about broad inference from local studies, research is needed on habitat characteristics associated with Golden-winged Warbler breeding density and population productivity throughout the western Great Lakes region, especially in northern Minnesota where populations are dense and highly productive, and in south-central Canada where the breeding distribution is expanding northward. In the Appalachian Mountains region, research is needed to determine the efficacy of current conservation and management actions intended to benefit Golden-winged Warblers, and to identify additional forested landscapes or landscapes that could be converted to forest where management efforts will have the greatest impact. Research on habitat associations and survival during the postfledging period is an immediate need in the Appalachian Mountains region; results from that research will be important for adapting management actions already underway and plans to be implemented in the near future.

Future Research on Golden-winged Warblers and Blue-winged Warblers

Over at least the past two decades, Vermivora conservation attention and action has primarily been driven by the decline and vulnerability of Golden-winged Warblers. Hybridization with Blue-winged Warblers has been viewed primarily as a threat to the continued persistence of Golden-winged Warblers, and as such the research agenda has largely focused on Golden-winged Warblers. We are not aware of any research that has focused on Blue-winged Warblers in the core of their breeding distribution where Golden-winged Warblers are rare or absent. In addition, no studies have considered the Vermivora complex as an adaptive evolutionary system in its own right. However, given their propensity to hybridize whenever sympatric, attempts to unravel individual life histories for Golden-winged and Blue-winged Warblers in isolation are unlikely to be successful.

Several studies have examined the genetics of hybridization between Golden-winged and Blue-winged Warblers with the goal of understanding the impact and implications of their intermixing (Gill 1997; Shapiro et al. 2004; Dabrowski et al. 2005; Vallender et al. 2007b, 2009). However, most genetic sampling to date has been done either in the zone of breeding distribution overlap between Golden-winged and Blue-winged Warblers or in regions where phenotypic Golden-winged Warblers are more common (Vallender et al. 2009). Our understanding of the genetic and ecological effects of hybridization and introgression on Golden-winged Warblers is growing, but far less is known about the impact on Blue-winged Warbler populations. Moreover, basic information is lacking about breeding, mate choice, wintering-ground ecology, migration, and many other aspects of Blue-winged Warbler ecology. It has therefore been recommended that general research be conducted on life history characteristics of Blue-winged Warblers, particularly in the distribution of phenotypically "pure" Blue-winged Warblers (Chapter 4, this volume). Such work could include behavioral and genetic analyses of mate choice in allopatry and sympatry with Golden-winged Warblers and hybrids, realized

reproductive success, and blood-parasite load (Hartman et al. 2012; Vallender et al. 2007a,b, 2012). An important first step will be additional genetic sampling and purity analyses of Blue-winged Warblers, especially from regions where they have been poorly sampled to date, such as Missouri and both peninsulas of Michigan.

NONBREEDING ECOLOGY

Key Findings and Knowledge Gaps

Until recently, little was known about the status, distribution, or ecology of Golden-winged Warblers on their nonbreeding grounds, and much of what has been observed is reported in the gray literature (Chavarría and Duriaux 2009, Chandler 2013). Chapters 1, 2, and 11 (this volume) are among the first peer-reviewed publications focused on Golden-winged Warblers during the nonbreeding period. We know from the work reported herein that wintering Golden-winged Warblers are concentrated in the highlands and Caribbean slopes from Guatemala and Belize to northwestern Nicaragua, at middle elevations of both Caribbean and Pacific slopes in Costa Rica and western Panama, and in an arc of the northern Andes from central Colombia to northern Venezuela. Little is known about patterns of habitat use across the full nonbreeding distribution (Chapter 2, this volume), but abundance appears to be greatest in mid-elevation primary and secondary forest (Bennett 2013, Chandler 2013). Golden-winged Warblers on the nonbreeding grounds select microhabitat features associated with intermediate forest disturbance, likely due to their use of hanging dead leaves as foraging substrates (Chandler and King 2011). Golden-winged Warbler use of agricultural cover types, such as shade coffee, appears to be contingent on adjacency of those cover types to intact primary or secondary forest (Chavarría and Duriaux 2009, Chandler 2010).

In Costa Rica and Nicaragua, Golden-winged Warblers maintain larger nonbreeding territories and home ranges, and may have larger area requirements than other Neotropical migrants. They display a high propensity to join mixed-species flocks where they forage by probing dead-leaf clusters and epiphytes (Chapter 11, this volume). A high degree of sociality may have important conservation implications. For example, territoriality and group size can clearly affect population density, which is a primary determinant of carrying capacity, and dependence upon mixed-species flocks may be a liability because forest fragmentation can disrupt cohesion within mixed-species flocks (Rappole and Morton 1985, Stouffer and Bierregaard 1995).

Non-Breeding-Grounds Research Needs

The first, and primary, hurdle in non-breeding-grounds research is gaining a better understanding of Golden-winged Warbler nonbreeding distribution. Even more so than on the breeding grounds, knowledge of non-breeding-grounds ecology in Golden-winged Warblers is limited geographically, with the published research coming from a few sites in the northwestern region of the nonbreeding distribution. Analyses of distribution-wide survey data yielded a coarse-scale indication of Golden-winged Warbler nonbreeding distribution and general cover-type associations (Chandler 2013); however, sampling intensity was modest relative to the area covered and was not evenly dispersed across the nonbreeding distribution. Therefore, a primary wintering-ecology research need is additional surveys in poorly sampled portions of the distribution, such as Belize, and in a greater range of land-use and cover-type associations.

Knowledge of the regions and cover types occupied by Golden-winged Warblers on the nonbreeding grounds is important to their conservation; however, despotic interactions among conspecifics may cause individuals to be forced into marginal sites where survival or other elements of overwinter performance can be compromised. The relationship between land use and cover-type-specific survival and body condition in Golden-winged Warblers is poorly understood. Therefore, broad-scale surveys of geographic and cover-type distribution should be followed up with detailed studies of demographic parameters. Last, replication of previous studies on space requirements and behavioral ecology should be undertaken in unstudied areas of the nonbreeding distribution. Better knowledge of the nonbreeding distribution, variation in cover-type associations, behavioral ecology throughout that distribution, and cover-type-specific survival will facilitate additional needed studies on correlations between nonbreeding habitat associations and subsequent survival and reproductive success, migratory connectivity between wintering and breeding locations, and winter-range shifts owing to climate change or other factors. A better understanding of resource selection and demography would permit the identification of the factors that limit population growth during the annual cycle, which can enable conservationists to target resources most efficiently and effectively.

MIGRATORY CONNECTIVITY AND MIGRATION ROUTES

Key Findings and Knowledge Gaps

The most difficult challenge for informing full life-cycle conservation plans for small migratory songbirds is connecting geographically distinct areas in the absence of adequate tools to track individual birds (e.g., GPS tags). Chapter 12 (this volume) used stable-isotope analysis to take a first step in understanding migratory connectivity across Golden-winged Warbler breeding and nonbreeding distributions and made reasonable speculation about likely migration routes. The authors used stable isotopes in tissues generated on the breeding grounds and collected on the nonbreeding grounds to create a coarse, but informative map linking Golden-winged Warblers from the Appalachian population segment to more southern wintering grounds, and Golden-winged Warblers from the Great Lakes breeding-distribution segment generally to more northwestern wintering grounds. It is possible that Chapter 12 (this volume) has provided an explanation, aside from insufficient non-breeding-ground sampling effort, for why thousands of Golden-winged Warblers banded on the breeding grounds have yielded no recaptures on the nonbreeding grounds. The majority of breeding-grounds research that involved banding was in the Appalachian breeding-distribution segment and the majority of non-breeding-grounds research was in northwestern Central America, which are likely two, largely unconnected populations.

Migratory Connectivity and Migration Research Needs

Establishing fine-resolution, distribution-wide migratory connectivity is a research challenge in Golden-winged Warblers and in many other migratory species (Hobson 2003; Faaborg et al. 2010a,b). Building upon the framework established by Chapter 12 (this volume) is necessary to allow conservation efforts to target individual populations on their breeding and wintering grounds and along their migration routes. Until recently, the only feasible step toward that goal was to increase the geographic coverage of non-breeding-grounds sampling locations and to increase the number of birds sampled across the nonbreeding distribution to improve accuracy of stable-isotope-based connectivity maps (Chapter 12, this volume). However, light-level geolocators have now been carried by Golden-winged Warblers with no deleterious effects (Peterson et al. 2015; Streby et al. 2015b,c) and preliminary results suggest geolocator tags will considerably improve knowledge of Golden-winged Warbler migratory connectivity in the next few years (H. M. Streby, unpubl. data).

The only information available about Golden-winged Warbler migration routes is speculation in this volume based on coarse migratory connectivity (Chapter 12, this volume) and incidental reports from eBird (www.ebird.org) or banding stations (Chapter 1, this volume). Chapters 1 and 12 (this volume) referenced eBird records, which suggest that Golden-winged Warblers use a trans-Gulf of Mexico migration pathway during spring and fall migrations. However, eBird observations of Golden-winged Warblers during migration are sparse (<0.5% of checklists) and the distribution of detections is likely biased by highly nonrepresentative spatial sampling of perceived migration hotspots such as the Gulf of Mexico Coast of North America. Current research using light-level geolocation has provided promising preliminary data about Golden-winged Warbler migration routes and stopover locations and duration (H. M. Streby, unpubl. data). Questions about when and where Golden-winged Warblers rest and refuel during migration, and about their diet, habitat associations, and survival along migration routes, and about how these factors affect their ability to reproduce, remain urgent research needs.

CONCLUSIONS

A common limiting factor in conservation of migratory birds is a lack of adequate knowledge from all stages of the annual cycle to inform full life-cycle models and identify conservation priorities for individual populations (Faaborg et al. 2010a). Despite the recent increase in knowledge about Golden-winged Warblers, their ecology during most of the annual cycle, especially during spring and fall migration, has not been studied thoroughly enough to inform such models.

Migratory connectivity and cover-type-specific survival on the wintering grounds are likely the most urgent problems to address with respect to full life-cycle research in Golden-winged Warblers. Until individual populations can be studied throughout their annual cycle, and factors most limiting growth for a population of interest can be identified, conservation plans will be limited to focusing on the apparent needs of different populations during different times of the year.

While acknowledging that many substantial research gaps remain and that some previously held truths about Golden-winged Warblers are being revised, the current information must be translated into conservation and management actions within areas of appropriate geographic inference sooner rather than later. Otherwise, we risk current knowledge being useful only for describing eventual extinction events (Lindenmayer et al. 2013), especially for the Appalachian breeding-distribution segment if long-term trends continue. Forest management efforts designed to benefit Golden-winged Warblers and other disturbance-dependent species are underway in many areas across the Golden-winged Warbler breeding distribution, and some progress is being made in reversing deforestation in areas of their nonbreeding distribution (Chapter 2, this volume). New research is needed not only to address remaining information gaps, but also to study the effects of conservation and management actions and to inform the adaptive nature of long-term management plans. Our volume of Studies in Avian Biology provides a solid foundation on which future research in Golden-winged Warblers can be built, but the recent acceleration in knowledge about Golden-winged Warblers indicates above all else that there is still much to be learned.

LITERATURE CITED

Aldinger, K. R., and P. B. Wood. 2014. Reproductive success and habitat characteristics of Golden-winged Warblers in high-elevation pasturelands. Wilson Journal of Ornithology 127:279–287.

Anders, A. D., D. C. Dearborn, J. Faaborg, and F. R. Thompson III. 1997. Juvenile survival in a population of migrant birds. Conservation Biology 11:698–707.

Anders, A. D., J. Faaborg, and F. R. Thompson III. 1998. Postfledging dispersal, habitat use, and home-range size of juvenile Wood Thrushes. Auk 115:349–358.

Bennett, R. E. 2013. Habitat associations of the Golden-winged Warbler in Honduras. M.S. thesis, Michigan Technological University, Houghton, MI.

Bulluck, L. P., and D. A. Buehler. 2008. Factors influencing Golden-winged Warbler (Vermivora chrysoptera) nest-site selection and nest survival in the Cumberland Mountains of Tennessee. Auk 125:551–559.

Chandler, R. B. 2010. Avian ecology and conservation in tropical agricultural landscapes with emphasis on Vermivora chrysoptera. Ph.D. thesis, University of Massachusetts, Amherst, MA.

Chandler, R. B. 2013. Analysis of Golden-winged Warbler winter survey data. Report prepared for the Cornell Lab of Ornithology, Ithaca, NY.

Chandler, R. B., and D. I. King. 2011. Habitat quality and habitat selection of Golden-winged Warblers in Costa Rica: an application of hierarchical models for open populations. Journal of Avian Ecology 48:1037–1048.

Chavarría, L., and G. Duriaux. [online]. 2009. Informe preliminar del primer censo de Vermivora chrysoptera: realizado en Marzo 2009 en la zona norcentral de Nicaragua. <http://www.bio-nica.info/Biblioteca/Chavarria2009FinalGWWA.pdf> (19 April 2014).

Confer, J. L., K. W. Barnes, and E. C. Alvey. 2010. Golden- and Blue-winged Warblers: distribution, nesting success, and genetic differences in two habitats. Wilson Journal of Ornithology 122:273–278.

Confer, J. L., P. Hartman, and A. Roth. 2011. Golden-winged Warbler (Vermivora chrystoptera). In A. Poole (editor), The birds of North America online. Cornell Lab of Ornithology, Ithaca, NY.

Confer, J. L., and K. Knapp. 1981. Golden-winged Warblers and Blue-winged Warblers: the relative success of a habitat specialist and a generalist. Auk 98:108–114.

Confer, J. L., J. L. Larkin, and P. E. Allen. 2003. Effects of vegetation, interspecific competition, and brood parasitism on Golden-winged Warbler (Vermivora chrysoptera) nesting success. Auk 120:138–144.

Dabrowski, A., R. Fraser, J. L. Confer, and I. J. Lovette. 2005. Geographic variability in mitochondrial introgression among hybridizing populations of Golden-winged (Vermivora chrysoptera) and Blue-winged Warblers (V. pinus). Conservation Genetics 6:843–853.

Faaborg, J., R. T. Holmes, A. D. Anders, K. L. Bildstein, K. M. Dugger, S. A. Gauthreaux, P. Heglund, K. A. Hobson, A. E. Jahn, D. H. Johnson, S. C. Latta, D. J. Levey, P. P. Marra, C. L. Merkord, E. Nol, S. I. Rothstein, T. W. Sherry, T. S. Sillett, F. R. Thompson, and N. Warnock. 2010a. Conserving migratory land birds in the New World: do we know enough? Ecological Applications 20:398–418.

Faaborg, J., R. T. Holmes, A. D. Anders, K. L. Bildstein, K. M. Dugger, S. A. Gauthreaux, P. Heglund, K. A. Hobson, A. E. Jahn, D. H. Johnson, S. C. Latta, D. J. Levey, P. P. Marra, C. L. Merkord, E. Nol, S. I. Rothstein, T. W. Sherry, T. S. Sillett, F. R. Thompson, and N. Warnock. 2010b. Recent advances in understanding migration systems of New World land birds. Ecological Monographs 80:3–48.

Ficken, M. S., and R. W. Ficken. 1968. Territorial relationships of Blue-winged Warblers, Golden-winged Warblers, and their hybrids. Wilson Bulletin 80:442–451.

Gill, F. B. 1980. Historical aspects of hybridization between Blue-winged and Golden-winged Warblers. Auk 97:1–18.

Gill, F. B. 1987. Allozymes and genetic similarity of Blue-winged and Golden-winged Warblers. Auk 104:444–449.

Gill, F. B. 1997. Local cytonuclear extinction of the Golden-winged Warbler. Evolution 51:519–525.

Gill, F. B., and B. G. Murray. 1972. Discrimination behavior and hybridization of the Blue-winged and Golden-winged Warblers. Evolution 26:282–293.

Harper, S. L., R. Vallender, and R. J. Robertson. 2010. Male song variation and mate choice in the Golden-winged Warbler. Condor 112:105–114.

Hartman, P. J., D. P. Wetzel, P. H. Crowley, and D. F. Westneat. 2012. The impact of extra-pair mating behavior on hybridization and genetic introgression. Theoretical Ecology 5:219–229.

Hobson, K. A. 2003. Making migratory connections with stable isotopes. Pp. 379–391 in P. Berthold, E. Gwinner, and E. Sonnenschein (editors), Avian migration. Springer-Verlag, Berlin, Germany.

Klaus, N. A., and D. A. Buehler. 2001. Golden-winged Warbler breeding habitat characteristics and nest success in clearcuts in the southern Appalachian Mountains. Wilson Bulletin 113:297–301.

Kubel, J. E., and R. H. Yahner. 2008. Quality of anthropogenic habitats for Golden-winged Warblers in central Pennsylvania. Wilson Journal of Ornithology 120:801–812.

Lindenmayer, D. B., M. P. Piggot, and B. A. Wintle. 2013. Counting the books while the library burns: why conservation monitoring programs need a plan for action. Frontiers in Ecology and the Environment 11:549–555.

Marshall, M. R., J. A. DeCecco, A. B. Williams, G. A. Gale, and R. J. Cooper. 2003. Use of regenerating clearcuts by late-successional bird species and their young during the post-fledging period. Forest Ecology and Management 183:127–135.

Martin, K. J., R. S. Lutz, and M. Worland. 2007. Golden-winged Warbler habitat use and abundance in northern Wisconsin. Wilson Journal of Ornithology 119:523–532.

Morrison, M. L., B. A. Collier, H. A. Mathewson, J. E. Groce, and R. N. Wilkins. 2012. The prevailing paradigm as a hindrance to conservation. Wildlife Society Bulletin 36:408–414.

Pagen, R. W., F. R. Thompson III, and D. E. Burhans. 2000. Breeding and post-breeding habitat use by forest migrant songbirds in the Missouri Ozarks. Condor 102:738–747.

Peterson, S. M., G. R. Kramer, H. M. Streby, J. A. Lehman, D. A. Buehler, and D. E. Andersen. 2015. Geolocators on Golden-winged Warblers do not affect migratory ecology. Condor 117:256–261.

Rappole, J. H., and E. S. Morton. 1985. Effects of habitat alteration on a tropical avian forest community. Ornithological Monographs 36:1013–1021.

Roth, A. M., D. J. Flaspohler, and C. R. Webster. 2014. Legacy tree retention in young aspen forest improves nesting habitat quality for Golden-winged Warbler (*Vermivora chrysoptera*). Forest Ecology and Management 321:61–70.

Sauer, J. R., J. E. Hines, J. E. Fallon, K. L. Pardieck, D. J. Ziolkowski, Jr., and W. A. Link. 2012. The North American Breeding Bird Survey, results and analysis 1966–2011, version 07.03.2013. USGS Patuxent Wildlife Research Center, Laurel, MD.

Shapiro, L. H., R. A. Canterbury, D. M. Stover, and R. C. Fleischer. 2004. Reciprocal introgression between Golden-winged Warblers (*Vermivora chrysoptera*) and Blue-winged Warblers (*V. pinus*) in eastern North America. Auk 121:1019–1030.

Stouffer, P. C., and R. O. Bierregaard. 1995. Use of Amazonian forest fragments by understory insectivorous birds. Ecology 76:2429–2445.

Stoleson, S. H. 2013. Condition varies with habitat choice in postbreeding forest birds. Auk 130:417–428.

Streby, H. M., and D. E. Andersen. 2011. Seasonal productivity in a population of migratory songbirds: why nest data are not enough. Ecosphere 2:78.

Streby, H. M., and D. E. Andersen. 2013a. Movements, cover-type selection, and survival of fledgling Ovenbirds in managed deciduous and mixed-coniferous forests. Forest Ecology and Management 287:9–16.

Streby, H. M., and D. E. Andersen. 2013b. Testing common assumptions in studies of songbird nest success. Ibis 155:327–337.

Streby, H. M., G. R. Kramer, S. M. Peterson, J. A. Lehman, D. A. Buehler, and D. E. Andersen. 2015c. Tornadic storm avoidance behavior in breeding songbirds. Current Biology 25:98–102.

Streby, H. M., J. P. Loegering, and D. E. Andersen. 2012. Spot mapping underestimates song-territory size and use of mature forest by breeding Golden-winged Warblers in Minnesota, USA. Wildlife Society Bulletin 36:40–46.

Streby, H. M., T. L. McAllister, G. R. Kramer, S. M. Peterson, J. A. Lehman, and D. E. Andersen. 2015b. Minimizing marker mass and handling time when attaching radio transmitters and geolocators to small songbirds. Condor 117:249–255.

Streby, H. M., S. M. Peterson, G. R. Kramer, and D. E. Andersen. 2015a. Post-independence fledgling ecology in a migratory songbird: implications for breeding-grounds conservation. Animal Conservation 18:228–235.

Streby, H. M., S. M. Peterson, J. A. Lehman, G. R. Kramer, B. J. Vernasco, and D. E. Andersen. 2014b. Do digestive contents confound body mass as a measure of relative condition in nestling songbirds? Wildlife Society Bulletin 38:305–310.

Streby, H. M., S. M. Peterson, T. L. McAllister, and D. E. Andersen. 2011. Use of early-successional managed northern forest by mature-forest species during the post-fledging period. Condor 113:817–824.

Streby, H. M., J. M. Refsnider, S. M. Peterson, and D. E. Andersen. 2014a. Retirement investment theory explains patterns in songbird nest-site choice. Proceedings of the Royal Society of London B 281:20131834.

Thogmartin, W. E. 2010. Modeling and mapping Golden-winged Warbler abundance to improve regional conservation strategies. Avian Conservation and Ecology 5:12.

Vallender, R., R. D. Bull, L. L. Moulton, and R. J. Robertson. 2012. Blood parasite infection and heterozygosity in pure and genetic-hybrid Golden-winged Warblers (*Vermivora chrysoptera*) across Canada. Auk 129:716–724.

Vallender, R., V. L. Friesen, and R. J. Robertson. 2007a. Paternity and performance of Golden-winged Warblers (*Vermivora chrysoptera*) and Golden-winged × Blue-winged Warbler (*V. pinus*) hybrids at the leading edge of a hybrid zone. Behavioral Ecology and Sociobiology 61:1797–1807.

Vallender, R., R. J. Robertson, V. L. Friesen, and I. J. Lovette. 2007b. Complex hybridization dynamics between Golden-winged and Blue-winged

Warblers (*Vermivora chrysoptera* and *V. pinus*) revealed by AFLP, microsatellite, intron and mtDNA markers. Molecular Ecology 16:2017–2029.

Vallender, R., S. Van Wilgenburg, L. Bulluck, A. Roth, R. Canterbury, J. Larkin, R. M. Fowlds, and I. J. Lovette. 2009. Extensive rangewide mitochondrial introgression indicates substantial cryptic hybridization in the Golden-winged Warbler. Avian Conservation and Ecology 4:4.

Vitz, A. C., and A. D. Rodewald. 2007. Vegetative and fruit resources as determinants of habitat use by mature forest birds during the postbreeding period. Auk 124:494–507.

Vitz, A. C., and A. D. Rodewald. 2013. Behavioral and demographic consequences of access to early-successional habitat in juvenile Ovenbirds (*Seiurus aurocapilla*): an experimental approach. Auk 130:21–29.

INDEX

A

Acer rubrum. See maple, red
Acer saccharum. See maple, sugar
ADEHABITAT package, 99, 133, 177
aerial photographs, 133, 143, 149 (figure), 150 (figure)
age-specific movement capability, 133–134
agriculture, 5, 31, 41, 49 (table), 50 (table), 52, 53 (table), 55 (table), 57 (table), 59–60, 113, 129, 207, 209, 211, 221
Akaike
    Information Criterion (AIC), 114–115, 116 (table), 119 (table), 120 (table), 132, 135, 136 (figure), 145, 196–197, 198 (table)
    weight, 114, 120 (table), 132, 135, 198 (table)
alerting mechanisms, predator, 69
allopatric populations, 70, 73–74, 221
Amplified Fragment Length Polymorphisms, 73
Anas platyrhynchos. See Mallard
Anas rubripes. See Black Duck, American
annual cycle, 4, 30, 193, 213, 223–224
Appalachian Mountains Bird Conservation Region, 82
apple, naturalized (*Malus* spp.), 83
ArcMap, 85–86, 131, 133
ash (*Fraxinus* spp.), 50 (figure), 51 (figure), 52–53, 54 (figure), 59, 82, 111
ash, black (*Fraxinus nigra*), 131
ash, green (*Fraxinus pennsylvanica*), 129
ash, white (*Fraxinus americana*), 83
aspen (*Populus* spp.), 41, 48, 50 (figure), 51 (figure), 52–53, 54 (figure), 59, 111, 129
aspen, quaking (*Populus tremuloides*), 49, 143
autumn-olive (*Elaeagnus umbellata*), 82–83

B

bare ground, 84, 89 (table), 90 (table), 98, 100, 102 (table), 113, 115–116, 117 (figure), 120, 121 (table), 122 (table), 123
Basileuterus culcivorous, 185 (table)
Basileuterus rufifrons, 185 (table)
Basileuterus tristriatus, 185 (table)
basswood (*Tilia* spp.), 51 (figure), 52, 54 (figure), 55, 56 (figure), 58 (figure)
basswood, American (*Tilia americana*), 129
beech (*Fagus* spp.), 50 (figure), 51 (figure), 52, 54 (figure), 55, 56 (figure), 58 (figure), 59
behavior, breeding adult
    feeding, 69, 83–84, 105, 137
    feeding/provisioning fledglings, 131, 162, 164
    foraging, 81, 91, 97–100, 101 (figure), 103–105, 131, 137, 175–178, 183–184, 186 (figure), 187 (figure), 188–189, 211, 219–220, 222
    nest-site choice/selection, 110, 111 (figure), 114, 116 (table), 119 (table), 120 (table), 121–123, 136, 169
    perching, 83–84, 98
    feeding/provisioning nestlings, 96, 161
    singing, 11, 70, 83–84, 91, 98, 131, 169, 178
behavior, fledgling
    age-specific movement capability, 133 (figure), 134
    begging, 131
    flying, 137
    foraging, 131, 137
    sitting quietly, 131
    travelling, 131
behavior, non-breeding adult
    energy expenditure, 175, 185
    foraging, specialized, 188
    gleans, 178, 184, 186 (figure), 188
    defense, flock and territory , 185, 188–189, 211
    mixed-species flocks, 176, 178, 183, 185, 188–189, 211–212, 222–223
    nuclear species, 183, 188
    predator avoidance, 176, 189
    probes, 178, 184
    response to broadcast vocalizations, 177–178, 180–181
    space use, 175, 177, 185
Belize, 17–18, 21–22, 201, 222–223
Betula allegheniensis. See birch, yellow
Betula lenta. See birch, sweet
Betula papyrifera. See birch, paper
Betula spp. See birch
BIOMOD, 46–47, 52–53, 61, 63

birch (*Betula* spp.), 52–53, 54 (figure), 83
birch, paper (*Betula papyrifera*), 50 (figure), 51 (figure), 129
birch, sweet (*Betula lenta*), 97, 99, 101 (table), 102, 103 (figure)
birch, yellow (*Betula alleghaniensis*), 50 (figure), 51 (figure), 52, 54 (figure), 56 (figure), 58 (figure), 59
Black Duck, American (*Anas rubripes*), 68
blackberry (*Rubus* spp.), 82–83, 89 (table), 90 (table), 97, 99–101, 102 (table), 103 (figure), 104–105, 113 (table), 115–116, 120 (table), 121 (table), 122 (table), 123
blueberry (*Vaccinium* spp.), 83, 97, 99–101, 102 (table), 103 (figure), 105
Bluebird, Eastern (*Sialia sialis*), 137
bootstrapping, 114, 117 (figure), 118 (figure)
Bunting, Indigo (*Passerina cyanea*), 156
Bush-tanager, Common (*Chlorospingus ophthalmicus*), 183

C

Calcarius lapponicus. See Longspur, Lapland
Cannabis sativa. See marijuana
canopy cover, 84, 86, 89 (table), 90 (table), 91, 129, 131–132, 135
Capitonidae, 188
capture methods
    decoys, 83, 177–178, 181, 185
    mist-nets, 143, 163
    passive netting, 143, 163
    target netting, 83, 112
    vocalization playback, 5, 83, 112, 180, 185, 195
Cardellina canadensis. See Warbler, Canada
Cardellina pusilla. See Warbler, Wilson's
Caribbean slope, 18, 20, 222
caterpillars, 96–97, 99–100, 102, 103 (figure), 104
    leaf-roller, 103, 137, 219
Catharus bicknelli. See Thrush, Bicknell's
Catharus ustulatus. See Thrush, Swainson's
Cecropia spp., 184
cherry, black (*Prunus serotina*), 50 (figure), 83, 99, 101 (figure), 102, 103 (figure), 104
cherry, pin (*Prunus pensylvanica*), 97, 99–100, 101 (figure), 102, 103 (figure), 104–105
chi-square test of independence, 86
Chickadee, Black-capped (*Poecile carolinensis*), 68
Chickadee, Carolina (*Poecile atricapillus*), 68, 75
Chickadee, Mountain (*Poecile gambeli*), 68
Chlorospingus ophthalmicus. See Bush-tanager, Common
chokecherry (*Prunus virginiana*), 97, 102
cinquefoil, common (*Potentilla simplex*), 83
classification tree, 43, 46–47, 48 (figure), 60–61
climate change, 22, 36, 209, 223
Colombia, 6, 17–18, 20–22, 194–195, 197 (figure), 199, 200 (figure), 201, 208, 211, 213, 222
compositional analysis, 99, 101 (figure), 133–134
Comptonia peregrina. See fern, sweet
Connecticut, 8–9, 11, 22
conservation
    breeding grounds, 15, 22, 59–60, 74, 76, 82, 123, 136, 158, 209–210, 220, 223
    migration stopover sites, 4, 193, 212
    non-breeding grounds, 4, 23, 30–36, 175–176, 184, 211–212, 223
conservation recommendations, 104, 209–211
core use areas, 91

Cornus racemosa. See dogwood, gray
Cornus sericea. See dogwood, red osier
Corylus spp. See hazel
Costa Rica, 6, 17–18, 20–22, 30–33, 36, 177–178, 179 (figure), 180 (figure), 181 (figure), 182 (figure), 183 (figure), 184 (figure), 185 (table), 186 (figure), 187 (figure), 188, 195, 197 (figure), 198 (figure), 199, 200 (figure), 201, 211, 222
cover-type selection, 128, 130 (table), 133 (figure), 134, 136–137, 219
Cowbird, Brown-headed (*Molothrus ater*), 96
Crataegus spp. See hawthorn
cross validation, 46, 157–158
cryptic hybrid, 13, 70, 71 (figure), 72 (figure), 73, 76
curvilinear model, 114

D

daily distance
    from edge of natal patch, 164, 165 (figure)
    from nest, 164, 165 (figure)
    maximum, 133
    minimum, 163
daily estimated availability, 133
darning needles, devil's (*Clematis virginiana*), 83
DeLorme atlas page sample area, 5, 14 (figure)
diet
    central Pennsylvania, 96, 99, 103 (figure)
    prey abundance, 96, 103 (figure), 105, 219
    prey availability, 95–96, 103
    western Great Lakes, 105, 219
digital orthophoto quadrangles, 133, 143, 163
distribution model, 5–6, 43–44, 46–47, 49, 61
distribution, breeding
    abundance, 6, 9, 13, 16, 23
    Appalachian breeding segment, 10 (figure), 12, 15–16, 22, 96, 104–105, 110, 114, 123, 138, 199–200, 208–211, 220, 223–224
    Canada, 11, 70, 72 (figure), 221
    density, 82, 219
    distribution shift and contraction, 4, 11–13, 22
    geographically imbalanced abundance, 210
    Golden-winged Warbler Atlas Project, 5
    Great Lakes breeding segment, 10 (figure), 12, 16, 22, 110, 123, 138, 208, 210, 223
    high-elevation populations, 3, 12, 15, 60, 210
    historic distribution expansion and contraction, 7, 8 (figure), 9, 17 (table), 22
    Michigan Breeding Bird Atlas, 11
    modeled future distribution, 45 (figure), 46, 48 (figure)
    Ontario Breeding Bird Survey, 11
    phenotypically pure and hybrid *Vermivora*, 4, 13, 14 (figure), 69, 72 (figure), 76, 222
    United States of America, 4
distribution, non-breeding
    abundance, 7, 30, 36, 176, 222
    density, 189
    inferring distribution from citizen reports, 18, 19 (figure)
    modeling abundance, 17 (figure)
dogwood, gray (*Cornus racemosa*), 82
dogwood, red osier (*Cornus sericea*), 82

deciduous, 12, 44, 49 (table), 50 (table), 52 (figure), 53 (table), 55 (table), 57 (table), 59–60, 62 (table), 143, 144 (table), 145, 146 (table), 147–148, 151, 152 (figure), 153 (figure), 155 (figure), 156, 157 (figure)

early successional, 15, 42, 60, 104, 128, 136, 142, 148, 162, 210, 218–220

mature, 110, 113, 120–121, 129, 130 (table), 133–135, 136 (figure), 137–138, 142, 144 (table), 148, 162–164, 169

swamp, 52, 54 (figure), 56 (figure), 110, 142, 154

upland, 128–129, 142–143

wetland, 129

young, 42, 60, 96–98, 105, 110–112, 209–210

Forest Inventory Analysis, 44, 210

Formicaridae, 188

Fragaria virginiana. See strawberry, Virginia

Fraxinus americana. See ash, white

Fraxinus nigra. See ash, black

Fraxinus pennsylvanica. See ash, green

Fraxinus spp. See ash

Furnariidae, 188

G

gene flow, 68, 70, 74–75, 221

generalized additive model, 46

generalized boosted regression, 46

genetic differentiation, 68, 75

genetic diversity, 22

genetic markers, 68, 221

genetic structure, 73

genetic swamping, 70

genetic variation. See genetic diversity

genetics, 75–76, 222

Geographic Information System, 85, 143, 163

Georgia, 8, 10–13, 15, 22, 194

grass, velvet (*Holcus lanatus*), 83

grasses, 83, 85, 110, 113, 129

grassland, 42, 44, 49 (table), 50 (table), 52 (figure), 56 (figure), 59, 62 (table), 112–116, 117 (figure), 118 (figure), 119 (table), 120 (table), 121 (table), 122 (table), 123, 130 (table), 131, 133, 136, 143, 144 (table), 146 (table), 148, 149 (figure), 151, 152 (figure), 153 (figure), 154 (figure), 155 (figure), 156, 157 (figure)

Guatemala, 17–18, 21, 23, 194, 201, 211, 222

H

habitat

Appalachian population, 13, 48, 53, 56 (figure), 57 (table), 58 (figure), 60, 82, 84, 87–88, 89 (table), 90 (table), 91, 96, 104, 111, 113–116, 120–123, 138, 142, 208, 210, 220–221

lateral vegetation cover, 164–166, 167 (figure)

nest-habitat structure, 98, 115–116, 123, 128, 132, 148, 162, 209, 217–218

non-breeding grounds, 30, 33–34, 36, 176–177, 188–189

postfledging, 122–123, 128, 136–138, 162, 168–169, 209

vegetation characteristics, 82–83, 90 (table), 98, 112, 114–116, 120 (table), 128, 163–167, 169, 218

vegetation strata, 163–164, 166

western Great Lakes population, 128–129, 132–133, 136–138, 152, 157–158, 163, 168–169, 210, 218–220

Haldane's Rule, 68, 74

hawthorn (*Crataegus* spp.), 82–83, 111

hay field, 162

hazel (*Corylus* spp.), 129, 143

hemlock, eastern (*Tsuga canadensis*), 51 (figure), 54 (figure), 55, 57 (figure), 58 (figure)

herbaceous vegetation, 83, 98, 114, 218

high-throughput sequencing, 75

Holcus lanatus. See grass, velvet

Honduras, 6, 17–18, 20–23, 31, 33 (figure), 34, 35 (figure), 179 (figure), 194–195, 197 (figure), 198 (figure), 199 (figure), 201, 211

honeysuckle, bush (*Lonicera* spp.), 82

human development, 56 (figure), 59, 97, 129, 130 (table), 133, 209, 211

human disturbance, 55, 60

hybridization, 4, 13, 14 (figure), 15, 23, 42, 59, 67–70, 72–76, 96, 142, 208–210, 220–222

hydrogen isotope, 194, 196

Hylocichla mustelina. See Thrush, Wood

Hylophilus decurtatus, 185 (table)

Hypericum prolificum. See St. John's wort, shrubby

I

Illinois, 7, 9–11, 13, 104

ImageJ, 131, 164

impact radius, 144–147, 150, 156

Indiana, 7, 9–11

indicator species analysis, 100, 102

insects, herbivorous, 103–105

Institute for Forest Conservation, 35

introgression, 13, 15, 68–75, 222

Iowa, 7

isotope-ratio mass spectrometry, 196

J

Juglans nigra. See walnut, black

K

k-fold cross validation, 148, 158

Kalmia latifolia. See laurel, mountain

Kentucky, 10, 12, 15, 22, 104

Kingbird, Eastern (*Tyrannus tyrannus*), 176

known fates module, 165

Kruskal–Wallis rank sum test, 100, 102–103

L

land cover, 42, 44, 46, 48–49, 52 (figure), 55, 57, 59–60, 61 (table), 112

Larix laricina. See tamarack

laurel, mountain (*Kalmia latifolia*), 83, 97, 99–101, 102 (table), 103 (figure), 104–105

life cycle, 23, 207–208, 212–214, 223–224

Limnothlypis swainsonii. See Warbler, Swainson's

linear models, 46, 60, 165, 196–197, 198 (table)

Liriodendron tulipifera. See poplar, yellow

locust, black (*Robinia pseudoacacia*), 83, 97, 99–100, 101 (table), 102, 103 (figure), 104–105

Thrush, Swainson's (*Catharus ustulatus*), 176
Thrush, Wood (*Hylocichla mustelina*), 35, 185
Tilia americana. See basswood, American
Tilia spp. See basswood
Troglodytidae, 188
Tsuga canadensis. See hemlock, eastern
Turdus assimilis. See Robin, White-throated
Turdus grayi, 185 (table)
Turdus migratorius. See Robin, American
Tyrannus tyrannus. See Kingbird, Eastern

U

U.S. Geological Survey, 83, 97, 131, 143, 163, 177, 197
Ulmus spp. See elm
unsustainable harvest, 68
U.S. Department of Agriculture, 208

V

Vaccinum spp. See blueberry
variable quantity map, 146
Venezuela, 6, 17–18, 20–22, 194–195, 197 (figure), 200 (figure), 201, 211, 222
Vermivora cyanoptera. See Warbler, Blue-winged
Vermivora hybrid. See Warbler, Brewster's
Vermivora hybrid. See Warbler, Lawrence's
Vermivora spp., 83, 88
Vermont, 8
    Lake Champlain Valley, 11
Viburnum dentatum. See viburnum, arrowwood
viburnum, arrowwood (*Viburnum dentatum*), 82
vine cover, 113 (table), 115–116, 120 (table), 121 (table), 122 (table), 123
Vireo, Blue-headed (*Vireo solitarius*), 188
Vireo flavifrons, 185 (table)
Vireo philadelphicus. See Vireo, Philadelphia
Vireo, Gray (*Vireo vicinior*), 188
Vireo, Philadelphia (*Vireo philadelphicus*), 188
Virginia, 8, 12, 15

W

Wagtail, White (*Motacilla alba*), 188
walnut, black (*Juglans nigra*), 82
Warbler, Black-throated Blue (*Setophaga caerulescens*), 176, 201
Warbler, Blue-winged (*Vermivora cyanoptera*), 48 (figure), 50 (figure), 51 (figure), 52 (figure), 53 (table), 54 (figure), 55 (table), 56 (figure), 57 (table), 58 (figure), 62 (table), 76, 83, 87, 96, 208, 221–222

X

STUDIES IN AVIAN BIOLOGY
Series Editor: Kate Huyvaert

34. *Beyond Mayfield: Measurements of Nest-Survival Data.* Jones, S. L., and G. R. Geupel, editors. 2007.

35. *Foraging Dynamics of Seabirds in the Eastern Tropical Pacific Ocean.* Spear, L. B., D. G. Ainley, and W. A. Walker. 2007.

36. *Status of the Red Knot (Calidris canutus rufa) in the Western Hemisphere.* Niles, L. J., H. P. Sitters, A. D. Dey, P. W. Atkinson, A. J. Baker, K. A. Bennett, R. Carmona, K. E. Clark, N. A. Clark, C. Espoz, P. M. González, B. A. Harrington, D. E. Hernández, K. S. Kalasz, R. G. Lathrop, R. N. Matus, C. D. T. Minton, R. I. G. Morrison, M. K. Peck, W. Pitts, R. A. Robinson, and I. L. Serrano. 2008.

37. *Birds of the US–Mexico Borderland: Distribution, Ecology, and Conservation.* Ruth, J. M., T. Brush, and D. J. Krueper, editors. 2008.

38. *Greater Sage-Grouse: Ecology and Conservation of a Landscape Species and Its Habitats.* Knick, S. T., and J. W. Connelly, editors. 2011.

39. *Ecology, Conservation, and Management of Grouse.* Sandercock, B. K., K. Martin, and G. Segelbacher, editors. 2011.

40. *Population Demography of Northern Spotted Owls.* Forsman, E. D., et al. 2011.

41. *Boreal Birds of North America: A Hemispheric View of Their Conservation Links and Significance.* Wells, J. V., editor. 2011.

42. *Emerging Avian Disease.* Paul, E., editor. 2012.

43. *Video Surveillance of Nesting Birds.* Ribic, C. A., F. R. Thompson III, and P. J. Pietz, editors. 2012.

44. *Arctic Shorebirds in North America: A Decade of Monitoring.* Bart, J. R., and V. H. Johnston, editors. 2012.

45. *Urban Bird Ecology and Conservation.* Lepczyk, C. A., and P. S. Warren, editors. 2012.

46. *Ecology and Conservation of North American Sea Ducks.* Savard, J.-P. L., D. V. Derksen, D. Esler, and J. M. Eadie, editors. 2015.

47. *Phenological Synchrony and Bird Migration: Changing Climate and Seasonal Resources in North America.* Wood, E. M., and J. L. Kellermann, editors. 2015.

48. *Ecology and Conservation of Lesser Prairie-Chickens.* Haukos, D. A. and C. W. Boal, editors. 2016.

49. *Golden-winged Warbler Ecology, Conservation, and Habitat Management.* Henry M. Streby, David E. Andersen, David A. Buehler, editors. 2016.

Printed and bound by CPI Group (UK) Ltd, Croydon, CR0 4YY

24/10/2024

01778298-0004